Dynamic System Reliability

Wiley Series in Quality and Reliability Engineering

Dr. Andre Kleyner

Series Editor

The Wiley Series in Quality and Reliability Engineering aims to provide a solid educational foundation for both practitioners and researchers in Q&R field and to expand the reader's knowledge base to include the latest developments in this field. The series will provide contribution to the teaching and practice of engineering.

The series coverage will contain, but is not exclusive to,

- statistical methods;
- physics of failure;
- reliability modeling;
- functional safety;
- Six Sigma methods;
- lead-free electronics;
- warranty analysis/management; and
- risk and safety analysis.

Wiley Series in Quality and Reliability Engineering

Design for Safety
by Louis J. Gullo, Jack Dixon
February 2018

Next Generation HALT and HASS: Robust Design of Electronics and Systems
by Krik A. Gray, John J. Paschkewitz
May 2016

Reliability and Risk Models: Setting Reliability Requirements, 2nd Edition
by Michael Todinov
September 2015

Applied Reliability Engineering and Risk Analysis: Probabilistic Models and Statistical Inference
By Ilia B. Frenkel, Alex Karagrigoriou, Anatoly Lisnianski, Andre V. Kleyner
September 2013

Design for Reliability
by Dev G. Raheja (Editor), Louis J. Gullo (Editor)
July 2012

Effective FMEAs: Achieving Safe, Reliable, and Economical Products and Process using Failure Mode and Effects Analysis
by Carl Carlson
April 2012

Failure Analysis: A Practical Guide for Manufactures of Electronic Components and Systems
by Marius Bazu, Titu Bajenescu
April 2011

Reliability Technology: Principles and Practice of Failure Prevention in Electronic Systems
by Norman Pascoe
April 2011

Improving Product Reliability: Strategies and Implementation
by Mark A. Levin, Ted T. Kalal
March 2003

Test Engineering: A Concise Guide to Cost-Effective Design, Development and Manufacture
by Patrick O'Connor
April 2001

Integrated Circuit Failure Analysis: A Guide to Preparation Techniques
by Friedrich Beck
January 1998

Measurement and Calibration Requirements for Quality Assurance to ISO 9000
by Alan S. Morris
October 1997

Electronic Component Reliability: Fundamentals, Modeling, Evaluation, and Assurance
by Finn Jensen
November 1995

Dynamic System Reliability

Modeling and Analysis of Dynamic and Dependent Behaviors

Liudong Xing
University of Massachusetts Dartmouth, USA

Gregory Levitin
The Israel Electric Corporation

Chaonan Wang
Jinan University, China

This edition first published 2019

Registered Offices
John Wiley & Sons, Inc., 111 River Street, Hoboken, NJ 07030, USA
John Wiley & Sons Ltd, The Atrium, Southern Gate, Chichester, West Sussex, PO19 8SQ, UK

Editorial Office
The Atrium, Southern Gate, Chichester, West Sussex, PO19 8SQ, UK

For details of our global editorial offices, customer services, and more information about Wiley products visit us at www.wiley.com.

Wiley also publishes its books in a variety of electronic formats and by print-on-demand. Some content that appears in standard print versions of this book may not be available in other formats.

Library of Congress Cataloging-in-Publication Data applied for
Hardback: 9781119507635

Cover design: Wiley
Cover image: © sakkmesterke/Shutterstock

Set in 10/12pt WarnockPro by SPi Global, Chennai, India
Printed in Singapore by C.O.S. Printers Pte Ltd

10 9 8 7 6 5 4 3 2 1

Contents

Foreword *ix*
Preface *xi*
Nomenclature *xv*

1 Introduction *1*
References *4*

2 Fundamental Reliability Theory *7*
2.1 Basic Probability Concepts *7*
2.1.1 Axioms of Probability *7*
2.1.2 Conditional Probability *7*
2.1.3 Total Probability Law *8*
2.1.4 Bayes' Theorem *9*
2.1.5 Random Variables *9*
2.2 Reliability Measures *10*
2.2.1 Time to Failure *11*
2.2.2 Failure Function *11*
2.2.3 Reliability Function *11*
2.2.4 Failure Rate *11*
2.2.5 Mean Time to Failure *11*
2.2.6 Mean Residual Life *12*
2.3 Fault Tree Modeling *12*
2.3.1 Static Fault Tree *13*
2.3.2 Dynamic Fault Tree *13*
2.3.3 Phased-Mission Fault Tree *14*
2.3.4 Multi-State Fault Tree *15*
2.4 Binary Decision Diagram *16*
2.4.1 Basic Concept *17*
2.4.2 ROBDD Generation *17*
2.4.3 ROBDD Evaluation *18*
2.4.4 Illustrative Example *19*
2.5 Markov Process *20*
2.6 Reliability Software *22*
References *22*

3 Imperfect Fault Coverage *27*
3.1 Different Types of IPC *27*
3.2 ELC Modeling *28*
3.3 Binary-State System *29*
3.3.1 BDD Expansion Method *29*
3.3.2 Simple and Efficient Algorithm *32*
3.4 Multi-State System *34*
3.4.1 MMDD-Based Method for MSS Analysis *35*
3.4.2 Illustrative Example *36*
3.5 Phased-Mission System *37*
3.5.1 Mini-Component Concept *37*
3.5.2 PMS SEA *38*
3.5.3 PMS BDD Method *40*
3.5.4 Summary of PMS SEA *42*
3.5.5 Illustrative Example *42*
3.6 Summary *43*
References *45*

4 Modular Imperfect Coverage *49*
4.1 Modular Imperfect Coverage Model *49*
4.2 Nonrepairable Hierarchical System *51*
4.3 Repairable Hierarchical System *55*
4.4 Summary *58*
References *58*

5 Functional Dependence *61*
5.1 Logic OR Replacement Method *61*
5.2 Combinatorial Algorithm *63*
5.2.1 Task 1: Addressing UFs of Independent Trigger Components *63*
5.2.2 Task 2: Generating Reduced Problems Without FDEP *63*
5.2.3 Task 3: Solving Reduced Reliability Problems *64*
5.2.3.1 Expansion Process *64*
5.2.3.2 Reduced FT Generation Procedure *65*
5.2.3.3 Dual Trigger-Basic Event Handling *65*
5.2.3.4 Evaluation of $P(\text{system fails}|ITE_i)$ *65*
5.2.4 Task 4: Integrating to Obtain Final System Unreliability *66*
5.2.5 Algorithm Summary *66*
5.2.6 Algorithm Complexity *66*
5.3 Case Study 1: Combined Trigger Event *67*
5.4 Case Study 2: Shared Dependent Event *70*
5.5 Case Study 3: Cascading FDEP *73*
5.5.1 Evaluation of $P(\text{system fails}|ITE_1)$ *74*
5.5.2 Evaluation of $P(\text{system fails}|ITE_2)$ *75*
5.5.3 Evaluation of UR_{system} *76*
5.6 Case Study 4: Dual Event and Cascading FDEP *76*
5.6.1 Evaluation of $P(\text{system fails}|ITE_1)$ *78*
5.6.2 Evaluation of UR_{system} *79*

5.7 Summary *79*
References *80*

6 Deterministic Common-Cause Failure *83*
6.1 Explicit Method *84*
6.1.1 Two-Step Method *84*
6.1.2 Illustrative Example *84*
6.2 Efficient Decomposition and Aggregation Approach *85*
6.2.1 Three-Step Method *86*
6.2.2 Illustrative Example *87*
6.3 Decision Diagram–Based Aggregation Method *89*
6.3.1 Three-Step Method *89*
6.3.2 Illustrative Example *91*
6.4 Universal Generating Function–Based Method *94*
6.4.1 System Model *94*
6.4.2 *u*-Function Method for Series-Parallel Systems *95*
6.4.3 *u*-Function Method for CCFs *97*
6.4.4 Illustrative Example *99*
6.5 Summary *104*
References *104*

7 Probabilistic Common-Cause Failure *107*
7.1 Single-Phase System *107*
7.1.1 Explicit Method *108*
7.1.2 Implicit Method *110*
7.1.3 Comparisons and Discussions *115*
7.2 Multi-Phase System *115*
7.2.1 Explicit Method *115*
7.2.2 Implicit Method *119*
7.2.3 Comparisons and Discussions *123*
7.3 Impact of PCCF *124*
7.4 Summary *125*
References *125*

8 Deterministic Competing Failure *127*
8.1 Overview *127*
8.2 PFGE Method *128*
8.2.1 *s*-Independent LF and PFGE *128*
8.2.2 *s*-Dependent LF and PFGE *128*
8.2.3 Disjoint LF and PFGE *129*
8.3 Single-Phase System with Single FDEP Group *129*
8.3.1 Combinatorial Method *129*
8.3.2 Case Study *131*
8.4 Single-Phase System with Multiple FDEP Groups *135*
8.4.1 Combinatorial Method *135*
8.4.2 Case Study *137*

8.5 Single-Phase System with PFs Having Global and Selective Effects *141*
8.5.1 Combinatorial Method *141*
8.5.2 Case Study *144*
8.6 Multi-Phase System with Single FDEP Group *150*
8.6.1 Combinatorial Method *150*
8.6.2 Case Study *153*
8.7 Multi-Phase System with Multiple FDEP Groups *158*
8.7.1 CTMC-Based Method *158*
8.7.2 Case Study *159*
8.8 Summary *166*
References *167*

9 Probabilistic Competing Failure *169*
9.1 Overview *169*
9.2 System with Single Type of Component Local Failures *170*
9.2.1 Combinatorial Method *170*
9.2.2 Case Study *172*
9.3 System with Multiple Types of Component Local Failures *181*
9.3.1 Combinatorial Method *181*
9.3.2 Case Study *182*
9.4 System with Random Failure Propagation Time *190*
9.4.1 Combinatorial Method *190*
9.4.2 Case Study: WSN System *192*
9.5 Summary *198*
References *199*

10 Dynamic Standby Sparing *201*
10.1 Types of Standby Systems *201*
10.2 CTMC-Based Method *202*
10.2.1 Cold Standby System *203*
10.2.2 Warm Standby System *204*
10.3 Decision Diagram–Based Method *205*
10.3.1 Cold Standby System *205*
10.3.2 Warm Standby System *208*
10.4 Approximation Method *211*
10.4.1 Homogeneous Cold Standby System *212*
10.4.2 Heterogeneous Cold Standby System *214*
10.5 Event Transition Method *216*
10.5.1 State-Space Representation of System Behavior *217*
10.5.2 Basic Steps *218*
10.5.3 Warm Standby System *218*
10.6 Overview of Optimization Problems *220*
10.7 Summary *222*
References *222*

Index *229*

Foreword by Dr. Andre Kleyner, Series Editor

"Dynamic System Reliability: Modeling and Analysis of Dynamic and Dependent Behaviors"

by Xing, Levitin and Wang

The importance of quality and reliability to a system can hardly be disputed. Product failures in the field inevitably lead to losses in the form of repair cost, warranty claims, customer dissatisfaction, product recalls, loss of sale, and in extreme cases, loss of life.

Engineering systems are becoming more and more complex with added functions and capabilities. Modeling of such complex systems, assessment of their performance, risk analysis and reliability prediction present an increasingly challenging task. Functional dependency, fault detection and coverage, common cause failures, redundancies, standby modes and other interactions among system components further complicate the modeling process requiring new methods and approaches to address the dynamic system reliability.

This book has been written by the leading experts in the field of dynamic reliability and multi-state systems. It discusses many technical aspects of modeling the reliability of complex systems when the reliabilities of their components change with time due to various types of interactions and state changes.

This book will be a great addition to the Wiley Series in Quality and Reliability Engineering, which aims to provide a solid educational foundation for researchers and practitioners in the field of quality and reliability engineering and to expand the knowledge base by including the latest developments in these disciplines.

Despite its obvious importance, quality and reliability education is paradoxically lacking in today's engineering curriculum. Few engineering schools offer degree programs or even a sufficient variety of courses in quality or reliability methods. Therefore, the majority of quality and reliability practitioners receive their professional training from colleagues, professional seminars, publications and technical books. The lack of formal education opportunities in this field greatly emphasizes the importance of technical publications for professional development.

We hope that this book, as well as the whole series, will continue Wiley's tradition of excellence in technical publishing and provide a lasting and positive contribution to the teaching and practice of engineering.

Preface

Dynamic behavior and dependence are typical characteristics of modern engineering and computing systems and products. Specifically, system load, stress levels, redundancy levels, and other operating environment parameters can be changing with time, causing dynamics in failure behavior of system components and in reliability requirements of the entire system. In addition, system components may have significant dependencies or correlations in time or function during the mission process. Modeling effects of these dynamic and dependent behaviors is crucial for accurate system reliability modeling and analysis, and further design optimization and maintenance activities.

Traditional system reliability models can define only the static logical structure of a system, but not the dynamic and dependent behaviors of the system and its components. Thus, reliability analysis results obtained using the traditional reliability models often deviate from the actual system reliability performance significantly, misleading system design, operation, and maintenance efforts. Therefore, the traditional reliability theory must be extended and enhanced for addressing the dependent and dynamic behaviors. This book presents recent developments of such extensions involving dynamic system reliability modeling theory, reliability evaluation methods, and case studies based on real-world examples.

The topic of the book "Dynamic System Reliability" has gained increasing attention in the reliability and safety community in the past few decades. Research articles on this subject are continuously being published in peer-reviewed journals and conference proceedings. However, to the best of the authors' knowledge, the subject has never been adequately or systematically included in any reliability book. Therefore, there is a great need for such a book covering recent developments on the dynamic system reliability modeling and analysis techniques. With an increased and sustained interest in this subject, it is the right time to publish this book.

This book particularly focuses on hot issues of dynamic system reliability, systematically introducing the reliability modeling and analysis methods for systems with imperfect fault coverage, systems with functional dependence, systems subject to deterministic or probabilistic common-cause failures, systems subject to deterministic or probabilistic competing failures, and dynamic standby sparing systems.

In the Introduction, the book describes the evolution from the traditional static reliability theory to the dynamic system reliability theory, and provides an overview description of dynamic and dependent behaviors addressed in the subsequent chapters of the book.

In Chapter 2, the book reviews basic probability and reliability concepts, various reliability measures, different types of fault trees, fundamentals of binary decision diagrams (a combinatorial model for system reliability analysis), and Markov processes. Some reliability analysis software tools are also introduced.

Chapter 3 introduces an inherent behavior of fault-tolerant systems called imperfect fault coverage. Just like any system component, the recovery mechanism of a system is hard to be perfect; it can fail such that the system cannot adequately detect, locate, isolate, or recover from a fault occurring in the system. The uncovered component fault may propagate through the system, causing extensive damage to the system. Reliability models and evaluation methods for addressing the imperfect fault coverage in binary-state systems, multi-state systems, and phased-mission systems are discussed in this chapter.

Chapter 4 discusses an extension of the traditional imperfect fault coverage concept to the modular imperfect fault coverage for systems with hierarchical structures. Due to the layered recovery of hierarchical systems, the extent of the damage from an uncovered component fault may exhibit multiple levels. This chapter introduces the modeling of such a modular imperfect fault coverage behavior as well as methods for considering the behavior in the reliability analysis of nonrepairable and repairable hierarchical systems.

Chapter 5 focuses on the functional dependence (Functional DEPendence, FDEP) behavior of complex systems, where the failure of one component (or in general the occurrence of a certain trigger event) causes other components (referred to as dependent components) within the same system to become unusable or inaccessible. The OR-gate replacement method is discussed for systems with perfect fault coverage. The combinatorial algorithm is discussed for systems with imperfect fault coverage. Case studies involving combined trigger events, cascading effects, dual-role events, and shared dependent events are also presented in this chapter.

Chapter 6 focuses on the reliability modeling of traditional deterministic common-cause failures, where the occurrence of a root cause results in deterministic failures of multiple system components simultaneously or in a short time interval. Methods based on Decomposition and Aggregation, Decision Diagrams, and Universal Generating Functions are discussed.

Chapter 7 discusses the extension of the traditional common-cause failures to the probabilistic common-cause failures, where the occurrence of a root cause results in failures of multiple system components with different probabilities. Both explicit and implicit methods are discussed for single-phase and multi-phase systems.

Chapter 8 presents the deterministic competing failure behavior in systems with the FDEP. This behavior is concerned with competitions in the time domain between the failure isolation and failure propagation effects, causing distinct system statuses. Reliability modeling of the deterministic competing effects is discussed for different types of systems, including single-phase systems with a single FDEP group, single-phase systems with multiple FDEP groups, single-phase systems with both global and selective effects, multi-phase systems with a single FDEP group, and multi-phase systems with multiple FDEP groups.

Chapter 9 focuses on probabilistic competing failures, which extend the deterministic competing failure behavior by considering probabilistic or uncertain failure isolation effects (commonly found in systems involving relayed wireless communications). Systems with a single type of local component failures, multiple different types of local component failures, and random propagation times are modeled and illustrated with real-world examples from wireless sensor networks, body sensor systems, and smart homes.

Chapter 10 presents diverse methods for the reliability analysis of standby sparing systems, including the traditional Markov-based method, the decision diagrams–based method, the approximation method based on the central limit theorem, and the recently developed event transition method.

The book has the following distinct features:

- It is the first book systematically focusing on dynamic system reliability modeling and analysis theory.
- It provides a comprehensive treatment on imperfect fault coverage (single-level or multi-level/modular), functional dependence, common-cause failures (deterministic or probabilistic), competing failures (deterministic or probabilistic), and dynamic standby sparing.
- It includes abundant illustrative examples and case studies based on real-world systems.
- It covers recent advances in combinatorial models and algorithms for dynamic system reliability analysis.
- It has a rich set of references, providing helpful resources for readers to pursue further study and research of the subjects.

The target audience of the book is undergraduate and graduate students, engineers and researchers in reliability and related disciplines. The readers should have a background in basic probability theory and stochastic processes. However, the book includes a chapter reviewing the fundamentals that the readers need to know for understanding the contents of the other chapters, covering advanced topics in reliability theory and case studies. The book can provide the readers with knowledge and insights on complex system reliability behaviors, as well as skills of modeling and analyzing these behaviors for guiding reliability design of real-world systems.

We would like to extend our sincere gratitude and appreciation to researchers who have developed some underlying concepts and models of this book, or have co-authored with us on some subjects of the book, to name a few, Professor Joanne Bechta Dugan and Professor Barry W. Johnson from the University of Virginia, Professor Kishor S. Trivedi from Duke University, Dr. Suprasad V. Amari from BAE Systems, USA, Dr. Akhilesh Shrestha from Autoliv Inc., USA, Dr. Ola Tannous from Illinois Institute of Technology, USA, Dr. Prashanthi Boddu from Global Prior Art Inc., USA, Dr. Yujie Wang from the University of Electronic Science and Technology of China, Ms. Guilin Zhao from the University of Massachusetts Dartmouth, USA, Professor Yuchang Mo from Huaqiao University, China, and Professor Rui Peng from the University of Science and Technology Beijing, China. There are many other researchers to mention. We have tried to recognize their contributions in the bibliographical references of this book.

Finally, it is our great pleasure to work with the editorial staff from Wiley, who have assisted in the publication of this book, their efforts and support are greatly appreciated.

June 8, 2018

Liudong Xing
Gregory Levitin
Chaonan Wang

Nomenclature

ACP	Application Communication Phase
BDD	Binary Decision Diagram
BEM	BDD Expansion Method
BSN	Body Sensor Network
CC	Common Cause
CCE	Common-Cause Event
CCF	Common-Cause Failure
CCG	Common-Cause Group
cdf	cumulative distribution function
CLT	Central Limit Theorem
CM	Computing Module
CPR	Combinatorial Phase Requirement
CPUC	CPU Chip
CSP	Cold SPare
CTE	Combined Trigger Event
CTMC	Continuous Time Markov Chain
DC	Dependent Component
DD	Decision Diagram
DFT	Dynamic Fault Tree
EDA	Efficient Decomposition and Aggregation
ELC	Element Level Coverage
EMB	External Memory Block
FCE	Failure Competition Event
FDEP	Functional DEPendence
FDG	Functional Dependence Group
FLC	Fault Level Coverage
FT	Fault Tree
FTS	Fault Tolerant System
HS	Hierarchical System
HSP	Hot SPare
IC	Interface Chip
ICP	Infrastructure Communication Phase
IFG	Isolation Factor Group
i.i.d.	independent and identically distributed
IoT	Internet of Things

IPC	ImPerfect Coverage
IPCM	IPC Model
ite	if-then-else
ITE	Independent Trigger Event
LF	Local Failure
MC	Memory Chip
MFT	Multi-state Fault Tree
MIPCM	Modular IPCM
MIU	Memory Interface Unit
MM	Memory Module
MMDD	Multi-state Multi-valued Decision Diagram
MRL	Mean Residual Life
MSS	Multi-State System
MTBF	Mean Time Between Failures
MTTF	Mean Time To Failure
MTTR	Mean Time To Repair
NDC	NonDependent Component
OBDD	Ordered BDD
PAND	Priority AND
PCCE	Probabilistic Common-Cause Event
PCCF	Probabilistic Common-Cause Failure
PCCG	Probabilistic Common-Cause Group
PDC	Performance Dependent Coverage
PDEP	Probabilistic-DEPendent
pdf	probability density function
PDO	Phase Dependent Operation
PF	Propagated Failure
PFD	Probabilistic Functional Dependence
PFDC	Probabilistic Functional Dependence Case
PFGE	Propagated Failure with Global Effect
PFSE	Propagated Failure with Selective Effect
pmf	probability mass function
PMS	Phased-Mission System
PTC	PorT Chip
RAP	Redundancy Allocation Problem
ROBDD	Reduced OBDD
r.v.	random variable
SBDD	Sequential BDD
SEA	Simple and Efficient Algorithm
SEQ	SEQquence enforcing
SESP	Standby Element Sequencing Problem
SFT	Static Fault Tree
ttf	time to failure
UF	Uncovered Failure
u-function	universal generating function
WSN	Wireless Sensor Network
WSP	Warm SPare

1

Introduction

The advances and interdisciplinary integration of science and technology are making modern engineering and computing systems more and more complex. For modern systems (especially those in, e.g. wireless sensor networks, Internet of Things (IoT), smart power systems, space explorations, and cloud computing industries), dynamic behavior and dependence are typical characteristics of the systems or products. System load, operating conditions, stress levels, redundancy levels, and other operating environment parameters are variables of time, causing dynamic failure behavior of the system components as well as dynamic system reliability requirements. In addition, components of these systems often have significant interactions or dependencies in time or functions. Effects of these dynamic and dependent behaviors must be addressed for accurate system reliability modeling and analysis, which is crucial for verifying whether a system satisfies desired reliability requirements and for determining optimal design and operation policies balancing different system parameters like cost and reliability. As a result, reliability modeling and analysis of modern dynamic systems become more challenging than ever.

Traditional reliability modeling methods, such as reliability block diagram [1] and fault tree analysis [2], can define the static logical structure of the system, but they lack the ability to describe dynamic state transfers of the system, and component fault dependencies and propagations. It is difficult or impossible to accurately reflect the actual behavior of modern complex fault-tolerant systems using the traditional reliability models. In other words, failure to address effects of dynamic behavior and dependencies of modem systems makes the reliability analysis results obtained using the traditional reliability models far from the actual system reliability performance, misleading the system design, operation, and maintenance efforts.

Different from the traditional static reliability modeling, the dynamic reliability theory considers that a system failure depends not only on the static logical combination of basic component failure events, but also on the timing of the occurrence of the events, correlations or interrelationship of the events, and impacts of operating environments. Therefore, the dynamic system reliability theory can provide a more accurate representation of actual complex system behavior, more effectively guiding the reliable design of real-world critical systems. The dynamic system reliability theory is the evolution and improvement of the traditional reliability modeling theory, and its research will promote the development and application of complex systems engineering.

This book focuses on dynamic reliability modeling of fault-tolerant systems with imperfect fault coverage, functional dependence, deterministic or probabilistic

Dynamic System Reliability: Modeling and Analysis of Dynamic and Dependent Behaviors,
First Edition. Liudong Xing, Gregory Levitin and Chaonan Wang.

common-cause failures, deterministic or probabilistic competing failures, as well as standby sparing.

Specifically, imperfect fault coverage is an inherent behavior of fault-tolerant systems designed with redundancies and automatic system recovery or reconfiguration mechanisms [3–5]. Just like any system component, the system recovery mechanisms involving fault detection, fault location, fault isolation, and fault recovery will likely not be perfect; they can fail such that the system cannot adequately detect, locate, isolate, or recover from a fault occurring in the system. The uncovered component fault may propagate through the system, causing an extensive damage to the system, sometimes failure of the entire system. Further, it is observed that the extent of the damage from an uncovered component fault occurring in a system with the hierarchical nature may exhibit multiple levels due to the layered recovery [6]. The traditional imperfect fault coverage concept has been extended to the modular imperfect fault coverage to model multiple levels of uncovered failure modes for components in hierarchical systems [7].

Functional dependence occurs in systems where the failure of one component (or, in general, the occurrence of a certain trigger event) causes other components (referred to as *dependent components*) within the same system to become unusable or inaccessible. A classic example is a computer network where computers can access the Internet through routers [8]. If the router fails, all computers connected to the router become inaccessible. It is said that these computers have functional dependence on the router.

In the case of systems with perfect fault coverage, the functional dependence behavior can be addressed as logic OR relationship [9]. However, for systems with imperfect fault coverage, the logic OR replacement method can lead to overestimation of system unreliability because it allows the disconnected dependent components (in the case of the trigger event occurring) to contribute to the system uncovered failure probability if they can fail uncovered. However, since these dependent components were disconnected or isolated, they could really not generate propagation effect or bring the system down [10]. New algorithms are required for addressing the coupled functional dependence and imperfect fault coverage behavior.

In addition to the imperfect fault coverage, common-cause failures are another class of behavior that can contribute significantly to the overall system unreliability [11–13]. Common-cause failures are defined as "A subset of dependent events in which two or more component fault states exist at the same time, or in a short time interval, and are direct results of a shared cause" [11]. Most of the traditional common-cause failure models assumed the deterministic failure of the multiple components affected by the shared root cause. Recent studies extended the concept to model probabilistic common-cause failures, where the occurrence of a root cause results in failures of multiple system components with different probabilities [14–16].

As one type of common-cause failures, a propagated failure with global effect (PFGE) originating from a system component can cause the failure of the entire system [17]. Such a failure can occur due to the imperfect fault coverage or destructive effect of a component failure on other system components (like overheating, explosion, etc.). However, PFGE may not always cause the overall system failure in systems with functional dependence behavior. Specifically, if the trigger event occurs before PFGEs of all the dependent components, these PFGEs can be isolated deterministically and thus cannot affect other parts of the system. On the other hand, if PFGE of any dependent component occurs before the trigger event, the failure propagation effect takes place, crashing the entire system. Therefore, there exist competitions in the time domain between the failure isolation and failure propagation effects, causing distinct system statuses [18, 19].

The pioneering works on addressing such competing failures in system with functional dependence have focused on deterministic competing failures, where the occurrence of the trigger event, as long as it happens first, can cause deterministic or certain isolation effect to any failures originating from the corresponding dependent components. Recent studies [20, 21] have revealed that in some real-world systems, e.g. systems involving relayed wireless communications, the failure isolation effect can be probabilistic or uncertain. Consider a specific example of a relay-assisted wireless sensor network where some sensors preferably deliver their sensed information to the sink device through a relay node due to wireless signal attenuation. These sensors have functional dependence on the relay node. However, unlike in the deterministic competing failure case, when the relay fails, each sensor is not necessarily isolated because it may increase transmission power to be wirelessly connected to the sink device with certain probability dependent on the percentage of remaining energy. A sensor is isolated only when its remaining energy is not sufficient to enable the direct transmission to the sink node. Similarly, there exist time-domain competitions between the probabilistic failure isolation effect and the failure propagation effect that can lead to dramatically different system statuses. The modeling of such probabilistic competing failures is naturally more complicated than modeling the deterministic competing failure behavior.

Another common dynamic behavior of modern systems, especially life or mission-critical systems requiring fault-tolerance and high-level of system reliability, is standby sparing. In the standby sparing systems, one or several units are online and operating while some redundant units serve as standby spares, which are activated to resume the system mission in the case of the online unit malfunction occurring [3]. Components in the standby sparing systems often exhibit dynamic failure behaviors; they have different failure rates before and after they are activated to replace the failed online component [22–26].

The above described dynamic behaviors abound in real-world systems, as detailed in case studies in subsequent chapters. Due to the existence of these dynamic behaviors, not only the system structure function is seriously affected, but also the system reliability modeling and analysis become more complicated. Ignoring the dynamic and dependence of failures, or simply performing system reliability analysis under the assumption that components behave independently of each other, often leads to excessive errors and even draws wrong conclusions. The following chapters present models and methods to address effects of the dynamic and dependent behaviors for different types of systems, covering binary-state and multi-state systems, single-phase, and multi-phase systems.

The traditional reliability models are mostly applicable to binary-state systems in which both the system and its components assume two and only two states (operation and failure). However, many practical systems are multi-state systems [27–30], such as those involving imperfect fault coverage, standby sparing, multiple failure modes [31], work sharing [32], load sharing [33], performance sharing [34, 35], performance degradation, and limited repair resources [36]. In these systems, both the system and its components can exhibit multiple states or performance levels varying from perfect function to complete failure. The nonbinary property and dependencies among different states of the same component must be addressed in modeling a multi-state system.

In addition to addressing effects of the dynamic behavior for reliability modeling and analysis of multi-state systems, this book considers multi-phase systems, also known as phased-mission systems. Traditional system reliability models generally assume that a system under study performs a single phased mission, during which the system does not change its task and configuration [37]. Due to an increased use of automation in diverse industries such as airborne weapon systems, aerospace, nuclear power, and communication networks, phased-mission systems have become a more appropriate and accurate model for many reliability problems since the 1970s [38, 39]. These systems perform a mission that involves multiple and consecutive phases with possibly different durations. During each phase, the system has to accomplish a specified and often different task. In addition, the system can be subject to different stress levels, environmental conditions, and reliability requirements. Thus, the system configuration, success criteria (structure function), and component behavior may vary from phase to phase [13, 40]. These dynamics as well as statistical dependence across different phases for a given component make reliability modeling and analysis of multi-phase systems more difficult than single-phase systems.

In summary, dynamic reliability models and methods are presented in this book to address effects of single-level or multi-level (modular) imperfect fault coverage, functional dependence, deterministic or probabilistic common-cause failures, deterministic or probabilistic competing failures, standby sparing, multi-state, and multi-phase behaviors.

References

1 Rausand, M. and Hoyland, A. (2003). *System Reliability Theory: Models, Statistical Methods, and Applications*, 2e. Wiley Inter-Science.

2 Dugan, J.B. and Doyle, S.A. (1996). New Results in Fault-Tree Analysis. In: *Tutorial Notes of Annual Reliability and Maintainability Symposium*, Las Vegas, Nevada, USA.

3 Johnson, B.W. (1989). *Design and Analysis of Fault Tolerant Digital Systems*. Addison-Wesley.

4 Arnold, T.F. (1973). The concept of coverage and its effect on the reliability model of a repairable system. *IEEE Transactions on Computers* C-22: 325–339.

5 Dugan, J.B. (1989). Fault trees and imperfect coverage. *IEEE Transactions on Reliability* 38 (2): 177–185.

6 Xing, L. and Dugan, J.B. (2001). Dependability analysis of hierarchical systems with modular imperfect coverage. In: *Proceedings of The 19th International System Safety Conference*, 347–356. Huntsville, AL.

7 Xing, L. (2005). Reliability modeling and analysis of complex hierarchical systems. *International Journal of Reliability, Quality and Safety Engineering* 12 (6): 477–492.

8 Xing, L., Levitin, G., Wang, C., and Dai, Y. (2013). Reliability of systems subject to failures with dependent propagation effect. *IEEE Transactions Systems, Man, and Cybernetics: Systems* 43 (2): 277–290.

9 Merle, G., Roussel, J.M., and Lesage, J.J. (2010). Improving the Efficiency of Dynamic Fault Tree Analysis by Considering Gates FDEP as Static. In: *Proceeding of European Safety and Reliability Conference*, Rhodes, Greece.

10 Xing, L., Morrissette, B.A., and Dugan, J.B. (2014). Combinatorial reliability analysis of imperfect coverage systems subject to functional dependence. *IEEE Transactions on Reliability* 63 (1): 367–382.
11 NUREG/CR-4780. (1988). Procedure for Treating Common-Cause Failures in Safety and Reliability Studies. *U.S. Nuclear Regulatory Commission*; vol. I and II, Washington DC, USA.
12 Fleming, K.N., Mosleh, A., and Kelly, A.P. (1983). On the analysis of dependent failures in risk assessment and reliability evaluation. *Nuclear Safety* 24: 637–657.
13 Xing, L. and Levitin, G. (2013). BDD-based reliability evaluation of phased-mission systems with internal/external common-cause failures. *Reliability Engineering & System Safety* 112: 145–153.
14 Xing, L. and Wang, W. (2008). Probabilistic common-cause failures analysis. In: *Proceedings of the Annual Reliability and Maintainability Symposium, Las Vagas, Nevada* 354–358.
15 Xing, L., Boddu, P., Sun, Y., and Wang, W. (2010). Reliability analysis of static and dynamic fault-tolerant systems subject to probabilistic common-cause failures. *Proc. IMechE, Part O: Journal of Risk and Reliability* 224 (1): 43–53.
16 Wang, C., Xing, L., and Levitin, G. (2014). Explicit and implicit methods for probabilistic common-cause failure analysis. *Reliability Engineering & System Safety* 131: 175–184.
17 Xing, L. and Levitin, G. (2010). Combinatorial analysis of systems with competing failures subject to failure isolation and propagation effects. *Reliability Engineering & System Safety* 95 (11): 1210–1215.
18 Xing, L., Wang, C., and Levitin, G. (2012). Competing failure analysis in non-repairable binary systems subject to functional dependence. *Proc IMechE, Part O: Journal of Risk and Reliability* 226 (4): 406–416.
19 Wang, C., Xing, L., and Levitin, G. (2012). Competing failure analysis in phased-mission systems with functional dependence in one of phases. *Reliability Engineering & System Safety* 108: 90–99.
20 Wang, Y., Xing, L., Wang, H., and Levitin, G. (2015). Combinatorial analysis of body sensor networks subject to probabilistic competing failures. *Reliability Engineering & System Safety* 142: 388–398.
21 Wang, Y., Xing, L., and Wang, H. (2017). Reliability of systems subject to competing failure propagation and probabilistic failure isolation. *International Journal of Systems Science: Operations & Logistics* 4 (3): 241–259.
22 Xing, L., Tannous, O., and Dugan, J.B. (2012). Reliability analysis of non-repairable cold-standby systems using sequential binary decision diagrams. *IEEE Transactions on Systems, Man, and Cybernetics, Part A: Systems and Humans* 42 (3): 715–726.
23 Zhai, Q., Peng, R., Xing, L., and Yang, J. (2013). BDD-based reliability evaluation of k-out-of-(n+k) warm standby systems subject to fault-level coverage. *Proc IMechE, Part O, Journal of Risk and Reliability* 227 (5): 540–548.
24 Levitin, G., Xing, L., and Dai, Y. (2013). Optimal sequencing of warm standby elements. *Computers & Industrial Engineering* 65 (4): 570–576.
25 Levitin, G., Xing, L., and Dai, Y. (2014). Cold vs. hot standby mission operation cost minimization for 1-out-of-N systems. *European Journal of Operational Research* 234 (1): 155–162.

26 Levitin, G., Xing, L., and Dai, Y. (2014). Mission cost and reliability of 1-out-of-N warm standby systems with imperfect switching mechanisms. *IEEE Transactions on Systems, Man, and Cybernetics: Systems* 44 (9): 1262–1271.
27 Zang, X., Wang, D., Sun, H., and Trivedi, K.S. (2003). A BDD-based algorithm for analysis of multistate systems with multistate components. *IEEE Transactions on Computers* 52 (12): 1608–1618.
28 Caldarola, L. (1980). Coherent systems with multistate components. *Nuclear Engineering and Design* 58 (1): 127–139.
29 Xing, L. and Dai, Y. (2009). A new decision diagram based method for efficient analysis on multi-state systems. *IEEE Transactions on Dependable and Secure Computing* 6 (3): 161–174.
30 Lisnianski, A. and Levitin, G. (2003). *Multi-state System Reliability: Assessment, Optimization and Applications*. World Scientific.
31 Mo, Y., Xing, L., and Dugan, J.B. (2014). MDD-based method for efficient analysis on phased-mission systems with multimode failures. *IEEE Transactions on Systems, Man, and Cybernetics: Systems* 44 (6): 757–769.
32 Levitin, G., Xing, L., Ben-Haim, H., and Dai, Y. (2016). Optimal task partition and state-dependent loading in heterogeneous two-element work sharing system. *Reliability Engineering & System Safety* 156: 97–108.
33 Kvam, P.H. and Pena, E.A. (2005). Estimating load-sharing properties in a dynamic reliability system. *Publications of the American Statistical Association* 100 (469): 262–272.
34 Levitin, G. (2011). Reliability of multi-state systems with common bus performance sharing. *IIE Transactions* 43 (7): 518–524.
35 Yu, H., Yang, J., and Mo, H. (2014). Reliability analysis of repairable multi-state system with common bus performance sharing. *Reliability Engineering & System Safety* 132: 90–96.
36 Amari, S.V., Xing, L., Shrestha, A. et al. (2010). Performability analysis of multi-state computing systems using multi-valued decision diagrams. *IEEE Transactions on Computers* 59 (10): 1419–1433.
37 Ma, Y. and Trivedi, K.S. (1999). An algorithm for reliability analysis of phased-mission systems. *Reliability Engineering & System Safety* 66 (2): 157–170.
38 Esary, J.D. and Ziehms, H. (1975). Reliability analysis of phased missions. In: *Reliability and Fault Tree Analysis: Theoretical and Applied Aspects of System Reliability and Safety Assessment*, 213–236. Philadelphia: SIAM.
39 Burdick, G.R., Fussell, J.B., Rasmuson, D.M., and Wilson, J.R. (1977). Phased mission analysis: a review of new developments and an application. *IEEE Transactions on Reliability* R-26 (1): 43–49.
40 Shrestha, A., Xing, L., and Dai, Y. (2011). Reliability analysis of multi-state phased-mission systems with unordered and ordered states. *IEEE Transactions on Systems, Man, and Cybernetics, Part A: Systems and Humans* 41 (4): 625–636.

2

Fundamental Reliability Theory

This chapter covers basic probability concepts, reliability measures, fault tree (FT) modeling, binary decision diagrams (BDDs), and Markov processes. Some reliability analysis software tools are also introduced.

2.1 Basic Probability Concepts

Random experiment is an experiment with its outcome being unknown ahead of time but all of its possible individual outcomes being known [1]. The set of all possible outcomes of a random experiment constitutes its *sample space*, denoted by Ω. Each individual outcome in the sample space is referred to as a *sample point*.

A subset of a sample space associated with a random experiment is defined as an *event* that occurs if the experiment is performed and the outcome observed is in the subset defining the event. There are two special events: a *certain event* (sample space itself Ω) that occurs with probability ONE (1) and an *impossible event* (empty set $\emptyset$) that occurs with probability ZERO (0). Because events are sets, the operations in the set theory like complement, union, and intersection can be applied to generate new events.

If two events A and B do not share any common sample points, i.e. $A \cap B = \emptyset$, then A and B are said to be *disjoint* or *mutually exclusive*.

2.1.1 Axioms of Probability

Let E denote a random event of interest. A real number $P(E)$ defined for E is refer to as the *probability* of event E if it satisfies the following three axioms [1]:

A1: $0 \leq P(E) \leq 1$,
A2: $P(\Omega) = 1$ (with probability ONE the outcome is a point in Ω),
A3: for any sequence of pair-wise disjoint events $E_1, E_2, \ldots$ (i.e. $E_i \cap E_k = \emptyset$ for any $i \neq k$), $P(\bigcup_{i=1}^{\infty} E_i) = \sum_{i=1}^{\infty} P(E_i)$ (the probability that at least one of them occurs is the sum of their individual probabilities).

2.1.2 Conditional Probability

Consider two events A and B. The *conditional probability* of event A given the occurrence of event B, denoted by $P(A|B)$, is a number such that $0 \leq P(A|B) \leq 1$ and $P(A \cap$

Dynamic System Reliability: Modeling and Analysis of Dynamic and Dependent Behaviors, First Edition. Liudong Xing, Gregory Levitin and Chaonan Wang.

$B) = P(B)P(A|B)$. In the case of $P(B) \neq 0$, Eq. (2.1) presents a commonly used formula for evaluating the conditional probability.

$$P(A|B) = \frac{P(A \cap B)}{P(B)}. \tag{2.1}$$

If $P(A \cap B) = P(A)P(B)$, implying $P(A|B) = P(A)$ and $P(B|A) = P(B)$ (neither event influences the occurrence of the other event), then events A and B are said to be independent. Note that independence does not mean that the two events cannot share common sample points. In other words, two independent events are not necessarily two mutually exclusive events. Mutually exclusive events are independent only when at least one of them has ZERO occurrence probability.

2.1.3 Total Probability Law

A collection of events $\{B_i\}_{i=1}^n$ forms a *partition* of sample place Ω if it satisfies the following three conditions:

1) $B_i \cap B_j = \emptyset$ for any $i \neq j$;
2) $P(B_i) > 0$ for $i = 1, 2, \ldots, n$;
3) $\bigcup_{i=1}^n B_i = \Omega$.

Based on partition $\{B_i\}_{i=1}^n$, for any event E defined on the same sample space Ω, its occurrence probability can be evaluated as

$$P(E) = \sum_{i=1}^{n} P(E|B_i)P(B_i). \tag{2.2}$$

Equation (2.3) presents a special case of the total probability law since for any event B, B and $\overline{B}$ form a partition of Ω.

$$P(E) = P(E|B)P(B) + P(E|\overline{B})P(\overline{B}). \tag{2.3}$$

Example 2.1 An on-line computer has three incoming communication lines with properties described in Table 2.1. Define E as an event that a chosen data is received without errors, B_i as an event that a chosen data is from line i ($i = 1, 2, 3$). Thus, $P(B_1) = 0.2$, $P(B_2) = 0.3$, $P(B_3) = 0.5$, $P(E|B_1) = 0.98$, $P(E|B_2) = 0.96$, $P(E|B_3) = 0.99$. Since B_1, B_2, and B_3 form a partition of the sample space, the total probability law (2.2) is applied to obtain the probability that a randomly selected data is free of error, $P(E)$, as:

$$P(E) = P(E|B_1)P(B_1) + P(E|B_2)P(B_2) + P(E|B_3)P(B_3) = 0.979.$$

Table 2.1 Properties of each communication line.

Line *i*	Probability of data from line *i*	Probability of data received without error given that it is from line *i*
1	0.2	0.98
2	0.3	0.96
3	0.5	0.99

2.1.4 Bayes' Theorem

Based on the total probability law and concept of condition probability, Bayes' theorem is

$$P(B_j|E) = \frac{P(E|B_j)P(B_j)}{P(E)} = \frac{P(E|B_j)P(B_j)}{\sum_{i=1}^{n} P(E|B_i)P(B_i)}. \tag{2.4}$$

Example 2.2 Consider the online computer system in Example 2.1. Suppose data selected at random are found to be free of error. To find the probability that the data came from communication line 2 (i.e. $P(B_2 \mid E)$), Eq. (2.4) is applied to obtain

$$P(B_2|E) = \frac{P(E|B_2)P(B_2)}{P(E)} = 0.294.$$

2.1.5 Random Variables

A random variable (*r.v.*) X is a real-valued function, which maps each individual outcome ω in some sample space Ω to a real number $X(\omega) \in R$, i.e. $X: \Omega \to R$.

A cumulative distribution function (*cdf*) F can be defined for X for a real number a as

$$F_X(a) = P(\omega : \omega \in \Omega\ AND\ X(\omega) \leq a) = P(X \leq a). \tag{2.5}$$

The *cdf* F is a nondecreasing function, that is, if $a < b$ then $F(a) \leq F(b)$. Also, for any $a < b$, $P(a < X \leq b) = F(b) - F(a)$.

A *r.v.* can be classified as *discrete* or *continuous.* A discrete *r.v.* can take on a countable number of possible values. In other words, the image of a discrete *r.v.* X is a finite or countably infinite subset of real numbers, denoted as $T = \{x_1, x_2, \ldots\}$. Besides *cdf*, a discrete *r.v.* X can be characterized by a function called probability mass function (*pmf*) defined as

$$p_X(a) = P(X = a) = P(\omega : \omega \in \Omega | X(\omega) = a). \tag{2.6}$$

The *pmf* has the following properties: $0 \leq p_X(a) \leq 1$ and $\sum_{a \in T} p_X(a) = 1$.

A continuous *r.v.* can take on a range of uncountable real values. In general, the time when a particular event happens is a continuous *r.v.* Besides *cdf*, a continuous *r.v.* X can be characterized by a function called probability density function (*pdf*) defined as

$$f_X(x) = F'_X(x) = \frac{dF_X(x)}{dx}. \tag{2.7}$$

The *pdf* $f_X(x)$ is integrable and for any real numbers $a < b$, $\int_a^b f_X(x)dx = P(a \leq X \leq b) = F_X(b) - F_X(a)$. For any real number a, $F_X(a) = P(X \leq a) = \int_{-\infty}^{a} f_X(x)dx$. Also, $F_X(\infty) = \int_{-\infty}^{\infty} f_X(x)dx = 1$.

The *mean* (also known as *expected value*) of a *r.v.* represents the long-run average of the variable or the expected average outcome over many observations. Equation (2.8) presents the formula calculating the mean of a discrete *r.v.* with *pmf* $p_X(a)$.

$$\mu = \mathrm{E}[X] = \sum_{a \in T} [a \cdot p_X(a)]. \tag{2.8}$$

Equation (2.9) presents the formula calculating the mean of a continuous *r.v.* with *pdf* $f_X(x)$.

$$\mu = \mathrm{E}[X] = \int_{-\infty}^{\infty} x f_X(x) dx. \tag{2.9}$$

The *variance* of a *r.v.* measures its statistical dispersion, indicating how far its values typically are from the expected value. For a *r.v.* X with mean μ, the variance of X is defined by

$$\mathrm{Var}(X) = \sigma^2 = \mathrm{E}[(X - \mu)^2]. \tag{2.10}$$

Equation (2.11) presents an alternative formula for computing the variance of X.

$$\begin{aligned}\mathrm{Var}[X] &= \mathrm{E}[(X - \mu)^2] = \mathrm{E}[X^2 - 2\mu X + \mu^2] \\ &= \mathrm{E}[X^2] - 2\mu \mathrm{E}[X] + \mathrm{E}[\mu^2] \\ &= \mathrm{E}[X^2] - 2\mu^2 + \mu^2 = \mathrm{E}[X^2] - \mu^2.\end{aligned} \tag{2.11}$$

The *standard deviation* of a *r.v.* is typically denoted by σ, which is simply the square root of the variance.

Example 2.3 As an illustration, one of the most widely used continuous *r.v.* in reliability engineering, the exponential *r.v.*, is explained.

A continuous *r.v.* X has an *exponential* distribution with parameter λ if its *pdf* has the form of

$$f_X(x) = \begin{cases} \lambda e^{-\lambda x} & x \geq 0, \\ 0 & x < 0. \end{cases} \tag{2.12}$$

Its *cdf* is thus

$$F_X(x) = \int_{-\infty}^{x} f_X(u) du = \begin{cases} 1 - e^{-\lambda x} & x \geq 0, \\ 0 & x < 0. \end{cases} \tag{2.13}$$

According to (2.9) and (2.10), the mean and variance of exponential *r.v.* X are, respectively,

$$\mathrm{E}[X] = 1/\lambda, \mathrm{Var}[X] = 1/\lambda^2. \tag{2.14}$$

The *exponential* distribution has the memoryless property defined by the following equation:

$$P(X > t + h | X > t) = P(X > h) \quad \forall t, h > 0. \tag{2.15}$$

2.2 Reliability Measures

Several quantitative reliability measures for a nonrepairable unit are presented, including the failure function $F(t)$, reliability function $R(t)$, failure rate function $z(t)$, mean time to failure (MTTF), and mean residual life (MRL). Their definitions are based on a continuous *r.v.* called *time to failure (ttf)*, which is defined in Section 2.2.1.

2.2.1 Time to Failure

The time to failure T is a continuous *r.v.* defined as the time elapsing from when the unit is first put into function until its first failure. It models the lifetime of a nonrepairable unit.

2.2.2 Failure Function

The failure function of the unit is given as the *cdf* of *r.v* T, that is

$$F(t) = P(T \leq t) = \int_0^t f(x)dx, \tag{2.16}$$

where $f(x)$ is the *pdf* of T. $F(t)$ gives the probability that the unit fails within time interval $(0, t]$.

Example 2.4 If the *ttf* of a unit follows an exponential distribution with parameter λ, then according to (2.13), its failure function (i.e. failure probability or unreliability) at time t is $F(t) = 1 - e^{-\lambda t}$.

2.2.3 Reliability Function

The reliability function (also known as the survival function) of a unit at time $t > 0$ is defined as

$$R(t) = 1 - F(t) = P(T > t) = \int_t^{\infty} f(x)dx, \tag{2.17}$$

It is the probability that the unit does not fail in time interval $(0, t]$, or the probability that the unit has survived interval $(0, t]$ and is still working at time t.

For the unit with the exponential *ttf* distribution in Example 2.4, its reliability function at time t is $R(t) = e^{-\lambda t}$.

2.2.4 Failure Rate

The failure rate function (also known as hazard rate function) of a unit measures the instantaneous speed of its failure, defined as

$$\begin{aligned} h(t) &= \lim_{\Delta t \to 0} \frac{P\{t < T \leq t + \Delta t | T > t\}}{\Delta t} \\ &= \lim_{\Delta t \to 0} \frac{F(t + \Delta t) - F(t)}{R(t)\Delta t} = \frac{f(t)}{R(t)}. \end{aligned} \tag{2.18}$$

For the unit with the exponential *ttf* distribution in Example 2.4, its failure rate function is constant, $h(t) = \lambda$.

2.2.5 Mean Time to Failure

MTTF is the expected or mean time that a unit will operate before its first failure, which is computed as

$$\text{MTTF} = \text{E}[T] = \int_0^{\infty} tf(t)dt. \tag{2.19}$$

Equation (2.20) gives another equivalent formula for computing MTTF:

$$\text{MTTF} = \int_0^\infty R(t)dt. \tag{2.20}$$

For the unit with the exponential *ttf* distribution in Example 2.4, $\text{MTTF} = \frac{1}{\lambda}$. The probability that the unit will survive its MTTF is $R(\text{MTTF}) = e^{-1} = 0.36788$.

For a repairable unit, the mean time between failures (MTBF) is typically of interest. It is computed as MTBF = MTTF + MTTR, where MTTR represents the mean time to repair the unit after it is failed. In case of a unit's MTTR being very short or negligible compared to its MTTF, the MTTF can be used to approximate the MTBF of the repairable unit.

2.2.6 Mean Residual Life

The MRL at time (or age) t is the mean remaining lifetime of a unit given that the unit has survived time interval $(0, t]$, which is evaluated as

$$\text{MRL}(t) = \int_0^\infty R(x|t)dx = \int_0^\infty \frac{R(x+t)}{R(t)}dx = \frac{1}{R(t)}\int_t^\infty R(x)dx. \tag{2.21}$$

When the unit is brand new (i.e. age is 0), the MRL at this age is simply MTTF, i.e., MRL(0) = MTTF.

Consider the unit with the exponential *ttf* distribution in Example 2.4. Because of the memoryless property of the exponential distribution, $\text{MRL}(t) = \text{MTTF}$ irrespective of the unit's age t, implying that the unit is statistically as good as new as long as the unit is still functioning.

2.3 Fault Tree Modeling

The FT technique was first introduced by Watson at Bell Telephone Laboratories in the 1960s for facilitating an analysis of a launch control system of the Minuteman intercontinental ballistic missile [2]. FT has evolved to be one of the most widely used techniques for system reliability and safety modeling and analysis.

FT is an analytical technique starting with identifying an undesired system event (typically the system considered being in a particular failure mode). Then the system is analyzed to identify all possible combinations of basic component failure events that can cause occurrence of the predefined undesired system event [3]. An FT can graphically represent the logical relationship between the undesired system event and the basic component failure events. It provides a logical framework for comprehending the possible ways in which a system can fail in a certain mode [4].

As a deductive technique, an FT analysis starts with a system failure scenario being considered (corresponding to the *TOP event* of FT). This failure symptom is then decomposed into its possible causes, each of which is further investigated and decomposed until basic causes of the failure (*basic events*) are understood. The FT model is completed in levels and constructed from top to bottom. During the top-down construction process, a set of intermediate events are usually involved. An *intermediate event* is a fault event that occurs due to one or more antecedent causes acting through logic gates [3].

There are two types of analysis based on the FT model. A qualitative analysis typically involves identifying minimal cutsets [4]. Each minimal cutset is a minimal combination of basic events whose occurrence results in the occurrence of the TOP event. A quantitative analysis determines the probability of the TOP event (corresponding to system unreliability or unavailability), given the probability of each basic event. Approaches for quantitative FT analysis include simulations (e.g. Monte Carlo simulations) [5] and analytical methods. The analytical methods can be further categorized into three types: state space-based methods [6–9], combinatorial methods [10–12], and modularization methods that integrate the former two methods as appropriate [13, 14]. Refer to [15] for a detailed discussion on these methods.

Based on the types of logic gates and events used for constructing the FT model, FT can be classified as static, dynamic, phased-mission, and multi-state FTs, which are described in the following sections.

2.3.1 Static Fault Tree

A static fault tree (SFT) represents the system failure criteria in terms of logic AND or OR combinations of fault events. Logical gates used for constructing an SFT are restricted to AND, OR, and VOTE (or K-out-of-N) gates, whose symbols are illustrated in Table 2.2.

2.3.2 Dynamic Fault Tree

A dynamic fault tree (DFT) models behavior of a dynamic system through logical gates in Table 2.2 as well as special dynamic gates, such as those illustrated in Table 2.3 [4].

A cold spare (CSP) gate has one primary basic event, corresponding to the component that is initially powered on and online. It also has one or several alternate basic events, corresponding to standby sparing components that are initially unpowered and serve as replacements for the primary online component. Since a CSP component is fully isolated from operational stresses, its failure rate before being activated can often be assumed as ZERO. The output of a CSP gate occurs when all of its input events have happened, i.e. when the primary component and all the standby sparing components have malfunctioned or become unavailable.

Table 2.2 SFT gate symbols and definitions.

Gate	Symbol	Definition
OR		Output event happens if at least one of the gate's input events happens.
AND		Output event happens if all of the gate's input events happen.
VOTE	K/N	Output event happens if at least K out of N input events happen.

Table 2.3 DFT gate symbols.

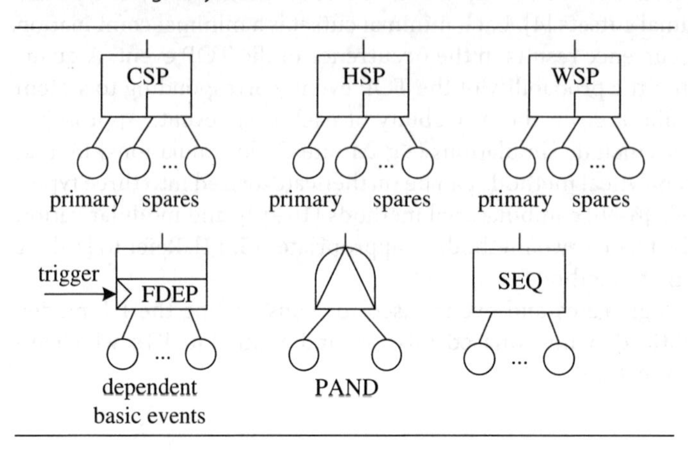

The CSP gate has two variations: hot spare (HSP) gate and warm spare (WSP) gate. Their graphical layouts are similar to that for CSP, with the change of word "CSP" to "HSP" and "WSP," respectively. An HSP component operates in parallel with the primary online component and is fully exposed to operational stresses; thus, it has the same failure rate before and after being switched into online for use. A WSP component is partially exposed to operational stresses and typically has a reduced failure rate before being switched into active use. Note that a basic event cannot be connected to different types of spare gates within the same system FT model. The reason is that the three spare gates not only model the standby sparing behavior but also affect failure rates of components associated with the gate's input basic events.

A functional dependence (FDEP) gate has a single trigger input event and one or several dependent basic events. The trigger event can be a basic event, or an intermediate event; its occurrence forces the corresponding dependent basic events to happen. An FDEP gate has no logical output; it is typically connected to the top gate of the system FT through a dashed line.

A priority AND (PAND) gate is logically equivalent to the logic AND gate requiring the occurrence of both input events to make the output event happen. However, there is an extra condition that the PAND gate's input events must occur in a predefined order (typically left-to-right order in which they appear under the PAND gate). In other words, the output event of a PAND gate will not occur if any of the gate's input events has not occurred or if a right input event occurs before a left input event.

A sequence enforcing (SEQ) gate forces all of the gate's input events to occur in a specified left-to-right order. The difference from the PAND gate is the SEQ gate only allows its input events to happen in a predefined order; the PAND gate detects whether its input events take place in the predefined order (these events may occur in any order in practice though).

2.3.3 Phased-Mission Fault Tree

Phased-mission FTs are used to model failure behavior of phased-mission systems (PMSs), which are systems involving multiple, nonoverlapping phases of operations or tasks to be accomplished in sequence [16, 17]. During different phases, the system configuration, reliability requirements, and component failure behavior can be different. A classic example is the flight of an aircraft, which involves taxi out, take-off, ascent, level-flight, descent, landing, and taxi in phases [18–20]. In the case of an aircraft with more than one engine, all the engines are typically required to be functioning during the take-off phase due to the enormous stress experienced by the airplane during this phase. However, while all the engines are desired to be functioning, only a subset of them is necessary during other phases. Moreover, the aircraft engines are more likely to fail during the take-off phase as they are generally under greater stress in this phase as compared to other phases of the flight profile [20]. These dynamic behaviors typically require a distinct FT model for each phase of a PMS in the system reliability analysis [21–25].

In a PMS with a phase-OR requirement, the entire mission fails if the system fails during any one phase. There are also PMSs with more general combinatorial phase requirements (CPRs) [26]. Specifically, their failure criteria can be expressed as a logical combination of phase failures in terms of logic AND, K-out-of-N, and OR. In a PMS with CPRs, a phase failure does not necessarily lead to failure of the entire mission; it may just cause degraded performance of the mission.

Example 2.5 A space data gathering system that has three consecutive phases; different combinations of phase FTs lead to different system performance levels [26]. For example, the system performs at an *Excellent* level if the data collection is successful in all of the three phases; the system performs at a *Good* level if the data collection is successful in one of the first two phases and in phase 3 at the same time. Figure 2.1 illustrates the FT models describing the CPR at these two performance levels. Refer to [26] for more details of this example PMS.

2.3.4 Multi-State Fault Tree

Multi-state fault trees (MFTs) are used to model systems, in which both the system and its components may exhibit multiple performance levels, corresponding to diverse states ranging from perfect function to complete failure [27–29]. These systems are referred to as multi-state systems (MSSs). MSSs can be used to model complex behaviors

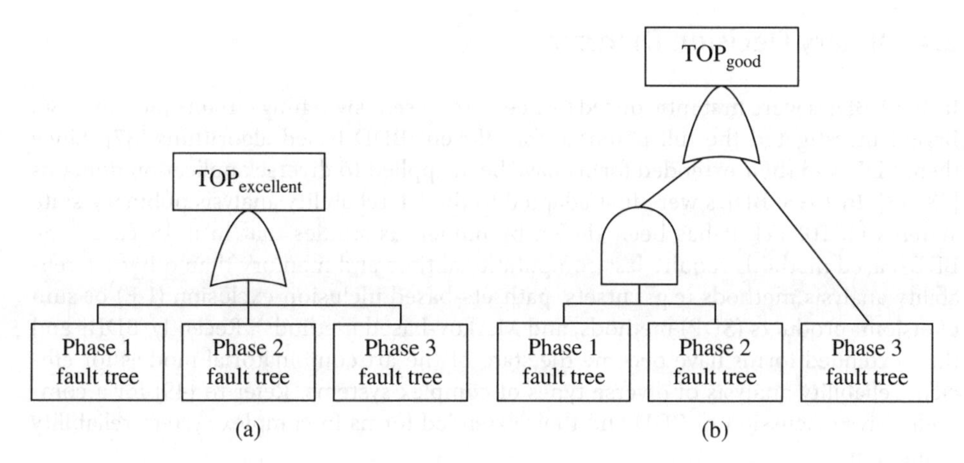

Figure 2.1 Examples of phased-mission FTs (a) $P(\text{excellent}) = 1 - P(\text{TOP}_{\text{excellent}})$. (b) $P(\text{good}) = 1 - P(\text{TOP}_{\text{good}})$.

such as performance degradation, shared load, imperfect fault coverage, and multiple failure modes in a wide range of real-world application systems, e.g. computer systems, sensor networks, circuits, power systems, communication networks, and transmission networks [30–33].

Similar to the traditional binary-state FT model, an MFT provides a graphical representation of combinations of component state events that can cause the entire system to occupy a specific state [34, 35]. For an MSS with n system states, n different MFTs must be constructed, one for each system state. Each MFT consists of a top event representing the system being in a specific state S_k, and a set of basic events each representing a multi-state component being in a particular state. The top event is resolved into a combination of basic and intermediate events that can cause the occurrence of system state S_k by means of OR, AND, and K-out-of-N logic gates. Given the occurrence probabilities of basic events, the quantitative analysis of an MFT for a particular system state can decide the probability of the MSS being in that state.

Example 2.6 A multi-state computer system with two boards (B_1, B_2) [34], each containing a processor and a memory module. The two memory modules (M_1, M_2) are shared by the two processors (P_1, P_2) through a common bus. Based on the status of its processor and memory module, each board B_i ($i = 1$ or 2) has four disjoint states: $B_{i,4}$ (both P and M are operational), $B_{i,3}$ (M is operational, but P is down), $B_{i,2}$ (P is operational, but M is down), and $B_{i,1}$ (both P and M are down). The entire computer system can assume three disjoint states: S_3 (at least one P and both M are operational), S_2 (at least one P and exactly one M are operational), and S_1 (no P or no M is operational). As an example, Figure 2.2 illustrates the MFT of the computer system being in state S_3. This MFT models the combinations of board states that make the entire computer system occupy state S_3: board B_1 is in state 4 and board B_2 is in state 3 or state 4; or B_1 is in state 3 and B_2 is in state 4.

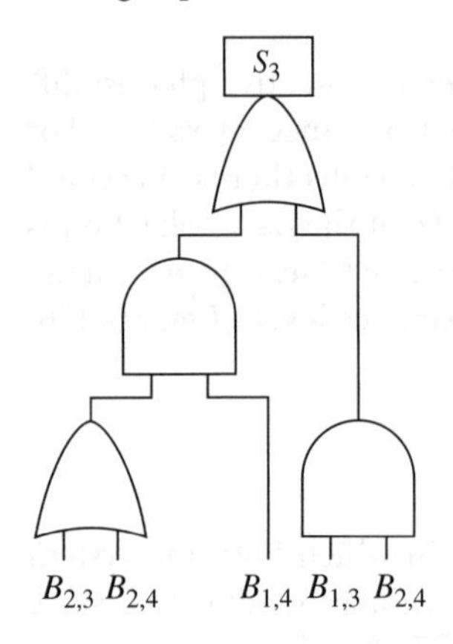

Figure 2.2 An example of MFTs [34].

2.4 Binary Decision Diagram

In 1959, BDDs were first introduced by Lee to represent switching circuits [36]. In 1986, Bryant investigated the full potential for efficient BDD-based algorithms [37]. Since then, BDD and their extended forms have been applied to diverse application domains [38–48]. In 1993, BDDs were first adapted to the FT reliability analysis of binary-state systems [4, 10, 11]. It has been shown by numerous studies that in most cases, the BDD-based methods require less computational time and memory than other FT reliability analysis methods (e.g. cutsets, pathsets-based inclusion-exclusion (I-E) or sum of disjoint products (SDP) methods, and Markov-based methods). Recently, BDDs and their extended forms have become the state-of-the-art combinatorial models for efficient reliability analysis of diverse types of complex systems. Refer to [49] for a comprehensive discussion of BDD and their extended forms in complex system reliability evaluation.

This section presents basics of BDDs and how to construct and evaluate a BDD model for system reliability analysis.

2.4.1 Basic Concept

BDDs are based on *Shannon's decomposition theorem* [37]: consider a Boolean expression F on a set of Boolean variables X, thus:

$$F = x \cdot F_{x=1} + \bar{x} \cdot F_{x=0} = x \cdot F_1 + \bar{x} \cdot F_0 = ite(x, F_1, F_0). \tag{2.22}$$

In (2.22), x is a Boolean variable belonging to the set X, $F_1 = F_{x=1}$ and $F_0 = F_{x=0}$ represent F evaluated at x being 1 and 0, respectively. To match the notation to the intuitive notion of binary tree induced by the Shannon's decomposition of a Boolean expression, the if-then-else (*ite*) format was introduced [10].

A BDD is a rooted, directed acyclic graph having two sink nodes and a set of non-sink nodes. The two sink nodes are labeled by a distinct logic value 0 and 1, representing the system being in a functional or a failed state, respectively. Each non-sink node is associated with a Boolean variable x, having two outgoing edges called 1-edge (or *then*-edge) and 0-edge (or *else*-edge), respectively (Figure 2.3). The 1-edge represents the failure of the corresponding component, and it leads to the child node $F_{x=1}$; the 0-edge represents the functioning of the component, and it leads to the child node $F_{x=0}$. Each non-sink node encodes a Boolean function in the *ite* format (e.g. the non-sink node in Figure 2.3 encodes (2.22)). A key characteristic of the BDD model is that $x \cdot F_{x=1}$ and $\bar{x} \cdot F_{x=0}$ are disjoint.

An ordered binary decision diagram (OBDD) can be obtained by applying the constraint that variables associated with non-sink nodes are ordered and every path from the root node to the sink node in the BDD model visits the variables in an ascending order. If each non-sink node in an OBDD encodes a different logic expression, then the OBDD is said to be a reduced ordered binary decision diagram (ROBDD).

ROBDDs can represent logical functions as graphs in a form that is both compact (any other graph representation contains more nodes) and canonical (the representation is unique for a certain variable ordering).

To perform a quantitative reliability analysis of an FT using the ROBDD, the FT is first converted to a ROBDD model (Section 2.4.2), and the resulted ROBDD is then evaluated to generate the system unreliability or reliability (Section 2.4.3).

2.4.2 ROBDD Generation

To construct a ROBDD from an FT, each input variable encoding the binary state of a system component is assigned a different index or order first. The size of the final ROBDD model can heavily rely on the order of these input variables. However, no exact procedure is currently available for determining the optimal ordering of input variables

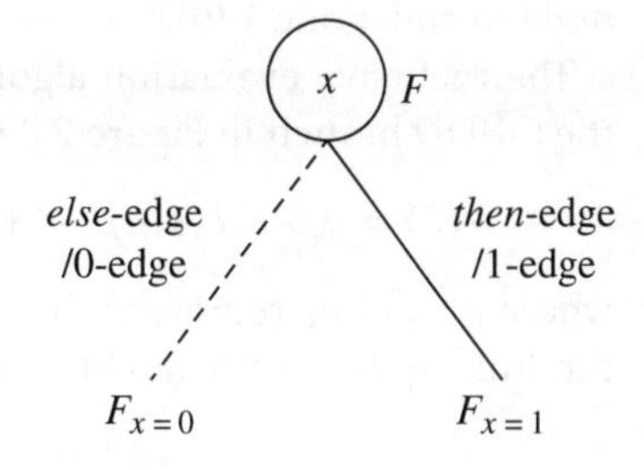

Figure 2.3 A non-sink node in the BDD model.

for a given FT structure. Heuristics are typically utilized to find a reasonably good ordering of input variables for ROBDD generation. Refer to [50–55] for several heuristics based on a depth-first search of the FT model.

After assigning the input variable ordering, OBDD is built in a bottom-up manner by applying the following operation rules in a recursive manner [37]:

$$\begin{aligned} G\Diamond H &= ite(x, G_{x=1}, G_{x=0})\Diamond ite(y, H_{y=1}, H_{y=0}) \\ &= ite(x, G_1, G_0)\Diamond ite(y, H_1, H_0) \\ &= \begin{cases} ite(x, G_1\Diamond H_1, G_0\Diamond H_0) & index(x) = index(y) \\ ite(x, G_1\Diamond H, G_0\Diamond H) & index(x) < index(y) \\ ite(y, G\Diamond H_1, G\Diamond H_0) & index(x) > index(y). \end{cases} \end{aligned} \tag{2.23}$$

In (2.23), G and H denote Boolean expressions corresponding to the traversed sub-FTs. G_i and H_i are subexpressions of G and H, respectively; $\Diamond$ denotes a logic operation (AND or OR).

Specifically, the rules of (2.23) are used to combine two sub-OBDD models representing logic expressions G and H into one OBDD model. In applying the rules, indexes of the two root nodes (x for G and y for H) are compared. If x and y have the same index (they belong to the same component), then the logic operation is applied to their children nodes (left child node of x with left child node of y, right child node of x with right child node of y); otherwise, the variable having a smaller index becomes the new root node of the combined OBDD model and the logic operation is applied to each child of the node having the smaller index and the other sub-OBDD as a whole. These rules are recursively applied until one of the sub-expressions becomes a constant 0 or 1. In this case, the Boolean algebra ($1 + x = 1$, $0 + x = x$, $1 \cdot x = x$, $0 \cdot x = 0$) is applied to simplify the model.

To generate a ROBDD, two reduction rules are applied: (i) merge isomorphic sub-OBDDs as one sub-OBDD because they encode the same Boolean function; and (ii) delete useless nodes, which are nodes with identical left and right child nodes.

2.4.3 ROBDD Evaluation

In the ROBDD generated, each path from the root node to a sink node denotes a disjoint combination of component failures or nonfailures. If the sink node of a path is labeled with 1, then the path leads to the entire system failure; if the sink node of a path is labeled with 0, then the path leads to the entire system function. The probability associated with each else-edge (or then-edge) on a path is reliability (or unreliability) of the corresponding component. Because all the paths in a ROBDD are disjoint, the system unreliability (reliability) can be simply evaluated by adding probabilities of all the paths from the root node to sink node 1 (0).

The recursive evaluation algorithm in the computer implementation with regard to the OBDD branch in Figure 2.3 can be represented as:

$$P(F) = q_x \cdot P(F_1) + p_x \cdot P(F_0),$$

where q_x and p_x represent the unreliability and reliability of component x. $P(F)$ gives the final system unreliability when x is the root node of the entire system ROBDD.

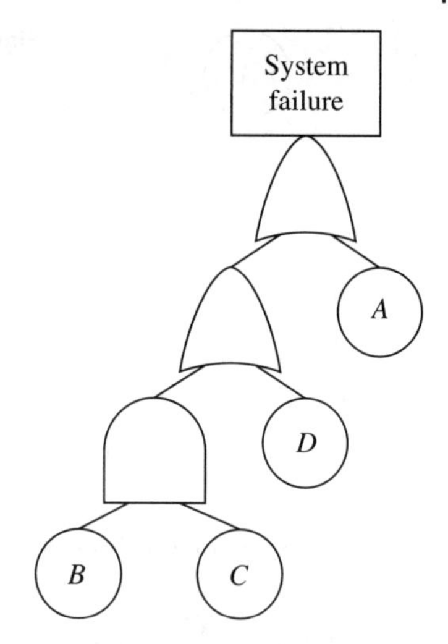

Figure 2.4 An example FT illustrating ROBDD generation.

The exit condition of this recursive algorithm is: If $F = 0$, then $P(F) = 0$; If $F = 1$, then $P(F) = 1$.

2.4.4 Illustrative Example

Example 2.7 Consider the FT in Figure 2.4. The variable ordering used for ROBDD generation is $A < B < C < D$. Figure 2.5 illustrates the final ROBDD model applying the generation procedure described in Section 2.4.2.

There are four disjoint paths from the root node A to sink node 1 in the ROBDD of Figure 2.5:

Path 1: A fails.
Path 2: A is operational, B and C fail.
Path 3: A and B are operational, D fails.
Path 4: A is operational, B fails, C is operational, and D fails.

Thus, by adding probabilities of these four paths, the system unreliability is obtained as

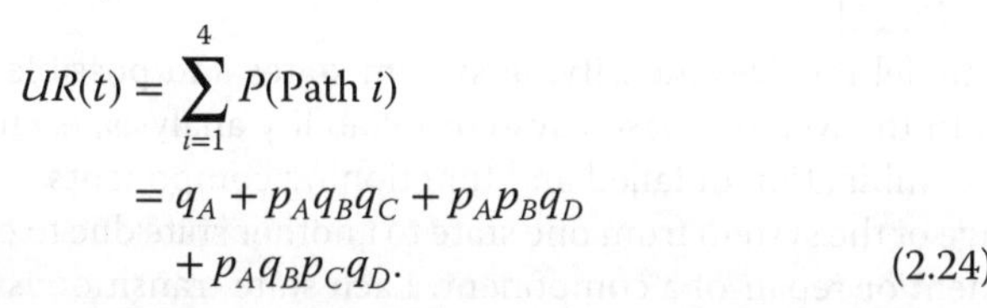

$$\begin{aligned} UR(t) &= \sum_{i=1}^{4} P(\text{Path } i) \\ &= q_A + p_A q_B q_C + p_A p_B q_D \\ &\quad + p_A q_B p_C q_D. \end{aligned} \tag{2.24}$$

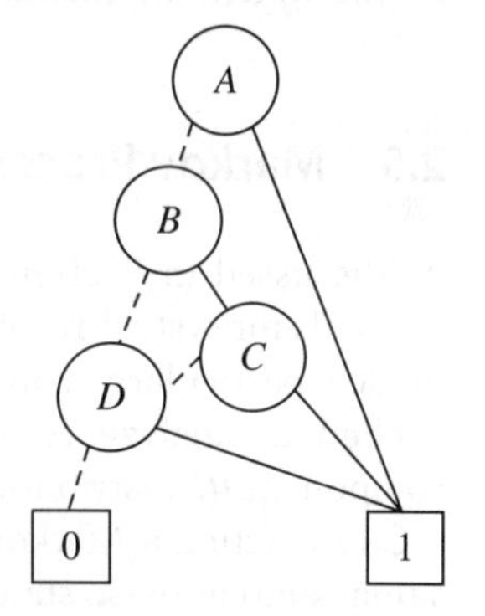

Figure 2.5 Final ROBDD for the example FT using $A < B < C < D$.

As mentioned in Section 2.4.2, the variable ordering can affect the generation of ROBDD. For illustration, Figure 2.6 shows the ROBDD generated from the FT of Figure 2.4 using a different ordering of $A < D < B < C$.

In the ROBDD of Figure 2.6, there are three disjoint paths from the root node A to sink node 1:

Path 1: A fails.
Path 2: A is operational, and D fails.
Path 3: A and D are operational, B and C fail.

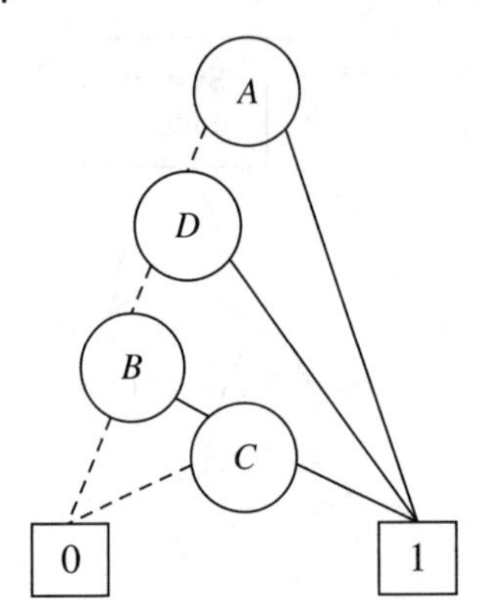

Figure 2.6 ROBDD for the example FT using $A < D < B < C$.

By adding probabilities of these three paths, the system unreliability is obtained as

$$\begin{aligned} UR(t) &= \sum_{i=1}^{3} P(\text{Path } i) \\ &= q_A + p_A q_D + p_A p_D q_B q_C. \end{aligned} \tag{2.25}$$

Note that though the unreliability expressions (2.24) and (2.25) generated based on different ROBDDs have different forms, they are essentially equivalent (i.e. the system unreliability results evaluated using them match exactly). Therefore, different variable orderings lead to different ROBDD models generated, but the same evaluation result for the system unreliability or reliability.

2.5 Markov Process

As discussed in Section 2.3, the state space-based methods belong to the analytical methods for system reliability analysis. Among the state space-based methods, Markov models particularly, continuous-time Markov chains (CTMCs) have commonly been applied to analyze reliability of systems with dynamic behavior and exponential component *ttf* distributions [7, 15].

Constructing a Markov model involves identifying system states and possible transitions among these states. In the Markov-based system reliability analysis, each state typically denotes a distinct combination of failed and functioning components. A state transition governs the change of the system from one state to another state due to events such as failure of a component or repair of a component. Each state transition is characterized by certain parameter(s), such as a component's failure rate or repair rate [56]. The state transition diagram starts with an initial state (typically a state in which all the system components are fault free). As time passes and component failures take place, the system goes from one state to another until one of the system failure states is reached.

Based on the state transition diagram generated, a set of differential equations $\boldsymbol{A}\boldsymbol{P}(t) = \boldsymbol{P}'(t)$ is constructed with $\boldsymbol{A}$ representing the state transition rate matrix, $\boldsymbol{P}(t)$ representing the vector of the system state probability at time t, and $\boldsymbol{P}'(t)$ representing the vector of the derivative of the system state probability at time t. Equation (2.26) shows the detailed form of the state equations for a system with n states.

$$\begin{bmatrix} -\alpha_{11} & \alpha_{21} & \alpha_{31} & \dots & \alpha_{n1} \\ \alpha_{12} & -\alpha_{22} & \alpha_{32} & \dots & \alpha_{n2} \\ \alpha_{13} & \alpha_{23} & -\alpha_{33} & \dots & \alpha_{n3} \\ \dots & \dots & \dots & \dots & \dots \\ \alpha_{1n} & \alpha_{2n} & \alpha_{3n} & \dots & -\alpha_{nn} \end{bmatrix} \bullet \begin{bmatrix} P_1(t) \\ P_2(t) \\ P_3(t) \\ \dots \\ P_n(t) \end{bmatrix} = \begin{bmatrix} P'_1(t) \\ P'_2(t) \\ P'_3(t) \\ \dots \\ P'_n(t) \end{bmatrix}. \tag{2.26}$$

In (2.26), $P_j(t)$ denotes the probability of the system being in state j at time t, $j = 1, 2, \dots, n$. $\alpha_{jk}(j \neq k)$ denotes the transition rate from state j to state k. The diagonal element is obtained as $\alpha_{jj} = \sum_{k=1,k\neq j}^{n} \alpha_{jk}$ (the sum of departure rates from state j). Hence, the sum of each column of the matrix $\boldsymbol{A}$ is zero.

Example 2.8 Consider the DFT in Figure 2.7. Figure 2.8 shows the state transition diagram for this dynamic system, where λ_A, λ_B, λ_C represent the failure rate of components A, B, and C (after being activated), respectively. The corresponding state equations are shown in (2.27).

$$\begin{bmatrix} -(\lambda_A + \lambda_B) & 0 & 0 \\ \lambda_B & -(\lambda_A + \lambda_C) & 0 \\ \lambda_A & \lambda_A + \lambda_C & 0 \end{bmatrix} \bullet \begin{bmatrix} P_1(t) \\ P_2(t) \\ P_3(t) \end{bmatrix} = \begin{bmatrix} P'_1(t) \\ P'_2(t) \\ P'_3(t) \end{bmatrix}. \tag{2.27}$$

To find all of the system state probabilities $P_j(t)$, the Laplace transform is typically applied to solve the differential equations in (2.26) with initial system state probabilities

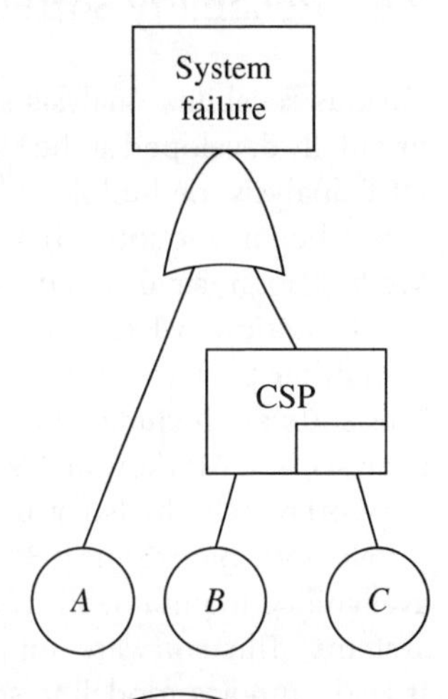

Figure 2.7 An example DFT illustrating the Markov analysis.

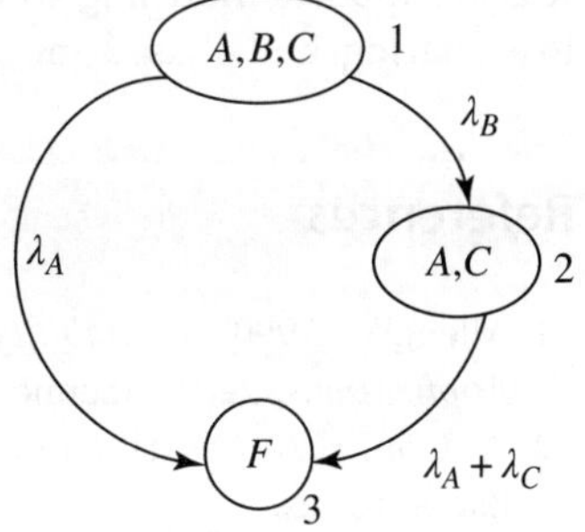

Figure 2.8 State transition diagram for the example DFT.

and $\sum_{j=1}^{n} P_j(t) = 1$ [1]. The system reliability (unreliability) can then be evaluated as the sum of all the system operational (failed) state probabilities.

The Markov model is more powerful in modeling dynamic and dependent behaviors than the traditional combinatorial methods (e.g. cutsets based I-E or SDP methods, BDD). However, the Markov model has the state space explosion problem (the model size can grow exponentially as the number of system components increases), leading to intractable models. To address this problem, the approximate bounding methods have been suggested, where only a portion of the complete system state space is generated and used for constructing state equations [57, 58]. Another solution suggested for addressing the state space explosion problem is the modular approach [13, 59], which combines the efficient BDD solution for static sub-FTs and the Markov solution for dynamic sub-FTs. Specifically, a system FT is decomposed into independent subtrees (sharing no input events). A fast and efficient modularization (subtree detection) algorithm can be found in [60]. The independent subtrees are then identified as either static or dynamic (based on relationships between input events of the subtree; refer to Section 2.3.1 and Section 2.3.2). The static subtrees are solved using the efficient combinatorial BDD-based method (Section 2.4); the dynamic subtrees are solved using the CTMC-based method (Section 2.5). Solutions of all the independent subtrees are finally integrated to obtain the solution to the entire system.

2.6 Reliability Software

Various reliability analysis software tools have been developed. For example, Galileo, originally developed at the University of Virginia, is a reliability analysis tool based on the DFT analysis methodology [61]. This software tool implements the modular approach (described in Section 2.5), where diverse time-to-failure distributions (e.g. exponential, Weibull, lognormal) and fixed failure probabilities are supported in the BDD-based solution to static sub-FTs. Galileo supports reliability analysis of PMSs.

Another FT-based reliability analysis software package is Isograph's FaultTree+ [62]. This software includes three modules, respectively, supporting FT analysis, event tree analysis, and Markov analysis. FaultTree+ has recently been incorporated into Reliability Workbench, the Isograph's flagship suite of reliability software.

ReliaSoft BlockSim [63] is a comprehensive software developed for reliability, availability, maintainability, and related analyses of both repairable and nonrepairable systems. This software supports modeling using FTs and reliability block diagrams. It also supports modeling some complex configurations such as standby redundancy, load sharing, and phased-missions.

References

1 Allen, A. (1990). *Probability, Statistics and Queuing Theory: with Computer Science Applications*, 2e. Academic Press.
2 Watson, H.A. (1961). *Launch Control Safety Study*. Murray Hill, NJ: Bell Telephone Laboratories.

3 Vesely, W.E., Goldberg, F.F., Roberts, N.H., and Haasl, D.F. (1981). *Fault Tree Handbook*. Washington, DC: U.S. Nuclear Regulatory Commission.
4 Dugan, J.B. and Doyle, S.A. (1996). New results in fault-tree analysis. In: *Tutorial Notes of Annual Reliability and Maintainability Symposium*, Las Vegas, NV, USA.
5 Ke, J., Su, Z., Wang, K., and Hsu, Y. (2010). Simulation inferences for an availability system with general repair distribution and imperfect fault coverage. *Simulation Modelling Practice and Theory* 18 (3): 338–347.
6 Bobbio, A., Franceschinis, G., Gaeta, R., and Portinale, L. (1999). Exploiting petri nets to support fault tree based dependability analysis. In: *Proceedings of the 8th International Workshop on Petri Nets and Performance Models*, 146–155.
7 Dugan, J.B., Bavuso, S.J., and Boyd, M.A. (1993). Fault trees and Markov models for reliability analysis of fault tolerant systems. *Reliability Engineering & System Safety* 39: 291–307.
8 Hura, G.S. and Atwood, J.W. (1988). The use of petri nets to analyze coherent fault trees. *IEEE Transactions on Reliability* 37 (5): 469–474.
9 Malhotra, M. and Trivedi, K.S. (1995). Dependability modeling using petri nets. *IEEE Transactions on Reliability* 44 (3): 428–440.
10 Rauzy, A. (1993). New algorithms for fault tree analysis. *Reliability Engineering & System Safety* 40: 203–211.
11 Coudert, O. and Madre, J.C. (1993). Fault tree analysis: 10^{20} prime implicants and beyond. In: *Proceedings of Annual Reliability and Maintainability Symposium*. Atlanta, GA, USA.
12 Sinnamon, R. and Andrews, J.D. (1996). Fault tree analysis and binary decision diagrams. In: *Proceedings of the Annual Reliability and Maintainability Symposium*. Las Vegas, NV, USA.
13 Gulati, R. and Dugan, J.B. (1997). A modular approach for analyzing static and dynamic fault trees. In: *Proceedings of the Annual Reliability and Maintainability Symposium*. Philadelphia, PA, USA.
14 Sahner, R., Trivedi, K.S., and Puliafito, A. (1996). *Performance and Reliability Analysis of Computer Systems: An Example-Based Approach Using the SHARPE Software Package*. Kluwer Academic Publisher.
15 Xing, L. and Amari, S.V. (2008). Fault tree analysis. In: *Handbook of Performability Engineering*. Springer-Verlag.
16 Xing, L. (2007). Reliability importance analysis of generalized phased-mission systems. *International Journal of Performability Engineering* 3 (3): 303–318.
17 Astapenko, D. and Bartlett, L.M. (2009). Phased mission system design optimisation using genetic algorithms. *International Journal of Performability Engineering* 5 (4): 313–324.
18 Dai, Y., Levitin, G., and Xing, L. (2014). Structure optimization of non-repairable phased mission systems. *IEEE Transactions on Systems, Man, and Cybernetics: Systems* 44 (1): 121–129.
19 Alam, M., Min, S., Hester, S.L., and Seliga, T.A. (2006). Reliability analysis of phased-mission systems: a practical approach. In: *Proceedings of Annual Reliability and Maintainability Symposium*. Newport Beach, CA. USA.
20 Murphy, K.E., Carter, C.M., and Malerich, A.W. (2007). Reliability analysis of phased-mission systems: a correct approach. In: *Proceedings of Annual Reliability and Maintainability Symposium*. Orlando, FL.

21 Dugan, J.B. (1991). Automated analysis of phased-mission reliability. *IEEE Transactions on Reliability* 40 (1): 45–52,55.

22 Smotherman, M.K. and Zemoudeh, K. (1989). A non-homogeneous Markov model for phased-mission reliability analysis. *IEEE Transactions on Reliability* 38 (5): 585–590.

23 Mura, I. and Bondavalli, A. (2001). Markov regenerative stochastic petri nets to model and evaluate phased mission systems dependability. *IEEE Transactions on Computers* 50 (12): 1337–1351.

24 Esary, J.D. and Ziehms, H. (1975). Reliability analysis of phased missions. In: *Reliability and fault tree analysis: theoretical and applied aspects of system reliability and safety assessment*, 213–236. Philadelphia, PA: SIAM.

25 Somani, A.K. and Trivedi, K.S. (1997). Boolean algebraic methods for phased-mission system analysis. *Technical Report NAS1–19480*. NASA Langley Research Center, Hampton, Virginia, USA.

26 Xing, L. and Dugan, J.B. (2002). Analysis of generalized phased mission system reliability, performance and sensitivity. *IEEE Transactions on Reliability* 51 (2): 199–211.

27 Xing, L. and Levitin, G. (2011). Combinatorial algorithm for reliability analysis of multi-state systems with propagated failures and failure isolation effect. *IEEE Transactions on Systems, Man, and Cybernetics, Part A: Systems and Humans* 41 (6): 1156–1165.

28 Shrestha, A., Xing, L., and Dai, Y.S. (2011). Reliability analysis of multi-state phased-mission systems with unordered and ordered states. *IEEE Transactions on Systems, Man, and Cybernetics, Part A: Systems and Humans* 41 (4): 625–636.

29 Levitin, G. and Xing, L. (2010). Reliability and performance of multi-state systems with propagated failures having selective effect. *Reliability Engineering & System Safety* 95 (6): 655–661.

30 Huang, J. and Zuo, M.J. (2004). Dominant multi-state systems. *IEEE Transactions on Reliability* 53 (3): 362–368.

31 Levitin, G. (2003). Reliability of multi-state systems with two failure-modes. *IEEE Transactions on Reliability* 52 (3): 340–348.

32 Chang, Y.-R., Amari, S.V., and Kuo, S.-Y. (2005). OBDD-based evaluation of reliability and importance measures for multistate systems subject to imperfect fault coverage. *IEEE Transactions on Dependable and Secure Computing* 2 (4): 336–347.

33 Li, W. and Pham, H. (2005). Reliability modeling of multi-state degraded systems with multi-competing failures and random shocks. *IEEE Transactions on Reliability* 54: 297–303.

34 Zang, X., Wang, D., Sun, H., and Trivedi, K.S. (2003). A BDD-based algorithm for analysis of multistate systems with multistate components. *IEEE Transactions on Computers* 52 (12): 1608–1618.

35 Amari, S.V., Xing, L., Shrestha, A. et al. (2010). Performability analysis of multi-state computing systems using multi-valued decision diagrams. *IEEE Transactions on Computers* 59 (10): 1419–1433.

36 Lee, C.Y. (1959). Representation of switching circuits by binary-decision programs. *Bell Systems Technical Journal* 38: 985–999.

37 Bryant, R.E. (1986). Graph-based algorithms for Boolean function manipulation. *IEEE Transactions on Computers* 35 (8): 677–691.

38 Miller, D.M. (1993). Multiple-valued logic design tools. In: *Proceedings of 23rd International Symposium on Multiple-Valued Logic (ISMVL)*, 2–11. Sacramento, CA, USA.
39 Miller, D.M. and Drechsler, R. (1998). Implementing a multiple-valued decision diagram package. In: *Proceedings of 23rd International Symposium on Multiple-Valued Logic (ISMVL)*, 52–57. Fukuoka, Japan.
40 Burch, J.R., Clarke, E.M., Long, D.E. et al. (1994). Symbolic model checking for sequential circuit verification. *IEEE Transactions on Computer-Aided Design of Integrated Circuits and Systems* 13 (4): 401–424.
41 Ciardo, G. and Siminiceanu, R. (2001). Saturation: an efficient iteration strategy for symbolic state space generation. In: *Tools and Algorithms for the Construction and Analysis of Systems*, 328–342.
42 Hermanns, H., Meyer-Kayser, J., and Siegle, M. (1999). Multi terminal binary decision diagrams to represent and analyse continuous time Markov chains. In: *Numerical Solution of Markov Chains*, 188–207.
43 Miner, A.S. and Cheng, S. (2004). Improving efficiency of implicit Markov chain state classification. In: *Proceedings of First International Conference on the Quantitative Evaluation of Systems (QEST '04)*, 262–271. Enschede, The Netherlands.
44 Ciardo, G. (2004). Reachability set generation for petri nets: can brute force be smart. In: *Proceedings of 25th International Conference on the Applications and Theory of Petri Nets (ICATPN '04)*, 17–34.
45 Miner, A.S. and Ciardo, G. (1999). Efficient reachability set generation and storage using decision diagrams. In: *Application and Theory of Petri Nets*, 6–25.
46 Burch, J.R., Clarke, E.M., McMillan, K.L. et al. (1990). Symbolic model checking: 10^{20} states and beyond. In: *Proceedings of Fifth Annual IEEE Symposium on the Logic in Computer Science (LICS' 90)*, 1–33. Philadelphia, PA, USA.
47 Chechik, M., Gurfinkel, A., Devereux, B. et al. (2006). Data structures for symbolic multi-valued model-checking. *Formal Methods in System Design* 29 (3): 295–344.
48 Corsini, M.-M. and Rauzy, A. (1994). Symbolic model checking and constraint logic programming: a cross-fertilization. In: *Proceedings of Fifth European Symp. Programming (ESOP'94)*, 180–194.
49 Xing, L. and Amari, S.V. (2015). *Binary Decision Diagrams and Extensions for System Reliability Analysis*. Salem, MA: Wiley-Scrivener.
50 Minato, S., Ishiura, N., and Yajima, S. (1990). Shared binary decision diagrams with attributed edges for efficient Boolean function manipulation. In: *Proceedings of the 27th ACM/IEEE Design Automation Conference*, 52–57. Orlando, FL, USA.
51 Fujita, M., Fujisawa, H., and Kawato, N. (1988). Evaluation and improvements of Boolean comparison method based on binary decision diagrams. In: *Proceedings of IEEE International Conference on Computer Aided Design*, 2–5. Santa Clara, CA, USA.
52 Fujita, M., Fujisawa, H., and Matsugana, Y. (1993). Variable ordering algorithm for ordered binary decision diagrams and their evaluation. *IEEE Transactions on Computer-Aided Design of Integrated Circuits and Systems* 12 (1): 6–12.
53 Bouissou, M., Bruyere, F., and Rauzy, A. (1997). BDD based fault-tree processing: a comparison of variable ordering heuristics. In: *Proceedings of ESREL Conference*, Lisbon, Portugal.

54 Bouissou, M. (1996). An ordering heuristics for building binary decision diagrams from fault-trees. In: *Proceedings of the Annual Reliability and Maintainability Symposium*. Las Vegas, NV, USA.

55 Butler, K.M., Ross, D.E., Kapur, R., and Mercer, M.R. (1991). Heuristics to compute variable orderings for efficient manipulation of ordered BDDs. In: *Proceedings of the 28th Design Automation Conference*. San Francisco, CA, USA.

56 Gulati, R. (1996). A modular approach to static and dynamic fault tree analysis. *M. S. Thesis, Electrical Engineering*, University of Virginia.

57 Sune, V. and Carrasco, J.A. (1997). A method for the computation of reliability bounds for non-repairable fault-tolerant systems. In: *Proceedings of the 5th IEEE International Symposium on Modeling, Analysis, and Simulation of Computers and Telecommunication System*, 221–228. Haifa, Israel.

58 Sune, V. and Carrasco, J.A. (2001). A failure-distance based method to bound the reliability of non-repairable fault-tolerant systems without the knowledge of minimal cutsets. *IEEE Transactions on Reliability* 50 (1): 60–74.

59 Manian, R., Dugan, J.B., Coppit, D., and Sullivan, K.J. (1998). Combining various solution techniques for dynamic fault tree analysis of computer systems. In: *Proceedings of the 3rd IEEE International High-Assurance Systems Engineering Symposium*, 21–28. Washington, DC, USA.

60 Dutuit, Y. and Rauzy, A. (1996). A linear time algorithm to find modules of fault trees. *IEEE Transactions on Reliability* 45 (3): 422–425.

61 Sullivan, K.J., Dugan, J.B., and Coppit, D. (1999). The Galileo fault tree analysis tool. In: *Proceedings of the 29th International Symposium on Fault-Tolerant Computing*. Madison, WI, USA.

62 FaultTree+, https://www.isograph.com/software/reliability-workbench/fault-tree-analysis-software, accessed in March 2018.

63 BlockSim, https://www.reliasoft.com/products/reliability-analysis/blocksim, accessed in March 2018.

3

Imperfect Fault Coverage

Many systems especially those used in life-critical or mission-critical applications such as aerospace, flight controls, nuclear plants, data storage systems and communication systems are fault-tolerant systems (FTSs) [1, 2]. An FTS can continue to perform its function correctly even in the presence of software errors or hardware failures [3, 4]. Its development typically requires using certain form of redundancy and an automatic reconfiguration and recovery mechanism to restore the system function in the case of the occurrence of a component failure. The mechanism itself (involving fault detection, fault location, fault isolation, and fault recovery) is often not perfect; it can fail such that the system cannot adequately detect, locate, isolate or recover from a component fault happening in the system. The uncovered component fault may propagate through the system and further cause the failure of the entire system or subsystem in spite of the presence of adequate redundancies. Such behavior is referred to as imperfect coverage (IPC) [5–7]. Since 1969, the IPC concept has been widely recognized as a significant concern in the reliability field.

Consider a hot standby server system with a primary server and a standby server. The standby server is switched online and operating upon the malfunction of the primary server. Under an ideal circumstance, the entire system functions correctly as long as one of the two servers functions correctly. However, in reality the failure of the primary server has to be detected and appropriately handled or isolated before the standby server can be activated and used to take over the system function. In other words, if the primary server's fault is not detected or isolated successfully, the entire system crashes even if the standby server is still available for use.

3.1 Different Types of IPC

Based on fault-tolerant techniques adopted, there are three types of IPC [8, 9]: element level coverage (ELC), fault level coverage (FLC), and performance dependent coverage (PDC).

Under ELC, also known as a single-fault model, each component has a particular fault coverage and is associated with a certain coverage probability or factor. This coverage probability does not depend on status of other components belonging to the same system. Under ELC, effectiveness of the system recovery and reconfiguration

Dynamic System Reliability: Modeling and Analysis of Dynamic and Dependent Behaviors, First Edition. Liudong Xing, Gregory Levitin and Chaonan Wang.

mechanism relies on the occurrence of each individual fault. For any particular component fault, the recovery mechanism's success or failure is independent of other component faults occurring in the same system. It is possible that a system can tolerate multiple coexisting single component faults.

Under FLC, also known as a multi-fault model, the fault coverage probability depends on the number of failed components within a particular set or group [10]. Effectiveness of the system recovery and reconfiguration mechanism is dependent on the occurrence of multiple faults within a certain recovery window. The FLC model has been applied to multi-processor computing systems with load-sharing [11] and aircraft computer control systems [8, 10].

Under PDC, effectiveness of the system recovery and reconfiguration mechanism relies on the condition or performance level of the overall system [12]. The PDC model is typically applicable to systems in which the system components perform the fault detection and recovery functions concurrently with their main functions. For instance, in a digital communication system, the same group of processors performs both fault detection and data exchange at the same time; the fault coverage probability in this case is relevant to the processor's loading and processing speed.

The remaining of this chapter focuses on the ELC modeling as well as reliability analysis methods considering ELC. Interesting readers may refer to [8–15] for detailed discussions on reliability modeling and analysis of FTSs subject to FLC and PDC.

3.2 ELC Modeling

In the seminal paper by Bouricius et al. [16], a fault coverage factor in ELC was defined as the conditional probability that a system can recover its function successfully given that a component fault has occurred. It measures a system's capability to perform fault detection, location, containment, and/or recovery when a component fault occurs in the system.

This section describes the imperfect coverage model (IPCM) introduced by Dugan and Trivedi, as illustrated in Figure 3.1 [7]. The subsequent sections present the incorporation of IPCM in reliability analysis of different types of systems including binary-state systems, multi-state systems, and multi-phase systems.

The IPCM is associated with a particular system component i and has a single entry point representing the occurrence of the component fault. The model also has three

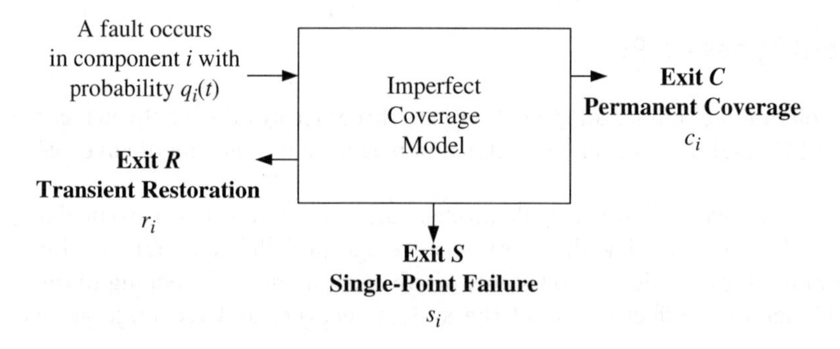

Figure 3.1 Structure of IPCM for component i [7].

disjoint exits *R*, *C*, *S*, representing all the possible outcomes of the fault recovery process triggered by the occurrence of the component fault event. Specifically,

- Exit *R*. Successful recovery from a transient component fault. The system function is restored without discarding the component.
- Exit *C*. Permanent nature of the fault is determined, and the faulty component is successfully isolated and removed. The system may or may not be functioning dependent on the remaining redundancy.
- Exit *S*. A single-point failure or an uncovered failure (UF) occurs, i.e. a single uncovered or undetected component fault causes the whole system to fail. The faulty component cannot be isolated or the system cannot be reconfigured successfully.

The three exits *R*, *C*, *S* for the IPCM of component *i* are associated with probabilities r_i, c_i, s_i (known as coverage factors), respectively, and $r_i + c_i + s_i = 1$. Their values can typically be estimated using techniques, e.g. fault injections [7, 17]. Let NF_i, CF_i, and UF_i denote event that component *i* is operational, failed covered, and failed uncovered, respectively. Let $q_i(t)$ represent the probability that component *i* experiences a fault by time *t*. Equation (3.1) gives the occurrence probabilities of those three events.

$$\begin{aligned} P(NF_i) &= n[i] = 1 - q_i(t) + q_i(t) \cdot r_i, \\ P(CF_i) &= c[i] = q_i(t) \cdot c_i, \\ P(UF_i) &= u[i] = q_i(t) \cdot s_i. \end{aligned} \tag{3.1}$$

3.3 Binary-State System

This section presents an explicit method, referred to as BDD expansion method (Section 3.3.1), and an implicit method named simple and efficient algorithm (SEA) (Section 3.3.2) for reliability analysis of binary-state systems considering effects of ELC modeled using the IPCM.

3.3.1 BDD Expansion Method

The BDD expansion method (BEM) addresses effects of ELC by explicitly inserting the IPCM in the path for components experiencing uncovered faults during traversal of the system BDD model (Section 2.4) [18]. Figure 3.2 illustrates the idea, where when node

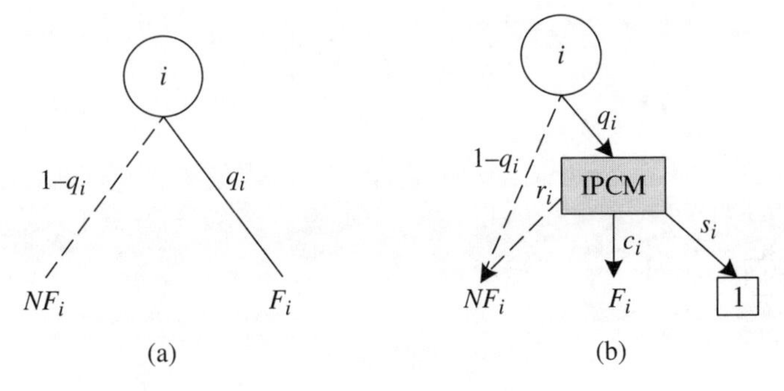

Figure 3.2 Inserting IPCM to a BDD path [18]. (a) Non-sink node without IPCM; (b) After inserting IPCM.

i can experience an uncovered fault and is traversed, the IPCM is inserted on the path led by the right branch (1-edge) of node i. Both the original left branch (representing no fault occurring) and the exit R from the IPCM point to NF_i; the exit C points to F_i; the exit S points to sink node 1 directly since an uncovered component fault leads to the failure of the entire system.

In the BDD generation (Section 2.4), some useless nodes can appear and should be removed for generating the ROBDD. In applying the BEM, these nodes, if they can experience uncovered faults, must be inserted back into the system BDD model. The reason is that their UFs can still make contributions to the entire system failure even though their covered failures make no contribution to the system unreliability for a particular path of the BDD. Figure 3.3 illustrates a scenario where node $(i+1)$ was useless and removed for generating the system ROBDD model. In applying the BEM, if the component corresponding to node $(i+1)$ can experience an uncovered fault, then node $(i+1)$ should be inserted back to the path along with its IPCM (Figure 3.4). The left branch (1-edge) as well as exits R and C of node $(i+1)$ lead to node $(i+2)$, while its exit S points to sink node 1 directly because the UF of component $(i+1)$ can cause the entire system to fail.

After IPCM is inserted for all nodes that can experience UFs, the system unreliability considering ELC can be obtained by summing probabilities of all the paths from the root to sink node 1, just like in the traditional BDD evaluation (Section 2.4).

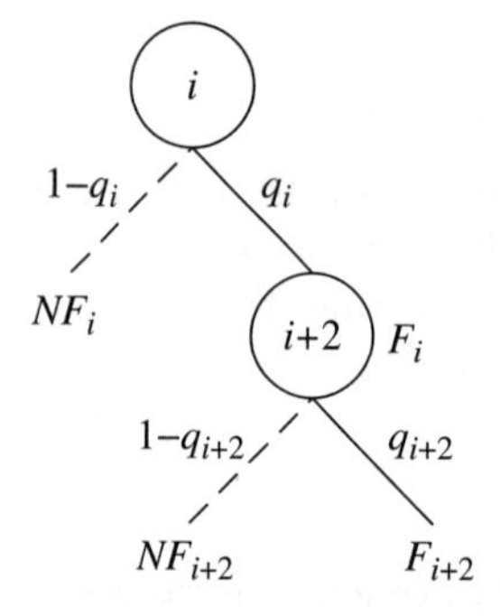

Figure 3.3 BDD with a useless node $i+1$ [19].

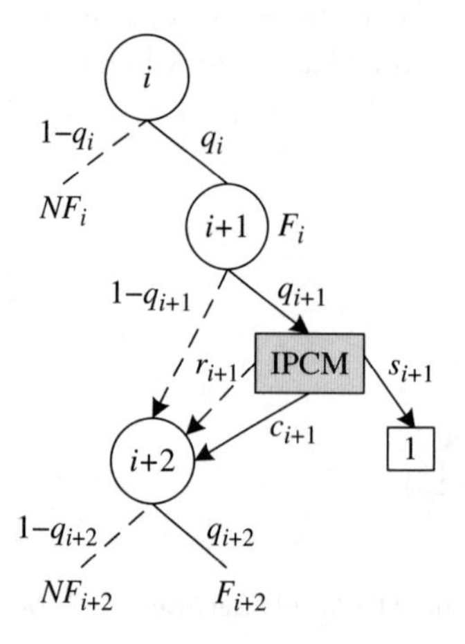

Figure 3.4 Insertion of useless node $i+1$ and its IPCM [19].

Example 3.1 Consider a two-component parallel system with FT illustrated in Figure 3.5. The BDD model without considering ELC is shown in Figure 3.6a. If both A and B can experience UFs, applying the BEM, the BDD model with IPCM is obtained as shown in Figure 3.6b. Thus, the system unreliability considering effects of ELC can be evaluated as the sum of probabilities of five paths from the root node to sink node 1 in the expanded BDD model.

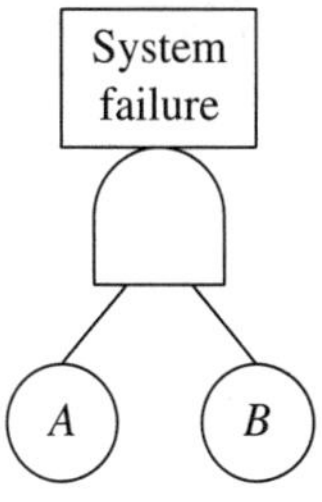

System failure

A B

Figure 3.5 FT of an example parallel system.

$$UR^{\mathrm{I}}(t) = q_A s_A + q_A c_A q_B c_B + q_A c_A q_B s_B + q_A r_A q_B s_B + (1 - q_A) q_B s_B. \tag{3.2}$$

Assume $q_A = q_B = 0.03$, $r_A = r_B = 0.9$, $c_A = c_B = 0.07$, and $s_A = s_B = 0.03$. According to (3.2), the unreliability of the example parallel system considering ELC is 0.001 803 6.

Figure 3.7 illustrates the unreliability of the example parallel system as the coverage factor c changes from 0 to 1. The r factor is assumed to be 0 for the two components for simplicity. Thus, the s factor is simply $1 - c$. When the c factor is ONE, implying that the system has perfect fault coverage, the system unreliability is the lowest (0.0009), which can be obtained by simply evaluating the BDD model in Figure 3.6a using $q_A = q_B = 0.03$. When the c factor is ZERO, implying that the component fault is always uncovered, the system actually becomes a series structure of components A and B with the highest system unreliability (0.0591). As the c factor decreases, the system becomes more unreliable since the probability of the component UF occurring increases and any component UF can bring down the entire system, increasing the system unreliability.

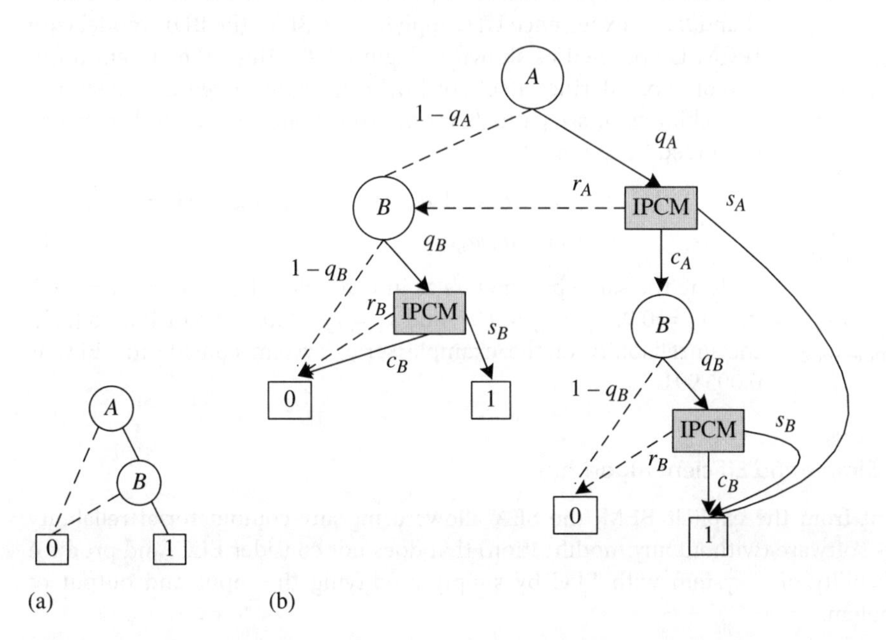

Figure 3.6 BDD for the example parallel system. (a) Without IPCM; (b) Expanded BDD with IPCM.

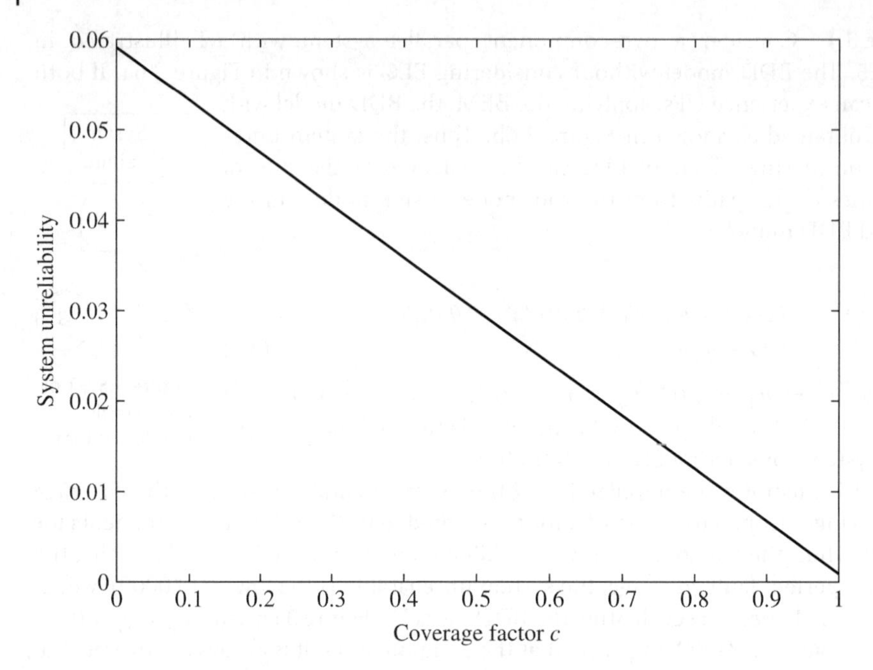

Figure 3.7 Parallel system unreliability vs. coverage factor *c*.

Example 3.2 Consider a two-component series system with FT illustrated in Figure 3.8. The BDD model without considering ELC is shown in Figure 3.9a. If both A and B can experience UFs, applying the BEM, the BDD model with IPCM is obtained as shown in Figure 3.9b. Thus, the system unreliability considering effects of ELC can be evaluated as the sum of probabilities of six paths from the root node to sink node 1 in the expanded BDD model.

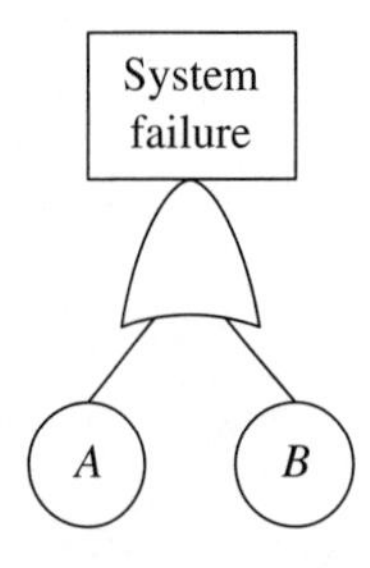

Figure 3.8 FT of an example series system.

$$UR^{\mathrm{I}}(t) = q_A s_A + q_A c_A + q_A r_A q_B s_B + q_A r_A q_B c_B + (1 - q_A) q_B s_B + (1 - q_A) q_B c_B. \tag{3.3}$$

Using the same parameters as in Example 3.1 (i.e. $q_A = q_B = 0.03$, $r_A = r_B = 0.9$, $c_A = c_B = 0.07$, and $s_A = s_B = 0.03$). According to (3.3), the unreliability of the example series system considering ELC is 0.005 991.

3.3.2 Simple and Efficient Algorithm

Different from the explicit BEM, the SEA allows using any combinatorial reliability analysis software (without any modification) that does not consider ELC, and produce the reliability of a system with ELC by simply modifying the input and output of the problem.

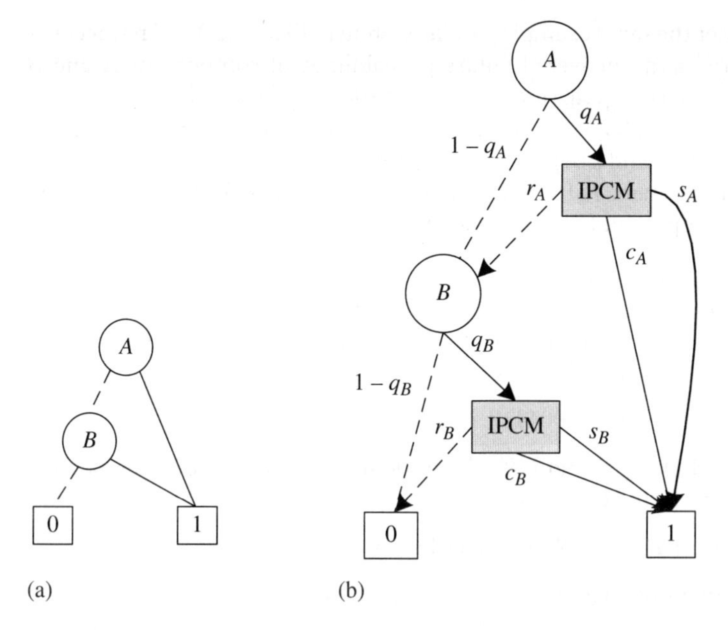

Figure 3.9 BDD for the example series system. (a) Without IPCM; (b) Expanded BDD with IPCM.

Based on the total probability theorem, the SEA [20] method evaluates the system unreliability with ELC modeled by the IPCM as:

$$\begin{aligned} UR^{\mathrm{I}}(t) &= P(\text{system fails} \mid \text{at least one UF}) \cdot P(\text{at least one UF}) \\ &\quad + P(\text{system fails} \mid \text{no UF}) \cdot P(\text{no UF}) \\ &= 1 \cdot [1 - P_u(t)] + UR^{\mathrm{C}}(t) \cdot P_u(t) \\ &= 1 - P_u(t) + P_u(t) \cdot UR^{\mathrm{C}}(t). \end{aligned} \tag{3.4}$$

$P_u(t)$ in (3.4) is the probability that no component in the system has a UF and can be computed based on (3.1) as

$$\begin{aligned} P_u(t) &= \prod_{\forall i} P(\text{no UF for component } i) \\ &= \prod_{\forall i} (1 - s_i \cdot q_i(t)) = \prod_{\forall i} (1 - u[i]). \end{aligned} \tag{3.5}$$

$UR^{\mathrm{C}}(t)$ in (3.4) is the conditional probability that the system fails given that no component experiences a UF. For evaluating $UR^{\mathrm{C}}(t)$ the fault occurrence probability $q_i(t)$ for each component i needs to be adjusted to a conditional probability $\widetilde{q}_i(t)$ denoted by (3.6) conditioned on no UF happening to this component.

$$\widetilde{q}_i(t) = \frac{c[i]}{1 - u[i]} = \frac{c_i \cdot q_i(t)}{(1 - s_i \cdot q_i(t))} = \frac{c[i]}{n[i] + c[i]}. \tag{3.6}$$

$UR^{\mathrm{C}}(t)$ can then be evaluated using any combinatorial method, e.g. the traditional ROBDD method (Section 2.4).

Example 3.3 Consider the same example parallel system in Example 3.1. First, according to (3.1), the covered and uncovered failure probabilities of components A and B are evaluated as $c[A] = q_A(t) \cdot c_A$, $u[A] = q_A(t) \cdot s_A$; $c[B] = q_B(t) \cdot c_B$, $u[B] = q_B(t) \cdot s_B$. Using parameters $q_A = q_B = 0.03$, $r_A = r_B = 0.9$, $c_A = c_B = 0.07$, and $s_A = s_B = 0.03$, one obtains $c[A] = c[B] = 0.0021$ and $u[A] = u[B] = 0.0009$.

According to (3.5), one obtains $P_u(t) = (1 - u[A])(1 - u[B]) = 0.998\,200\,81$. According to (3.6), the modified component failure probabilities are

$$\widetilde{q}_A(t) = \frac{c[A]}{1 - u[A]} = 0.00210189, \quad \widetilde{q}_B(t) = \frac{c[B]}{1 - u[B]} = 0.00210189.$$

Using the modified component failure probabilities obtained, the ROBDD model in Figure 3.6a is evaluated to obtain $UR^C(t)$ as

$$UR^C(t) = \widetilde{q}_A(t) \cdot \widetilde{q}_B(t) = 0.000004418.$$

Finally according to (3.4), $P_u(t)$ and $UR^C(t)$ can be integrated to obtain the parallel system unreliability considering effects of ELC as

$$UR^I(t) = 1 - P_u(t) + P_u(t) \cdot UR^C(t) = 0.0018036.$$

This result matches that obtained using the BEM in Example 3.1.

Example 3.4 Consider the same example series system in Example 3.2. The evaluation of $P_u(t)$, $\widetilde{q}_A(t)$, and $\widetilde{q}_B(t)$ is the same as that in Example 3.3. Using the modified component failure probabilities, the ROBDD model in Figure 3.9a is evaluated to obtain $UR^C(t)$ as

$$UR^C(t) = \widetilde{q}_A(t) + [1 - \widetilde{q}_A(t)]\widetilde{q}_B(t) = 0.0041994.$$

Finally according to (3.4), $P_u(t)$ and $UR^C(t)$ can be integrated to obtain the series system unreliability considering effects of ELC as

$$UR^I(t) = 1 - P_u(t) + P_u(t) \cdot UR^C(t) = 0.005991.$$

This result matches that obtained using the BEM in Example 3.2.

3.4 Multi-State System

MSSs are systems in which the system and/or its components can exhibit multiple states or performance levels [21]. Particularly, consider an MSS with n multi-state components and w different system states or performance levels. Each component j under the IPCM has $m_j + 1$ states ($j = 1, \ldots, n$). Among these states, state 0 corresponds to the UF of component j, state 1 corresponds to the covered failure of component j, states 2, …, m_j correspond to different performance levels of the component operational state. Let x_j denote the state indicator variable of component j, $p^I_{j,k}(t)$ denote the probability that component j is in state k at time t under the IPCM, i.e. $p^I_{j,k}(t) = P(x_j = k)$. Table 3.1 illustrates the state and probability space of multi-state component j under IPCM.

Similar to the SEA method for binary-state systems (Section 3.3.2), the total probability theorem is applied to reduce an MSS reliability problem with IPCM to a problem

Table 3.1 State space of multi-state component *j* [22].

Operational states		Failed states
different performance levels	$P(x_j = m_j) = p^{\mathrm{I}}_{j,m_j}(t)$	$P(x_j = 1) = p^{\mathrm{I}}_{j,1}(t)$
	……	covered failure
	$P(x_j = 3) = p^{\mathrm{I}}_{j,3}(t)$	$P(x_j = 0) = p^{\mathrm{I}}_{j,0}(t)$
	$P(x_j = 2) = p^{\mathrm{I}}_{j,2}(t)$	uncovered failure

with perfect fault coverage [23]. Particularly, the probability of an MSS being in a specific non failure state S_k can be evaluated as

$$\begin{aligned} P^{\mathrm{I}}_{S_k}(t) &= P(S_k) \\ &= P(S_k \mid \text{at least one UF}) \cdot P(\text{at least one UF}) + P(S_k \mid \text{no UF}) \cdot P(\text{no UF}) \\ &= 0 \cdot [1 - P_u(t)] + P^{\mathrm{C}}_{S_k}(t) \cdot P_u(t) = P_u(t) \cdot P^{\mathrm{C}}_{S_k}(t). \end{aligned} \tag{3.7}$$

$P_u(t)$ in (3.7) is the probability that no multi-state components undergo UFs, which is evaluated as (3.8).

$$P_u(t) = \prod_{\forall j} P(\text{no UF for component } j) = \prod_{\forall j} [1 - p^{\mathrm{I}}_{j,0}(t)]. \tag{3.8}$$

Before evaluating $P^{\mathrm{C}}_{S_k}(t) = P(S_k \mid \text{no UF})$ in (3.7), the component state probability $p^{\mathrm{I}}_{j,k}(t)$ needs to be adjusted to a conditional state probability $p^{\mathrm{C}}_{j,k}(t)$ conditioned on no UF occurring to this component *j* using (3.9).

$$p^{\mathrm{C}}_{j,k}(t) = P(x_j = k \mid x_j \neq 0) = p^{\mathrm{I}}_{j,k}(t)/(1 - p^{\mathrm{I}}_{j,0}(t)). \tag{3.9}$$

Using the modified component state probabilities, $P^{\mathrm{C}}_{S_k}(t)$ in (3.7) can be evaluated using any MSS reliability analysis method that does not consider IPCM [19]. An efficient multi-state multi-valued decision diagram (MMDD)-based method [21] is adopted, which is reviewed next.

3.4.1 MMDD-Based Method for MSS Analysis

MMDD is a straightforward extension of the traditional BDD model (Section 2.4) to the multi-valued logic. Each multi-state component A with r states is associated with a multi-valued variable $x_A \in \{1, 2, \ldots, r\}$, modeled by a non-sink node with r outgoing edges in the MMDD model as shown in Figure 3.10.

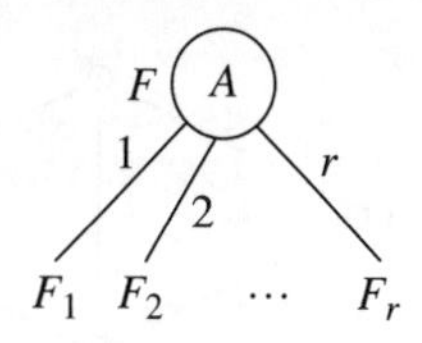

Figure 3.10 A non-sink node in MMDD.

The MMDD model in Figure 3.10 encodes a multi-valued logic expression F, which is decomposed into r subexpressions with regard to component A and can be represented using the *case* format as shown in (3.10).

$$\begin{aligned} F &= A_1 \cdot F_{x_A=1} + A_2 \cdot F_{x_A=2} + \ldots + A_r \cdot F_{x_A=r} \\ &= case\,(A, F_1, F_2, \ldots, F_r). \end{aligned} \tag{3.10}$$

A different MMDD may be constructed for each system state of the MSS. The MMDD for state S_k has two sink nodes 0 and 1, representing the MSS is not in or in the particular state S_k, respectively. To evaluate $P^C_{S_k}(t)$, the probability that the MSS is in state S_k under the perfect fault coverage, the following two steps are performed.

Step 1: Generate MMDD from MFT.

Based on the MFT representation of an MSS (Section 2.3.4), the MMDD model can be constructed in a bottom-up manner using manipulation rules of (3.11), which are straightforward extensions of the traditional BDD manipulation rules of (2.23) [21]. In (3.11), logic expressions $G = case(x, G_1, \ldots, G_r)$, and $H = case(y, H_1, \ldots, H_r)$ represent two sub-MMDD models to be combined using logic operation $\Diamond$ (AND or OR).

$$\begin{aligned} G \Diamond H &= case(x, G_1, \ldots G_r) \Diamond case(y, H_1, \ldots, H_r) \\ &= \begin{cases} case(x, G_1 \Diamond H_1, \ldots, G_r \Diamond H_r) & index(x) = index(y), \\ case(x, G_1 \Diamond H, \ldots, G_r \Diamond H) & index(x) < index(y), \\ case(y, G \Diamond H_1, \ldots, G \Diamond H_r) & index(x) > index(y). \end{cases} \end{aligned} \tag{3.11}$$

Step 2: Evaluate MMDD.

Each path from the root to sink node 1 in the final MMDD model indicates a disjoint combination of component states that can cause the MSS to be in a specific state S_k. If the path leads from a node i to its l-edge, then the state l of the component is considered for that path. In other words, the probability $p^C_{i,l}(t)$ should be used for calculating the path probability, which is the multiplication of probabilities of all edges appearing on the path. The system state probability $P^C_{S_k}(t)$ can be simply computed as the sum of probabilities for all paths from the root to sink node 1.

3.4.2 Illustrative Example

Example 3.5 Consider a multi-state bridge network system in Figure 3.11 [22, 23]. Each link in the network has six states (including a UF state) corresponding to different capacities or performance levels. The whole network can be in multiple states, among which the *acceptable* state is defined by

$$F_{\text{accept}} = l_{1:2}\, l_{3:2}\, l_{5:2} + l_{1:2}\, l_{4:2} + l_{2:2}\, l_{5:2} + l_{2:2}\, l_{3:2}\, l_{4:2},$$

where $l_{j:k}$ represents link j being at performance level k or above.

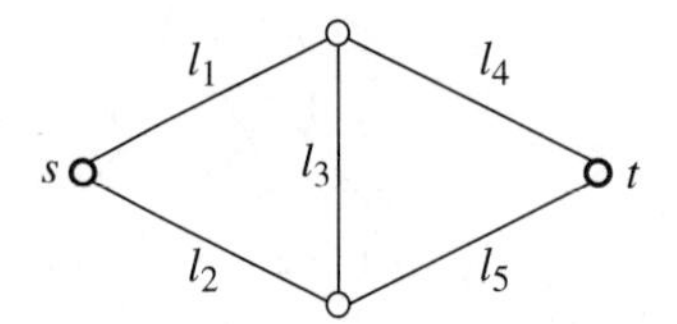

Figure 3.11 An example bridge network [22, 23].

Table 3.2 Link state probabilities.

State k	Perfect coverage $p^{I}_{j,k}(t)=p^{C}_{j,k}(t)$	Imperfect coverage $p^{I}_{j,k}(t)$	$p^{C}_{j,k}(t)$
0	—	0.1	—
1	0.1	0.1	0.111111
2	0.2	0.2	0.222222
3	0.3	0.3	0.333333
4	0.3	0.2	0.222222
5	0.1	0.1	0.111111

Table 3.3 Analysis results for the example network.

Perfect coverage	$P^{I}_{S_{accept}}(t) = P^{C}_{S_{accept}}(t) = 0.97848$	
Imperfect coverage	$P_u(t)$	0.590490
	$P^{C}_{S_{accept}}(t)$	0.973293
	$P^{I}_{S_{accept}}(t)$	0.57472

Assume all the links are *s*-identical. Table 3.2 gives the link state probabilities. Under perfect fault coverage, $p^{I}_{j,k}(t)$ and $p^{C}_{j,k}(t)$ are identical; under ELC modeled by the IPCM, $p^{I}_{j,k}(t)$ is given and $p^{C}_{j,k}(t)$ is obtained using (3.9) and given $p^{I}_{j,k}(t)$.

According to (3.8), $P_u(t)$ can be calculated for cases with IPCM. To evaluate the *accept* state probability given that no links experience UFs, i.e. $P^{C}_{S_{accept}}(t)$, the MMDD model of Figure 3.12 is generated (edges from a non-sink node leading to the same child node are grouped together to simplify the model representation). Evaluating the generated MMDD model using $p^{C}_{j,k}(t)$ of Table 3.2, $P^{C}_{S_{accept}}(t)$ can be obtained. The final system *accept* state probability $P^{I}_{S_{accept}}(t)$ is further obtained using (3.7). Table 3.3 summarizes the analysis results of the example bridge network system.

3.5 Phased-Mission System

Phased-mission systems (PMSs) involve multiple, consecutive, and nonoverlapping phases of operations during the mission (Section 2.3.3). Based on the mini-component concept (Section 3.5.1), this section presents an extension of the SEA method, called PMS SEA to consider effects of ELC in the reliability analysis of nonrepairable PMSs (Section 3.5.2).

3.5.1 Mini-Component Concept

Esary and Ziehms introduced the mini-component concept in 1975 for handling *s*-dependence across phases for a given component in the PMS [24]. The concept

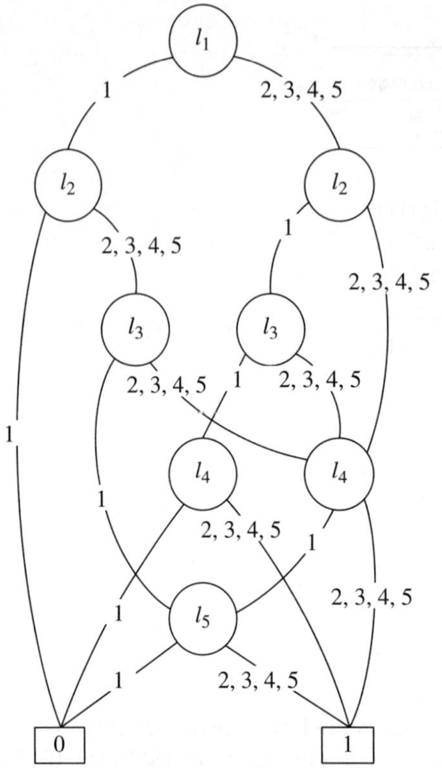

Figure 3.12 MMDD for the example network: F_{accept}.

is to replace a component in each phase of the PMS with a series of *s*-independent *mini-components*. Specifically, a component A in phase i is replaced by a series of mini-components a_1, a_2, ..., a_i performing *s*-independently with logic relation of $A_i = a_1 a_2 \ldots a_i$, meaning that A is functioning in phase i (i.e. $A_i = 1$) if and only if (*iff*) this component has operated correctly in all of the previous phases. The relation can also be translated as $\overline{A_i} = \overline{a_1} + \overline{a_2} + \ldots + \overline{a_i}$, meaning that A is failed in phase i ($A_i = 0$) *iff* it has failed in phase i or any of the previous phases.

According to (3.1) and given the failure function $q_{a_i}(t)$ and coverage factors $(r_{a_i}, c_{a_i}, s_{a_i})$ of each mini-component a_i, Eq. (3.12) gives the probabilities that a_i is operational (event NF_{a_i}), is failed covered (event CF_{a_i}), and is failed uncovered (event UF_{a_i}).

$$
\begin{aligned}
P(NF_{a_i}) &= n[a_i] = 1 - q_{a_i}(t) + q_{a_i}(t) \cdot r_{a_i}, \\
P(CF_{a_i}) &= c[a_i] = q_{a_i}(t) \cdot c_{a_i}, \\
P(UF_{a_i}) &= u[a_i] = q_{a_i}(t) \cdot s_{a_i}.
\end{aligned}
\tag{3.12}
$$

3.5.2 PMS SEA

Similar to the SEA method for single-phase systems (Section 3.3.2), the idea of PMS SEA is to apply the total probability theorem to separate UFs of all the system components from combinatorics of the PMS reliability solution [25]. According to (3.4), the unreliability of a PMS with n components and m phases subject to ELC can be evaluated as $UR^{I}(t) = 1 - P_u(t) + P_u(t)\, UR^{C}(t)$, where $P_u(t)$ is the probability of no mini-component experiencing a UF throughout the mission and can be evaluated using (3.13).

$$
\begin{aligned}
P_u(t) &= P(\overline{UF}_1 \cap \overline{UF}_2 \cap \ldots \cap \overline{UF}_n) \\
&= \prod_{A=1}^{n}(1 - P(UF_A)) = \prod_{A=1}^{n}(1 - u[A]) \\
&= \prod_{A=1}^{n}(1 - u[A_m]).
\end{aligned}
\tag{3.13}
$$

In (3.13), UF_A represents event that component A fails uncovered during the mission; UF_A for different components are s-independent. The occurrence probability of UF_A, denoted by $u[A]$ is equivalent to $u[A_m]$, which is the probability that component A has failed uncovered before the end of the last phase m.

For a nonrepairable PMS, if a component is failed uncovered in one phase, the component remains in the same state for the rest of the mission. Also, a component can fail uncovered in a particular phase j only if it has survived all the previous phases [24]. Thus, $u[A_j] = P$(component A fails uncovered before the end of phase j) can be evaluated as [25]:

$$
\begin{aligned}
u[A_j] &= P(UF_{A_j}) = P(\text{any mini-component } a_{i\in\{1\ldots j\}} \text{ is failed uncovered}) \\
&= P\{UF_{a_1} \cup (NF_{a_1} \cap UF_{a_2}) \cup \ldots \cup (NF_{a_1} \cap \ldots \cap NF_{a_{j-1}} \cap UF_{a_j})\} \\
&= u[a_1] + n[a_1]u[a_2] + \ldots + n[a_1]n[a_2] \ldots n[a_{j-1}]u[a_j] \\
&= u[a_1] + \sum_{i=2}^{j}\left\{\left(\prod_{k=1}^{i-1} n[a_k]\right) u[a_i]\right\}.
\end{aligned}
\tag{3.14}
$$

$u[A_m]$ in (3.13) can be evaluated using (3.14) by setting j as the index of the last phase m.

Based on the mini-component concept, $n[A_j] = P$(component A has not failed before the end of phase j) and $c[A_j] = P$(component A fails covered before the end of phase j) can be similarly calculated as:

$$
\begin{aligned}
n[A_j] &= P(NF_{A_j}) = P(\text{all mini-components } a_{i\in\{1\ldots j\}} \text{ are functioning}) \\
&= P\{NF_{a_1} \cap \ldots \cap NF_{a_{j-1}} \cap NF_{a_j}\} \\
&= n[a_1] \ldots n[a_{j-1}]n[a_j] = \prod_{i=1}^{j} n[a_i],
\end{aligned}
\tag{3.15}
$$

$$
\begin{aligned}
c[A_j] &= P(CF_{A_j}) = P(\text{any mini-component } a_{i\in\{1\ldots j\}} \text{ is failed covered}) \\
&= P\{CF_{a_1} \cup (NF_{a_1} \cap CF_{a_2}) \cup \ldots \cup (NF_{a_1} \cap \ldots \cap NF_{a_{j-1}} \cap CF_{a_j})\} \\
&= c[a_1] + n[a_1]c[a_2] + \ldots + n[a_1]n[a_2] \ldots n[a_{j-1}]c[a_j] \\
&= c[a_1] + \sum_{i=2}^{j}\left\{\left(\prod_{k=1}^{i-1} n[a_k]\right) c[a_i]\right\}.
\end{aligned}
\tag{3.16}
$$

When $j = 1$, $n[A_1] = n[a_1]$, $c[A_1] = c[a_1]$, $u[A_1] = u[a_1]$ by definition.

To evaluate $UR^C(t)$ in (3.4) for a PMS, the failure function of each component A in each phase j is modified to a conditional failure probability $P^C(A_j)$ conditioned on no UF occurring during the entire mission.

$$P^C(A_j) = P(CF_{A_j} \mid \overline{UF_A}) = \frac{c[A_j]}{1-u[A]} = \frac{c[A_j]}{1-u[A_m]}. \tag{3.17}$$

Using the modified component failure probabilities $P^C(A_j)$, $UR^C(t)$ can then be evaluated using the traditional PMS reliability analysis method that does not consider ELC, for example, the BDD-based methods [19, 26], recursive combinatorial methods [27, 28], state space-based methods [29, 30], and simulations [31]. In this chapter, the BDD-based method is adopted for evaluating $UR^C(t)$ and its main steps are described in Section 3.5.3.

3.5.3 PMS BDD Method

The BDD-based PMS reliability analysis method involves variable ordering, generating single-phase BDDs, combining single-phase BDDs to obtain the PMS BDD, and PMS BDD evaluation. The following summarizes those four steps. Refer to [19, 26] for more detailed explanation of the method.

Step 1: Input variable ordering.

Two types of variables need to be ordered for generating BDD for a PMS: variables of different system components and variables representing the same component in different mission phases. For the former, heuristics [32–37] can be used to determine the ordering; for the latter, forward or backward ordering can be applied [26]. The forward method orders variables of the same component using the same order as the phase order, e.g. $A_1 < A_2 < \ldots < A_m$ while the backward method orders them using the reverse of the phase order, e.g. $A_m < \ldots < A_2 < A_1$. Consider an example of a PMS with two components and three phases. For component ordering of $A < B$, the forward method gives the total ordering of $A_1 < A_2 < A_3 < B_1 < B_2 < B_3$; the backward method gives the total ordering of $A_3 < A_2 < A_1 < B_3 < B_2 < B_1$. The backward ordering was shown to outperform the forward method for PMS analysis [26]. Therefore, only the backward method is illustrated and used in the subsequent discussions.

Step 2: Generate single-phase BDDs.

In this step, the traditional BDD generation method (Section 2.4.2) is directly applicable to generate the BDD model from each single-phase fault tree model of the PMS.

Step 3: Generate PMS BDD.

In this step, single-phase BDDs generated in step 2 are combined to generate the BDD model of the entire PMS. Two cases must be treated differently: when a logical operation is performed on two variables of different components, the traditional BDD manipulation rules of (2.23) should be used; when the operation is performed on two variables of the same component but different mission phases (e.g. phases i, j, and $i<j$), the special phase dependent operation (PDO) of (3.18) [26] must be used to handle the s-dependence between the two variables ($A_j = 0 \rightarrow A_i = 0$ meaning that component A must be operational in phase i if it is operational in a later phase j).

$$\begin{aligned} G\Diamond H &= ite(A_i, G_{A_i=1}, G_{A_i=0})\Diamond ite(A_j, H_{A_j=1}, H_{A_j=0}) \\ &= ite(A_i, G_1, G_0)\Diamond ite(A_j, H_1, H_0) \\ &= ite(A_j, G\Diamond H_1, G_0\Diamond H_0). \end{aligned} \tag{3.18}$$

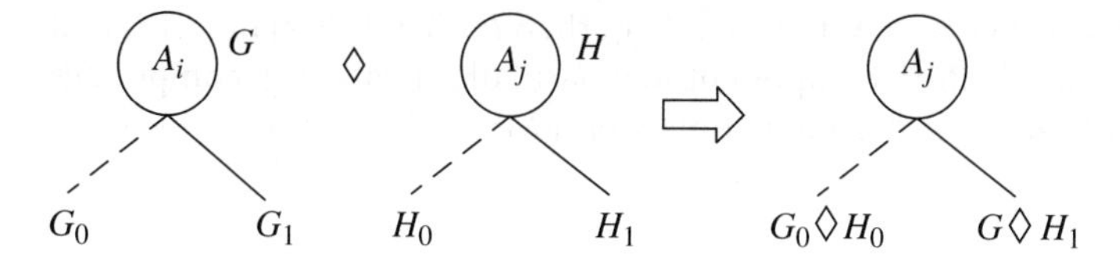

Figure 3.13 PDO for PMS.

As the backward ordering is used, A_j is ordered before A_i and is selected as the new root node of the combined BDD as shown in the *ite* format of (3.18) and illustrated in Figure 3.13. Then the logic operation (AND or OR) represented by $\Diamond$ is applied to the right child of A_j (H_1) and the other BDD encoding G as a whole; the same operation is applied to the left child of A_j (H_0) and the left child of A_i (G_0). Note that the following two stringent rules need to be followed to apply the PDO correctly:

1) Orderings used for generating single-phase BDDs in step 2 must be consistent for all the mission phases.
2) Orderings for variables of the same component but different mission phases must be grouped together.

Refer to [38] for procedures that can relax these two constraints, allowing any arbitrary ordering strategies for generating PMS BDDs.

Step 4: Evaluate PMS BDD.

The PMS BDD generated in step 3 is evaluated to find the PMS unreliability. The PMS BDD evaluation is recursive. Again two cases exist and must be treated differently. Consider the sub-PMS BDD in Figure 3.14. For 0-edge (left) or 1-edge (right) linking variables of different components, the following traditional BDD evaluation method (Section 2.4.3) should be used:

$$P(G) = P(x) \cdot P(G_1) + [1 - P(x)] \cdot P(G_0) = P(G_1) + [1 - P(x)] \cdot [P(G_0) - P(G_1)]. \tag{3.19}$$

For 1-edge linking two dependent variables belonging to the same component but different mission phases (e.g. x and y in Figure 3.14 are A_j and A_i, respectively), the evaluation method in (3.20) should be used to handle the s-dependence [26].

$$P(G) = P(G_1) + [1 - P(x)] \cdot [P(G_0) - P(H_0)]. \tag{3.20}$$

$P(G)$ gives the PMS unreliability if x is the root node of the PMS BDD model generated in step 3. The recursive evaluation has the following exit conditions: if $G = 0$ (system is

Figure 3.14 A sub-PMS BDD.

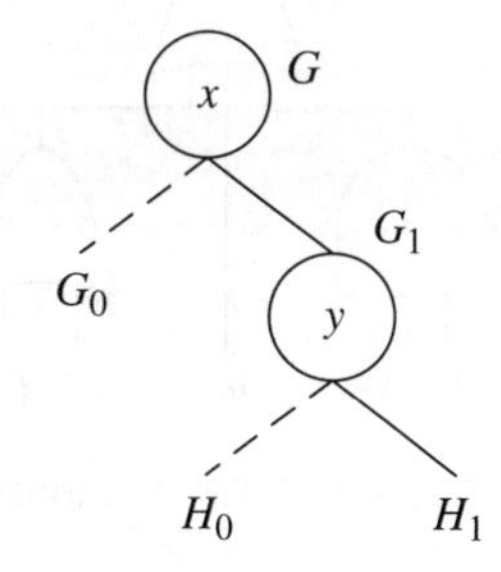

operational), then $P(G) = 0$; if $G = 1$ (system is failed), then $P(G) = 1$. $P(x)$ in (3.19) and (3.20) is the unreliability of x (a PMS component in a particular phase, e.g. component A in phase j). In the application of the above four-step PMS BDD method to evaluate $UR^C(t)$, $P(x)$ is evaluated as the modified component failure probability $P^C(A_j)$.

3.5.4 Summary of PMS SEA

Based on discussions in Sections 3.5.1–3.5.3, this section presents a step-by-step procedure for the reliability analysis of a nonrepairable PMS with ELC.

1) Based on (3.12), evaluate the mini-component event occurrence probabilities $u[a_j]$, $n[a_j]$, $c[a_j]$ for each component A at each phase j.
2) Based on (3.14), (3.15), (3.16) and the mini-component event probabilities computed in step 1, evaluate $u[A_j]$, $n[A_j]$, $c[A_j]$ for each component A at each phase j.
3) Based on (3.13), evaluate $P_u(t)$.
4) Based on (3.17), evaluate the conditional failure probability $P^C(A_j)$ for each component A at each phase j.
5) Generate the PMS BDD model using the generation procedure of Section 3.5.3.
6) Evaluate $UR^C(t)$ recursively from the PMS BDD using the evaluation algorithm in Section 3.5.3 and $P^C(A_j)$ obtained in step 4.
7) Use $UR^I(t) = 1 - P_u(t) + P_u(t)\ UR^C(t)$ (i.e. (3.4)) to integrate $P_u(t)$ evaluated in step 3 and $UR^C(t)$ evaluated in step 6 to obtain $UR^I(t)$, the PMS unreliability considering effects of ELC.

3.5.5 Illustrative Example

Example 3.6 Consider the data gathering PMS in Section 2.3.3. The excellent level is analyzed here for illustration. Based on the high-level FT in Figure 2.1, the detailed FT is shown in Figure 3.15, where four types of components (A_a, A_b), (B_a), (C_a, C_b),

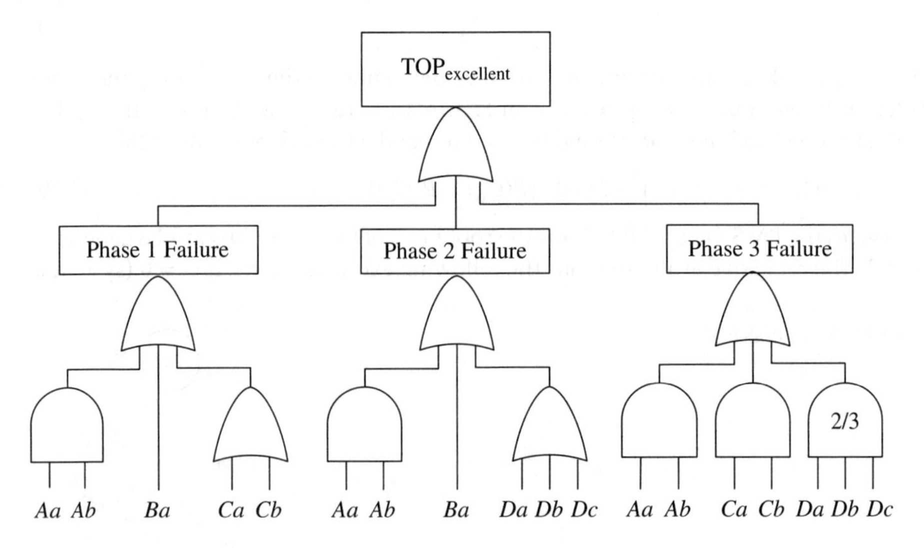

Figure 3.15 PMS FT model [25] (P(excellent) = 1 – P(TOP$_{excellent}$)).

Table 3.4 Failure parameters (λ and μ are in 10^{-6} hr^{-1}) and coverage factors.

Component Type		*A*	*B*	*C*	*D*
Phase 1	p/λ	$p = 1e-4$	$\lambda = 1.5$	$p = 2.5e-3$	$p = 1e-3$
	c	0.99	0.97	0.97	0.99
Phase 2	p/λ	$p = 1e-4$	$\lambda = 1.5$	$\lambda = 1$	$p = 2e-3$
	c	0.99	0.97	0.99	0.99
Phase 3	$p/\mu,\alpha$	$p = 1e-4$	$p = 1e-4$	$\mu = 1.6, \alpha = 2$	$p = 1e-4$
	c	0.99	0.97	1	0.97

(D_a, D_b, D_c) are involved in a three-phased mission. The probability that the data gathering PMS is at the excellent level is the complement of the occurrence probability of the top event in Figure 3.15.

Based on the PMS FT, one of the two type *A* components is required in all of the three mission phases, i.e. either A_a or A_b must be functioning all the time; the type *B* component, i.e. (B_a) is required and must be functioning in phase 1 and phase 2; both type *C* components, i.e. (C_a, C_b) are required in phase 1 and at least one of them must be functioning in phase 3; the type *D* components, i.e. (D_a, D_b, D_c) are not required in phase 1, but required in phase 2 (all must be functioning) and phase 3 (at least two must be functioning). Refer to [25] for more details on the example system.

To illustrate that the PMS SEA method is applicable to arbitrary types of component *ttf* distributions, components of the example PMS can experience faults either with a fixed probability p, or following an exponential distribution with constant rate λ, or following a Weibull distribution with parameters μ and α. Table 3.4 gives values of those failure parameters for the four types of components used in the analysis. The parameter values of the same component can vary with phases.

Table 3.4 also presents values of the coverage factor c. The r factor is assumed to be 0. Thus, the value of the s factor is simply obtained as $1 - c$. Durations of the three phases are 33 hours for phase 1, 100 hours for phase 2, and 67 hours for phase 3.

Following the seven-step procedure in Section 3.5.4, $P(\text{TOP}_{\text{excellent}})$ can be analyzed. Particularly, $P_u(t)$ is evaluated as 0.999 734 in step 3. Figure 3.16 shows the PMS BDD model [25] generated from the PMS FT of Figure 3.15 in step 5. In step 6, $UR^C(t)$ is calculated as 0.001 387 by recursively traversing the PMS BDD generated. In step 7, $UR^I(t) = 1 - P_u(t) + P_u(t)\,UR^C(t) = 0.014\,133$. Finally, the probability that the data gathering PMS is at the excellent level is $P_{\text{excellent}} = 1 - UR^I(t) = 0.985\,867$.

3.6 Summary

This chapter discusses imperfect fault coverage, an inherent behavior of FTSs. While there exist diverse types of imperfect coverage (e.g. ELC, FLC, and PDC), this chapter focuses on the modeling and analysis of ELC modeled by the IPCM. Under the IPCM, each component is associated with a set of three coverage factors, reflecting the effectiveness of the system recovery and reconfiguration mechanism in the case of a fault

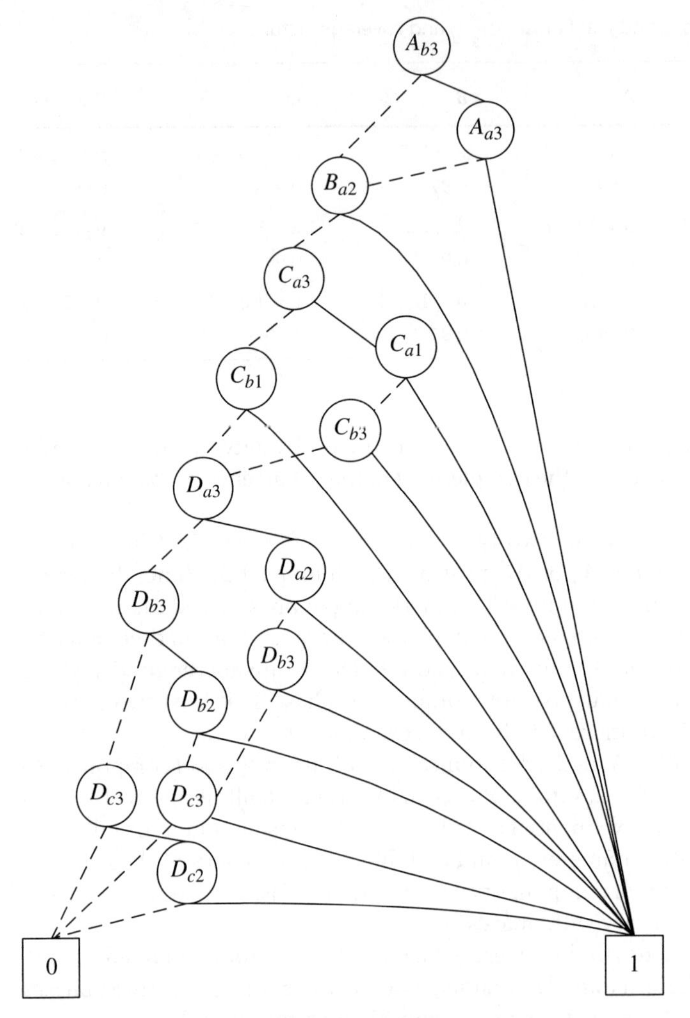

Figure 3.16 PMS BDD generated using $A_b < A_a < B_a < C_a < C_b < D_a < D_b < D_c$.

occurring to this component. Both an explicit method based on BDD (called BEM) and an implicit method based on the total probability law (called SEA) are discussed for the reliability analysis of binary-state systems subject to the ELC. Extensions of the SEA method for analyzing multi-state systems and phased-mission systems subject to the ELC are further presented. The SEA methodology enables the users to apply reliability analysis software that does not consider ELC (e.g. BDD for binary-state systems, MMDD for multi-state systems, and PMS BDD for phased-mission systems), and generate the final system reliability considering effects of ELC via modifying problem input and output.

References

1. Amari, S.V., Pham, H., and Dill, G. (2004). Optimal design of k-out-of-n: G subsystems subjected to imperfect fault-coverage. *IEEE Transactions on Reliability* 53 (4): 567–575.
2. Myers, A.F. (2006). *FCASE: Flight Critical Aircraft System Evaluation.*
3. Johnson, B.W. (1989). *Design and Analysis of Fault Tolerant Digital Systems.* Addison-Wesley.
4. Shooman, M.L. (2002). *Reliability of Computer Systems and Networks: Fault Tolerance, Analysis, and Design.* Wiley.
5. Arnold, T.F. (1973). The concept of coverage and its effect on the reliability model of a repairable system. *IEEE Transactions on Computers* C-22: 325–339.
6. Dugan, J.B. (1989). Fault trees and imperfect coverage. *IEEE Transactions on Reliability* 38 (2): 177–185.
7. Dugan, J.B. and Trivedi, K.S. (1989). Coverage modeling for dependability analysis of fault-tolerant systems. *IEEE Transactions on Computers* 38 (6): 775–787.
8. Myers, A.F. (2007). k-out-of-n: G system reliability with imperfect fault coverage. *IEEE Transactions on Reliability* 56 (3): 464–473.
9. Levitin, G. and Amari, S.V. (2009). Three types of fault coverage in multi-state systems. In: *Proceedings of the 8th International Conference on Reliability, Maintainability and Safety (ICRMS)*, 122–127. Chengdu, China.
10. Myers, A. and Rauzy, A. (2008). Assessment of redundant systems with imperfect coverage by means of binary decision diagrams. *Reliability Engineering & System Safety* 93 (7): 1025–1035.
11. Amari, S.V., Myers, A.F., Rauzy, A., and Trivedi, K.S. (2008). Imperfect coverage models: status and trends. In: *Handbook of Performability Engineering*. London: Springer.
12. Levitin, G. and Amari, S.V. (2008). Multi-state systems with static performance dependent fault coverage. *Proc. IMechE, PartO: Journal of Risk and Reliability* 222 (2): 95–103.
13. Chang, Y., Amari, S.V., and Kuo, S. (2004). Computing system failure frequencies and reliability importance measures using OBDD. *IEEE Transactions on Computers* 53 (1): 54–68.
14. Peng, R., Zhai, Q., Xing, L., and Yang, J. (2014). Reliability of demand-based phased-mission systems subject to fault level coverage. *Reliability Engineering & System Safety* 121: 18–25.
15. Levitin, G. and Amari, S.V. (2008). Multi-state systems with multi-fault coverage. *Reliability Engineering & System Safety* 93: 1730–1739.
16. Bouricius, W.G., Carter, W.C., and Schneider, P.R. (1969). Reliability modeling techniques for self-repairing computer systems. In: *Proceedings of 24th Ann. ACM., National Conference*, 295–309. New York, NY, USA.
17. Cukier, M., Powell, D., and Arlat, J. (1999). Coverage estimation methods for stratified fault-injection. *IEEE Transactions on Computers* 48 (7): 707–723.

18 Dugan, J.B. and Doyle, S.A. (1996). New results in fault-tree analysis. In: *Tutorial Notes of Annual Reliability and Maintainability Symposium*, Las Vegas, Nevada, USA.
19 Xing, L. and Amari, S.V. (2015). *Binary Decision Diagrams and Extensions for System Reliability Analysis*. MA: Wiley-Scrivener.
20 Amari, S.V., Dugan, J.B., and Misra, R.B. (1999). A separable method for incorporating imperfect coverage in combinatorial model. *IEEE Transactions on Reliability* 48 (3): 267–274.
21 Xing, L. and Dai, Y. (2009). A new decision diagram based method for efficient analysis on multi-state systems. *IEEE Transactions on Dependable and Secure Computing* 6 (3): 161–174.
22 Chang, Y.-R., Amari, S.V., and Kuo, S.-Y. (2005). OBDD-based evaluation of reliability and importance measures for multistate systems subject to imperfect fault coverage. *IEEE Transactions on Dependable and Secure Computing* 2 (4): 336–347.
23 Shrestha, A., Xing, L., and Amari, S.V. (2010). Reliability and sensitivity analysis of imperfect coverage multi-state systems. In: *Proceedings of the 56th Annual Reliability & Maintainability Symposium*. San Jose, CA, USA.
24 Esary, J.D. and Ziehms, H. (1975). Reliability analysis of phased missions. In: *Reliability and Fault Tree Analysis: Theoretical and Applied Aspects of System Reliability and Safety Assessment*. Philadelphia, PA: SIAM.
25 Xing, L. and Dugan, J.B. (2002). Analysis of generalized phased mission system reliability, performance and sensitivity. *IEEE Transactions on Reliability* 51 (2): 199–211.
26 Zang, X., Sun, H., and Trivedi, K.S. (1999). A BDD-based algorithm for reliability analysis of phased-mission systems. *IEEE Transactions on Reliability* 48 (1): 50–60.
27 Levitin, G., Xing, L., and Amari, S.V. (2012). Recursive algorithm for reliability evaluation of non-repairable phased mission systems with binary elements. *IEEE Transactions on Reliability* 61 (2): 533–542.
28 Levitin, G., Amari, S.V., and Xing, L. (2013). Algorithm for reliability evaluation of non-repairable phased-mission systems consisting of gradually deteriorating multi-state elements. *IEEE Transactions on Systems, Man, and Cybernetics: Systems* 43 (1): 63–73.
29 Mura, I. and Bondavalli, A. (2001). Markov regenerative stochastic petri nets to model and evaluate phased mission systems dependability. *IEEE Transactions on Computers* 50 (12): 1337–1351.
30 Dugan, J.B. (1991). Automated analysis of phased-mission reliability. *IEEE Transactions on Reliability* 40 (1): 45–52, 55.
31 Altschul, R.E. and Nagel, P.M. (1987). The efficient simulation of phased fault trees. In: *Proceedings of the Annual Reliability and Maintainability Symposium*, 292–296, Philadelphia, Pennsylvania, USA.
32 Xing, L. and Levitin, G. (2013). BDD-based reliability evaluation of phased-mission systems with internal/external common-cause failures. *Reliability Engineering & System Safety* 112: 145–153.
33 Xing, L., Tannous, O., and Dugan, J.B. (2012). Reliability analysis of non-repairable cold-standby systems using sequential binary decision diagrams. *IEEE Transactions on Systems, Man, and Cybernetics, Part A: Systems and Humans* 42 (3): 715–726.

34 Zhai, Q., Peng, R., Xing, L., and Yang, J. (2013). BDD-based reliability evaluation of k-out-of-(n+k) warm standby systems subject to fault-level coverage. *Proc IMechE, Part O, Journal of Risk and Reliability* 227 (5): 540–548.

35 Bouissou, M., Bruyere, F., and Rauzy, A. (1997). BDD based fault-tree processing: a comparison of variable ordering heuristics. In: *Proceedings of ESREL Conference*, Lisbon, Portugal.

36 Bouissou, M. (1996). An ordering heuristics for building binary decision diagrams from fault-trees. In: *Proceedings of the Annual Reliability and Maintainability Symposium*. Las Vegas, NV, USA.

37 Butler, K.M., Ross, D.E., Kapur, R., and Mercer, M.R. (1991). Heuristics to compute variable orderings for efficient manipulation of ordered BDDs. In: *Proceedings of the 28th Design Automation Conference*. San Francisco, California, USA.

38 Xing, L. and Dugan, J.B. (2004). Comments on PMS BDD generation. *IEEE Transactions on Reliability* 53 (2): 169–173.

4

Modular Imperfect Coverage

A hierarchical system (HS) is a system whose underlying architecture can be characterized by multiple layers, with each layer housing different modules and/or components. In an HS, the failure behavior of an upper level often relies on the failure behavior of its lower level(s) [1, 2].

Chapter 3 focuses on the traditional *kill-all* imperfect fault coverage, where an uncovered component fault can lead to the failure of the entire system even when adequate redundancies still remain [3, 4]. In an HS, however, the hierarchical nature of the system may aid in the fault coverage [2, 5]: if an undetected fault escapes from one level of the system, it may be tolerated at a higher level. In other words, the extent of the damage from an uncovered component fault in a certain layer of a system does not necessarily cause the entire system loss. On the other hand, a fault can cause the entire system to fail only if it remains uncovered through all levels of the system hierarchy.

In summary, the extent of the damage from an uncovered component fault occurring in an HS may exhibit multiple levels due to the hierarchical recovery: it can fail the current layer in which the fault occurs only, or it can fail some layer(s) above the current layer, or it can fail the entire system. This chapter introduces the model of representing such multilevel imperfect coverage behavior and methods of considering the behavior in the reliability analysis of nonrepairable and repairable HSs.

4.1 Modular Imperfect Coverage Model

The modular imperfect coverage model (MIPCM) was first introduced in [2] to analyze effects of the multiple levels of uncovered failure (UF) modes for components in HSs. Figure 4.1 illustrates the general structure of MIPCM for a component located at layer i of an HS with a total of L layers.

Similar to the one-level imperfect coverage model (IPCM) discussed in Chapter 3, the entry point of MIPCM represents that a component located at layer i experiences a fault. The $(L-i+3)$ exits represent all possible outcomes of the fault recovery process triggered by the occurrence of the component fault. The exits R and C have the same meaning as in the traditional IPCM in Figure 3.1. The remaining $(L-i+1)$ exits correspond to different levels of component UF modes. Particularly, if the component uncovered fault happening in layer i is covered at layer $i+1$, then the layer i UF exit $S-i$ is reached; if the uncovered fault remains uncovered at layer $i+1$, but is covered

Dynamic System Reliability: Modeling and Analysis of Dynamic and Dependent Behaviors, First Edition. Liudong Xing, Gregory Levitin and Chaonan Wang.

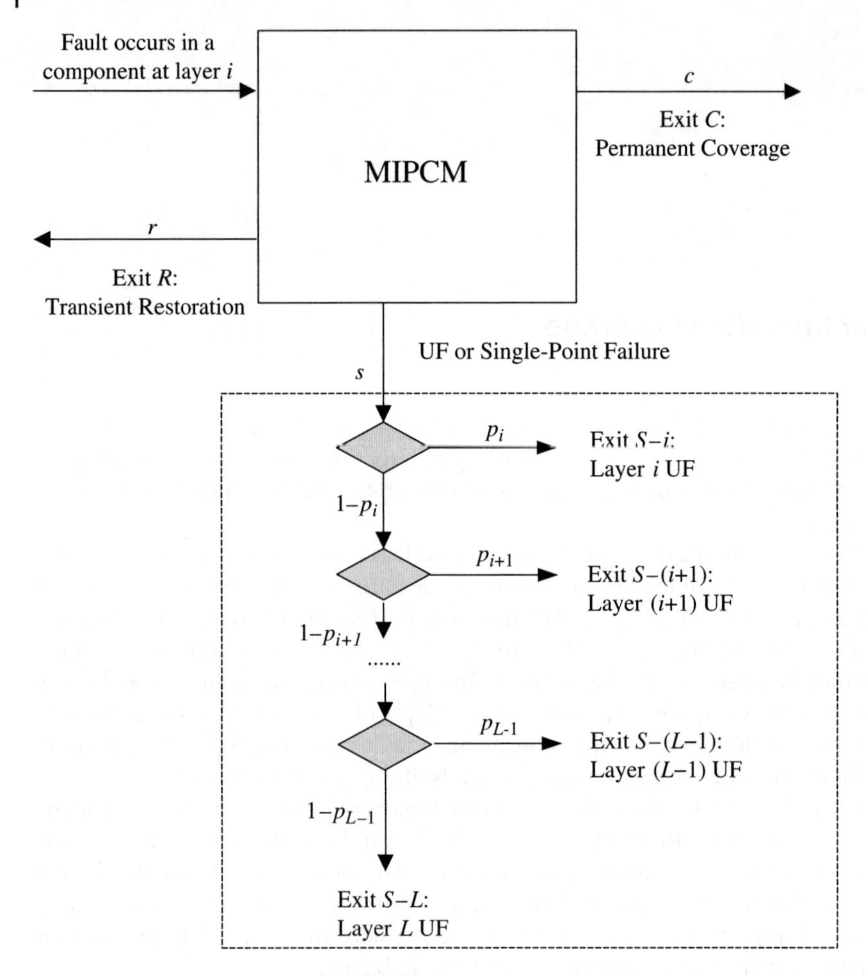

Figure 4.1 General structure of MIPCM [2].

at layer $i+2$, then layer $(i+1)$ UF exit $S-(i+1)$ is reached; and if the component fault remains uncovered through all layers of the system hierarchy and hence crashes the entire system, then the layer L UF exit $S-L$ is reached.

In Figure 4.1, p_k $(k = i \ldots L-1)$ represents the conditional probability that an uncovered component fault escapes from layer k but is covered at layer $k+1$ given that an uncovered fault takes place in layer i. Hence, the probabilities of taking the UF exits $S-i, S-(i+1), \ldots, S-L$ are respectively, $s{\cdot}p_i$, $s{\cdot}(1-p_i){\cdot}p_{i+1}, \ldots, s \cdot \prod_{j=i}^{L-1}(1-p_j)$.

Let $q_{Ai}(t)$ represent the cumulative failure function of component A located at layer i of the system under study. Let $n_{Ai}(t)$, $c_{Ai}(t)$, $u_{Ak}(t)$ denote the probability that A is functioning, failed covered, and layer k failed uncovered $(k = i \ldots L)$, respectively. Equation (4.1) gives their evaluation method:

$$n_{Ai}(t) = 1 - q_{Ai}(t) + q_{Ai}(t) \cdot r_{Ai},$$

$$c_{Ai}(t) = q_{Ai}(t) \cdot c_{Ai},$$

$$u_{Ak}(t) = \begin{cases} q_{Ai}(t) \cdot s_{Ai} \cdot p_{Ai} & k = i, \\ q_{Ai}(t) \cdot s_{Ai} \cdot \prod_{j=i}^{k-1} (1 - p_{Aj}) \cdot p_{Ak} & i < k < L, \\ q_{Ai}(t) \cdot s_{Ai} \cdot \prod_{j=i}^{L-1} (1 - p_{Aj}) & k = L. \end{cases} \quad (4.1)$$

In Section 4.2, a hierarchical and separable approach of incorporating MIPCM in the reliability analysis of nonrepairable HSs is presented. In Section 4.3, a separable Markov-based method is presented for incorporating effects of MIPCM in analyzing repairable HSs.

Phased-mission systems (PMSs) with combinatorial phase requirements can be considered as a special class of HSs, where two UF modes from an uncovered component fault are identified: a phase UF causing an immediate phase failure and a mission UF causing the entire system failure. Refer to [6] for a combinatorial ternary decision diagram-based method for the reliability analysis of PMSs with combinatorial phase requirements and modular imperfect coverage.

4.2 Nonrepairable Hierarchical System

Based on the simple and efficient algorithm (SEA) of [7, 8] for addressing the one-level IPCM, Figure 4.2 illustrates a general solution methodology of analyzing an HS subject

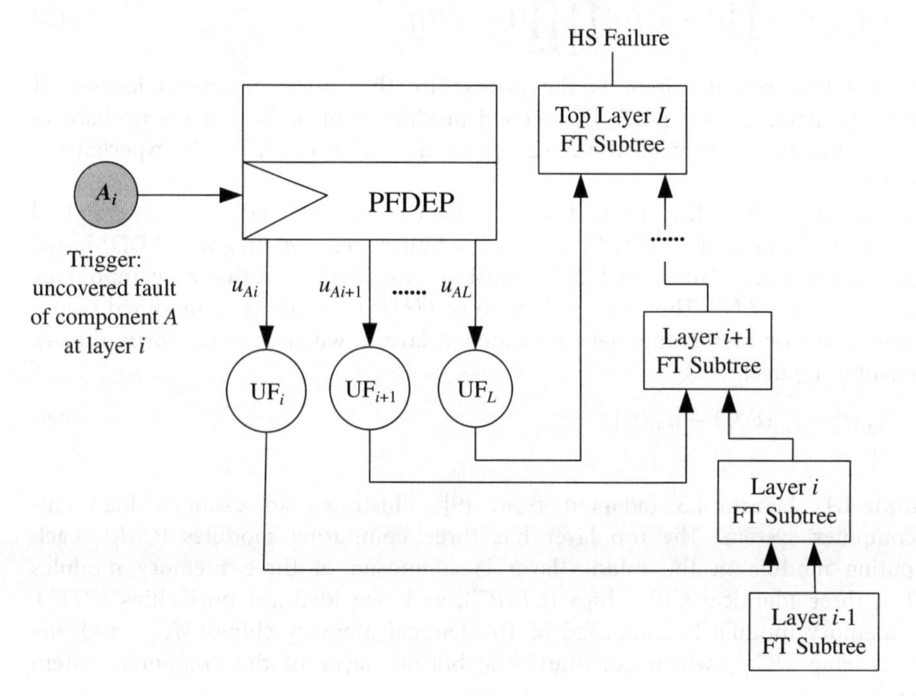

Figure 4.2 Hierarchical FT solution to consider MIPCM [5].

to the modular imperfect fault coverage. The HS fault tree (FT) is composed of independent fault subtrees for each layer. The layer subtrees are then solved in a hierarchical and bottom-up manner, where a subtree is replaced by a single component in its parent layer subtrees whose occurrence probability is the occurrence probability of the corresponding layer subtree.

The probabilistic functional dependency (PFDEP) gate in Figure 4.2 has a trigger input event (representing the occurrence of an uncovered component fault at layer i), and one or more dependent events (UF events of different layers). When the trigger event happens, the dependent events are then forced to occur with certain (different) probabilities given by $u_{Ak}(t)$ of (4.1). Each dependent event is distributed to a corresponding layer subtree and makes its contribution to the failure of that layer.

To analyze each layer subtree, a separable approach is adopted [2]. Consider two mutually exclusive events related to the failure of layer i: E_1: 1 or more components (including components belonging to layer i and components from lower layers (1 ... $i-1$) experience layer i UF; E_2: no components undergo layer i UF.

Let N_i represent the number of components in layer i and E represent an event that the system fails at layer i. According to the total probability law, the occurrence probability of failure of layer i can be evaluated as:

$$UR_{\text{layer } i} = P(E) = \sum_{i=1}^{2}[P(E_i) \cdot P(E \mid E_i)]. \tag{4.2}$$

In (4.2),

$$P(E_1) = 1 - \prod_{A=1}^{N_i}[1 - u_{Ai}(t)] \prod_{k=1}^{i-1}\prod_{B=1}^{N_k}[1 - u_{Bi}(t)], \tag{4.3}$$

where the first product item is the probability that no components located at layer i experience a layer i UF. The second product item in (4.3) is the probability that no components from all the lower layers (i.e. layer 1, …, i – 1) experience a layer i UF.

In (4.2), $P(E_2) = 1 - P(E_1)$, $P(E|E_1) = 1$, $P(E|E_2)$ can be evaluated using any standard approach that ignores the MIPCM such as the binary decision diagram (BDD)-based method (Section 2.4). However, $P(E|E_2)$ must be evaluated given that no components experience a layer i UF. Thus, the evaluation of $P(E|E_2)$ should use a modified failure function $\widetilde{q}_{Ai}(t)$ for each component A located at layer i, which is in the form of (4.4), instead of $q_{Ai}(t)$ itself.

$$\widetilde{q}_{Ai}(t) = c_{Ai}(t)/[1 - u_{Ai}(t)]. \tag{4.4}$$

Example 4.1 Figure 4.3 (adapted from [9]) illustrates an example hierarchical computer system. The top layer has three computing modules (CM_i). Each computing module in the middle layer is composed of three memory modules ($MM_{i,j}$), three identical CPU chips ($CPUC_i$), and two identical port chips (PTC_i). Each memory module is composed of 10 identical memory chips ($MC_{i,j}$) and one interface chip ($IC_{i,j}$), which constitute the bottom layer of the computer system hierarchy.

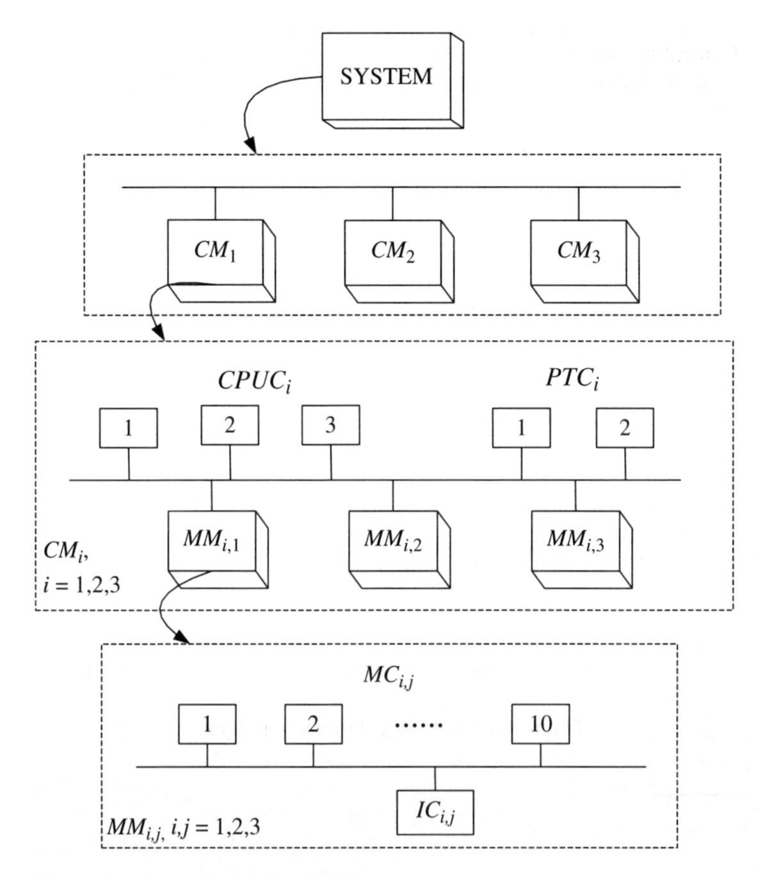

Figure 4.3 An example of an HS [2].

The following lists the operational criteria of each layer:

- *Bottom layer*. At least eight memory chips and the one interface chip have to be operational for the memory module to be functioning.
- *Middle layer*. At least two memory modules, two CPU chips, and the one port chip must be functioning to make each computing module operational.
- *Top layer*. At least two computing modules must be operational for the entire system to be functioning.

Figures 4.4–4.6 give FTs representing the failure criteria of each layer. Table 4.1 lists input parameter values for reliability analysis of the example HS subject to MIPCM. Mission time t considered for the analysis is 200 hours.

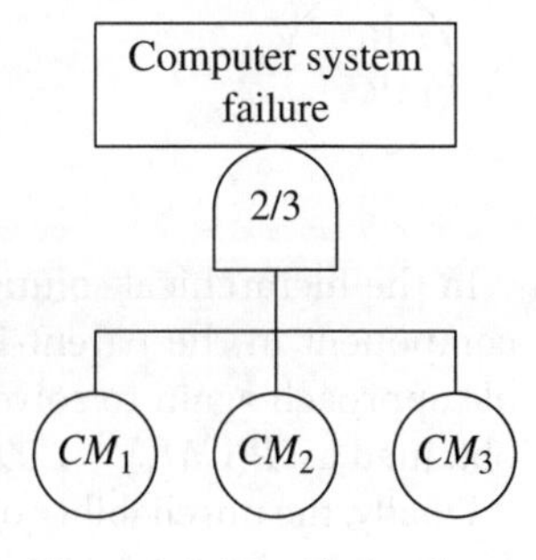

Figure 4.4 FT at the top layer.

The hierarchical analysis starts with the evaluation of the bottom-layer FT model in Figure 4.6 using the separable approach presented in this section, which gives the failure probability of each memory module (MM) in each of the three computing modules as: $P(MM_1) = 4.3799e-5$, $P(MM_2) = 8.3796e-5$, $P(MM_3) = 1.2379e-4$.

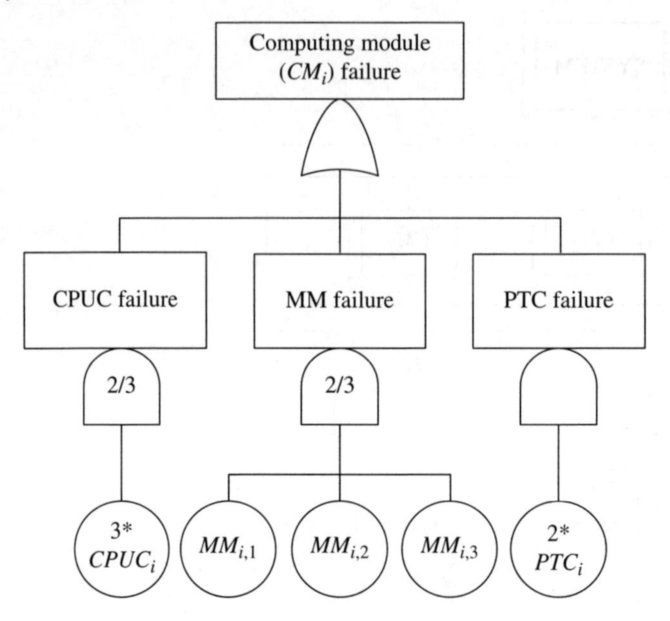

Figure 4.5 FT of CM_i at the middle layer.

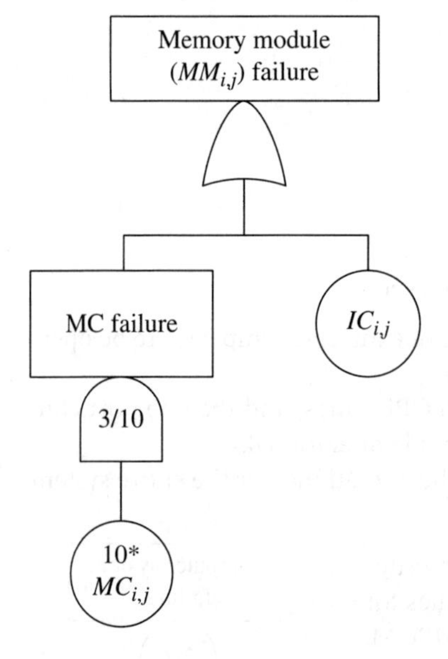

Figure 4.6 FT of $MM_{i,j}$ at the bottom layer.

In the hierarchical solution depicted by Figure 4.2, each MM is replaced by a single component in the parent-layer sub-FT as shown in Figure 4.5. Applying the separable approach again to solve the FT of Figure 4.5, the failure probability of each CM is obtained as: $P(CM_1) = 1.0915e-5$, $P(CM_2) = 2.1356e-5$, $P(CM_3) = 3.1917e-5$.

Finally, the unreliability of the entire computer HS is obtained by solving the top-layer FT model in Figure 4.4 as $UR_{\text{HS}} = 3.2588e-6$.

Table 4.1 Input failure parameters and coverage factors.

	Failure rate (10^{-6}/hr)			
HS component	CM_1	CM_2	CM_3	**Coverage factors**
CPUC	0.6	1.2	1.8	$r = 0, c = 0.99, p_2 = 0.95$
PTC	0.6	1.2	1.8	$r = 0, c = 0.97, p_2 = 0.95$
MC	0.2	0.2	0.2	$r = 0, c = 0.99, p_1 = 0.95, p_2 = 0.99$
IC	0.2	0.4	0.6	$r = 0, c = 1$

4.3 Repairable Hierarchical System

For HSs with components subject to independent repair, a separable approach similar to that in Section 4.2 is combined with the Markov method for the system availability analysis [2].

The possible states for each component A located at layer i of a repairable HS include: operational state (A_{iO}), covered failure state (A_{iC}), layer i UF state (A_{iUi}), layer $(i+1)$ UF state ($A_{iU_{i+1}}$), ... and layer L UF state (A_{iU_L}). Figure 4.7 shows the continuous time Markov chain (CTMC) modeling the failure and repair behavior of component A at layer i.

Solving the CTMC of Figure 4.7, the state occupation probability for each state $P(A_{iO})$, $P(A_{iC})$, $P(A_{iUi})$, $P(A_{iUi+1})$, ..., $P(A_{iUL})$ can be obtained. The separable approach presented in Section 4.2 can then be similarly applied to separate the consideration of layer i UF from the combinatorics of the solution to evaluating the unavailability of a particular layer.

Specifically, the layer i unavailability can be decomposed into two disjoint parts: layer i UF probability denoted by $UA_{\text{uncovered}}$ and layer i covered failure probability denoted by UA_{covered}. Based on the total probability law, the unavailability of layer i can be

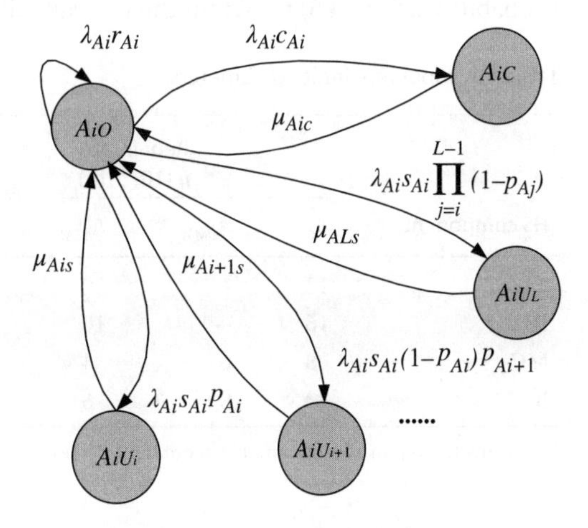

Figure 4.7 CTMC for component A at layer i [2].

obtained as:

$$UA_{\text{layer } i} = UA_{\text{uncovered}} + (1 - UA_{\text{uncovered}}) \cdot UA_{\text{covered}}, \tag{4.5}$$

where similar to (4.3), $UA_{\text{uncovered}} = 1 - \prod_{A=1}^{N_i} [1 - P(A_{iUi})] \cdot \prod_{k=1}^{i-1} \prod_{B=1}^{N_k} [1 - P(B_{kUi})]$. UA_{covered} in (4.5) should be evaluated given that no components experience layer i UF. Hence, before calculating UA_{covered}, the covered failure state occupation probability $P(A_{iC})$ should be modified to a conditional probability $\widetilde{P}(A_{iC})$ conditioned on no layer i UF occurring using (4.6).

$$\widetilde{P}(A_{iC}) = \frac{P(A_{iC})}{1 - P(A_{iUi})}. \tag{4.6}$$

Using these conditional covered failure state probabilities, UA_{covered} can then be evaluated using the BDD-based method.

Based on the hierarchical solution depicted in Figure 4.2, the solution to each layer's unavailability can be combined hierarchically to obtain the unavailability of the entire HS.

Example 4.2 Consider the hierarchical computer system in Example 4.1. Repair rates for each component upon reaching different failure states are given in Table 4.2, which are identical in the three CMs. A_C represents the covered failure state, A_{UMM}, A_{UCM}, A_{Usys} denote the UF state with respect to the MM, CM, and the entire system, respectively.

According to the method presented in this section, a CTMC is constructed for the memory and interface chips at the bottom layer, and a CTMC for the CPU and port chips at the middle layer as shown in Figures 4.8 and 4.9, respectively. Based on the parameters of Tables 4.1 and 4.2, the steady-state probabilities for each component can be obtained by solving balance equations of those CTMCs. Note that the steady-state probabilities for different CMs are different since the failure rates are different for each component in the three CMs, which are given in Tables 4.3–4.5.

An ordinary BDD (Section 2.4) is constructed for each layer using the covered failure state variables A_C, and then is evaluated using modified covered state probabilities calculated from (4.6). The BDD evaluation result is then combined with the layer UF probability using (4.5) to obtain the unavailability of this particular layer.

Table 4.2 Input repair rate parameters.

	Repair rate μ (10^{-4} /hr)			
HS component	A_C	A_{UMM}	A_{UCM}	A_{Usys}
CPUC	12	—	10	8
PTC	16	—	10	8
MC	8	6	4	2
IC	30	8	6	4

(— means the repair rate parameter is not applicable)

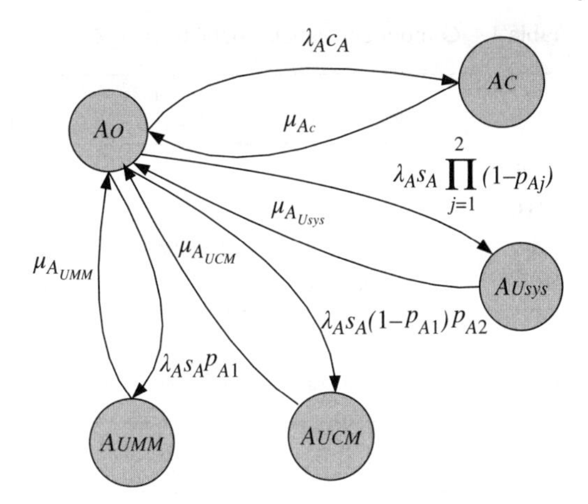

Figure 4.8 CTMC for MC, IC at the bottom layer.

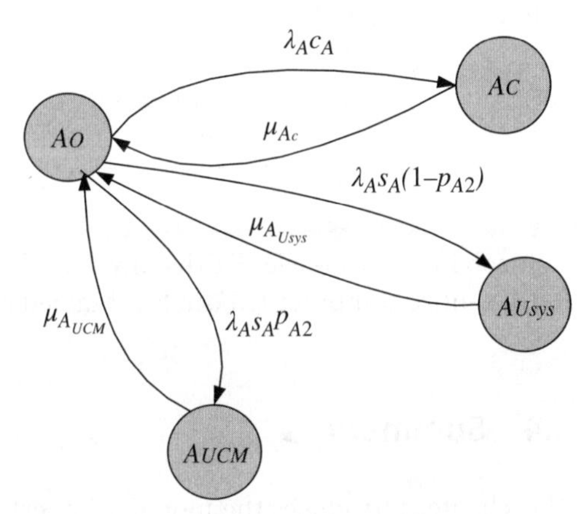

Figure 4.9 CTMC for CPUC, PTC at the middle layer.

Table 4.3 Component state probabilities in CM_1.

	$P(A_C)$	$P(A_{UMM})$	$P(A_{UCM})$	$P(A_{Usys})$
CPUC	$4.9475e-4$	—	$5.6971e-6$	$3.7481e-7$
PTC	$3.6361e-4$	—	$1.7093e-5$	$1.1246e-6$
MC	$2.4744e-4$	$3.1659e-6$	$2.4744e-7$	$4.9987e-9$
IC	$6.5996e-5$	0	0	0

By solving the bottom-layer FT using the separable approach, the steady-state unavailability for the MM at the middle layer for CM_1, CM_2, and CM_3 can be obtained as $9.7654e-5$, $1.6364e-4$, and $2.2961e-4$, respectively.

By solving the middle-layer FT containing MMs whose occurrence probabilities are the steady-state unavailabilities obtained in previous step, the steady-state unavailability for each of the three CMs can be obtained as $5.4645e-5$, $1.0853e-4$, and $1.6428e-4$, respectively.

Table 4.4 Component state probabilities in CM_2.

	$P(A_C)$	$P(A_{UMM})$	$P(A_{UCM})$	$P(A_{Usys})$
CPUC	$9.8901e-4$	—	$1.1389e-5$	$7.4925e-7$
PTC	$7.2694e-4$	—	$3.4174e-5$	$2.2483e-6$
MC	$2.4744e-4$	$3.1659e-6$	$2.4744e-7$	$4.9987e-9$
IC	$1.3198e-4$	0	0	0

Table 4.5 Component state probabilities in CM_3.

	$P(A_C)$	$P(A_{UMM})$	$P(A_{UCM})$	$P(A_{Usys})$
CPUC	0.0015	—	$1.7074e-5$	$1.1233e-6$
PTC	0.0011	—	$5.1241e-5$	$3.3711e-6$
MC	$2.4744e-4$	$3.1659e-6$	$2.4744e-7$	$4.9987e-9$
IC	$1.9796e-4$	0	0	0

By solving the top-layer FT that involves the three CMs, the steady-state unavailability for the entire computer HS can be obtained as $UA_{\mathrm{HS}} = 2.0413e-5$.

4.4 Summary

This chapter introduces the modular imperfect coverage for systems with hierarchical structures. The MIPCM extends the traditional one-level IPCM by considering multiple levels of UFs resulting from the layered recovery of HSs. Based on FT modeling, the hierarchical and separable methodology is presented for addressing effects of MIPCM in the reliability analysis of nonrepairable HSs and availability analysis of repairable HSs.

References

1 Xing, L. (2002). Dependability modeling and analysis of hierarchical computer-based systems. *Ph.D. Dissertation, Electrical and Computer Engineering, University of Virginia,* Charlottesville, VA, USA.

2 Xing, L. and Dugan, J.B. (2001). Dependability analysis of hierarchical systems with modular imperfect coverage. In: *Proceedings of the 19th International System Safety Conference (ISSC'2001),* 347–356. Huntsville, AL. The International System Safety Society.

3 Doyle, S.A., Dugan, J.B., and Patterson-Hine, A. (1995). A combinatorial approach to modeling imperfect coverage. *IEEE Transactions on Reliability* 44 (1): 87–94.

4 Dugan, J.B. (1989). Fault trees and imperfect coverage. *IEEE Transactions on Reliability* 38 (2): 177–185.

5 Xing, L. (2005). Reliability modeling and analysis of complex hierarchical systems. *International Journal of Reliability, Quality and Safety Engineering* 12 (6).
6 Xing, L. and Dugan, J.B. (2004). A separable ternary decision diagram based analysis of generalized phased-mission reliability. *IEEE Transactions on Reliability* 53 (2): 174–184.
7 Amari, S.V., Dugan, J.B., and Misra, R.B. (1999). A separable method for incorporating imperfect coverage in combinatorial model. *IEEE Transactions on Reliability* 48 (3): 267–274.
8 Xing, L. and Dugan, J.B. (2002). Analysis of generalized phased-mission systems reliability, performance and sensitivity. *IEEE Transaction on Reliability* 51 (2): 199–211.
9 Sune, V. and Carrasco, J.A. (2001). A failure-distance based method to bound the reliability of non-repairable fault-tolerant systems without the knowledge of minimal cutsets. *IEEE Transactions on Reliability* 50 (1): 60–74.

5

Functional Dependence

Functional dependence takes place when the malfunction of one system component (referred to as *the trigger*) causes other components (referred to as *dependent components*) within the same system to become unusable or inaccessible. For example, in a computer system, peripheral devices such as keyboards and monitors are accessed through I/O controllers. If the I/O controller malfunctions, the peripheral devices connected to it become unusable [1]. In other words, the peripheral devices have functional dependence on the I/O controller. Another example is a computer network system where a computer can access the Internet or communicate with other computers through routers [2]. The router is the trigger component, and computers connected to the router have functional dependence on the router.

The functional dependence behavior is modeled via Functional DEPendence (FDEP) gates in the dynamic fault tree (DFT) analysis [3, 4] (Section 2.3.2). Note that the FDEP gate has no logical output; it is connected to the top gate of the system DFT using a dashed line. In this chapter, a logic OR replacement method is discussed for the reliability analysis of systems with FDEP and perfect fault coverage. A combinatorial approach is presented for reliability analysis of systems subject to imperfect fault coverage.

5.1 Logic OR Replacement Method

For systems with perfect fault coverage, the FDEP behavior can be addressed by replacing each FDEP gate with logic OR gates (one for each dependent component) in the system DFT model [5].

Example 5.1 Consider the DFT in Figure 5.1a with one FDEP gate. The failure of the trigger component A causes the dependent components B, D, and F to become unusable. Figure 5.1b shows the equivalent DFT model after replacing the FDEP gate in Figure 5.1a with three OR gates, one for each of the three dependent components.

The logic OR replacement method works because each dependent component in an FDEP group fails when either it malfunctions intrinsically or its corresponding trigger component malfunctions. With the logic OR replacement process, the original system DFT model can actually be transformed into a traditional static fault tree (FT) model. The resultant model can thus be analyzed using efficient combinatorial approaches, e.g. the binary decision diagram (BDD)-based method presented in Section 2.4 [6–9].

Dynamic System Reliability: Modeling and Analysis of Dynamic and Dependent Behaviors, First Edition. Liudong Xing, Gregory Levitin and Chaonan Wang.

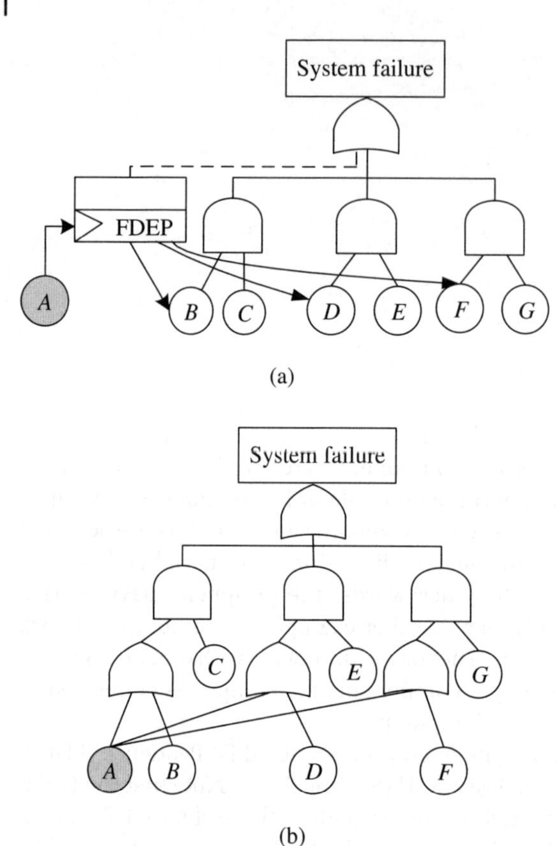

Figure 5.1 An illustrative example. (a) Original DFT model; (b) DFT after the replacement.

For systems with imperfect coverage (IPC) (Chapter 3) [10–12], however, the logic OR replacement method is not applicable. Consider the aforementioned computer system example. To achieve high reliability, the critical I/O controller can be backed up with a standby I/O controller. This standby controller takes over the task upon the failure of the primary controller. Under ideal circumstances, the computer operates correctly as long as one of the two I/O controllers functions correctly. However, it may happen in practice that the malfunction of the primary controller is not successfully detected by the system recovery mechanism, which fails the computer system even though the standby controller is still available. In this case, the computer system is said to be subject to the IPC behavior. The IPC behavior can make significant contributions to the system unreliability for highly reliable systems [13, 14].

Suppose the logic OR replacement method is used for addressing FDEP in reliability analysis of IPC systems. Consider a certain FDEP group where the dependent components are subject to uncovered failures (UFs). In the case of the corresponding trigger component failure, these dependent components are disconnected from the system, but are allowed to contribute to the system UF probability since they could still fail uncovered. However, since those dependent components were disconnected (or isolated), their UFs could not really bring the system down. Hence, the logic OR replacement method generates overestimated unreliability results when being applied to IPC systems.

To avoid overcounting UFs of dependent components in an FDEP group, the state space-based methods, in particular, the continuous time Markov chain (CTMC)-based method [15] can be applied to analyze the entire system or at least its parts with IPC through the modular approach [16–18]. As discussed in Section 2.5, the CTMC method suffers the state-space explosion problem [19, 20] and is typically limited to exponential component *ttf* distributions.

In Section 5.2, a combinatorial method is presented for reliability analysis of IPC systems with FDEP, which can overcome the problems of the logic OR replacement method and the CTMC-based method. The IPCM [3, 6] described in Section 3.2 is used to model the IPC behavior.

5.2 Combinatorial Algorithm

The combinatorial approach aims to successfully "turn off" the disconnected or isolated dependent components (when the corresponding trigger component fails or becomes unavailable) so that they could not contribute to the system UF probability. The approach involves four tasks that, respectively, address UFs of independent trigger components of FDEP, transform the system with FDEP into subsystems without FDEPs, evaluate resultant subsystems, and integrate for the final system unreliability.

5.2.1 Task 1: Addressing UFs of Independent Trigger Components

Based on the simple and efficient algorithm (SEA) (Section 3.3.2), particularly (3.4), the system unreliability is evaluated as:

$$UR_{system}(t) = 1 - P_{u,IT}(t) + P_{u,IT}(t) \cdot Q(t). \tag{5.1}$$

$P_{u,IT}(t)$ in (5.1) is the probability that no independent trigger components undergo UFs, and $Q(t)$ is the conditional system unreliability given that no independent trigger components experiences UFs. Note that in the case of cascading FDEP appearing in the system DFT, the trigger components that also serve as dependent components (referred to as dependent trigger components) are not included in the evaluation of $P_{u,IT}(t)$. According to (3.5), it is computed as

$$P_{u,IT}(t) = \prod_{\forall trigger-i} (1 - s_i \cdot q_i(t)). \tag{5.2}$$

As discussed in Section 3.3.2, for each independent trigger component a conditional failure probability $\widetilde{q}_{trigger-i}(t)$ given that the component undergoes no UFs should be calculated and used for evaluating $Q(t)$. According to (3.6), it is evaluated as $\widetilde{q}_{trigger-i}(t) = \frac{c_i \cdot q_i(t)}{(1-s_i \cdot q_i(t))}$.

5.2.2 Task 2: Generating Reduced Problems Without FDEP

Evaluating $Q(t)$ in (5.1) needs to address effects of FDEP. The transformation technique developed in [21, 22] is utilized to transform a reliability problem with FDEP into reduced problems without FDEP so that the SEA method can be safely applied to components that are still not disconnected or isolated.

Assume the system under analysis involves m elementary independent trigger events (ITEs) (denoted by T_i, $i = 1 \ldots m$). An ITE space is first built, containing 2^m mutually exclusive and collectively exhaustive combined events called ITEs. Each ITE is a distinct and disjoint combination of the m elementary ITEs, illustrated as follows:

$$\begin{aligned} ITE_1 &= \overline{T_1} \cap \overline{T_2} \cap \ldots \overline{T_{m-1}} \cap \overline{T_m}, \\ ITE_2 &= \overline{T}_1 \cap \overline{T_2} \cap \ldots \overline{T_{m-1}} \cap T_m, \\ &\ldots \\ ITE_{2^m-1} &= T_1 \cap T_2 \cap \ldots T_{m-1} \cap \overline{T_m}, \\ ITE_{2^m} &= T_1 \cap T_2 \cap \ldots T_{m-1} \cap T_m. \end{aligned} \tag{5.3}$$

Applying the total probability law, $Q(t)$ in (5.1) can be evaluated as:

$$Q(t) = \sum_{i=1}^{2^m} [P(\text{system fails}|ITE_i) \cdot P(ITE_i)]. \tag{5.4}$$

where $P(ITE_i)$ can be evaluated based on the definition of ITE_i in (5.3) and conditional occurrence probabilities of elementary trigger events involved in the definition of $\widetilde{q}_{trigger-i}(t)$. Evaluating $P(\text{system fails}|ITE_i)$ in (5.4) is a reduced reliability problem, where components affected by ITE_i do not appear in the system reliability model due to the occurrence of ITE_i. The evaluation is detailed in Section 5.2.3.

5.2.3 Task 3: Solving Reduced Reliability Problems

To evaluate $P(\text{system fails}|ITE_i)$, the set of components affected by ITE_i (denoted by S_{ITE_i}) needs to be determined. Define a functional dependence group (FDG) as a group of components that is functionally dependent on the same trigger event. For example, components B, D, and F in Figure 5.1a form the FDG for component A, denoted as $FDG_A = \{B, D, F\}$. The set S_{ITE_i} is determined as the union of all FDGs whose corresponding trigger event takes place when ITE_i happens.

For example, consider an FDEP system with two elementary ITEs T_1 and T_2. The ITE space constructed contains $ITE_1 = \overline{T_1} \cap \overline{T_2}$, $ITE_2 = \overline{T_1} \cap T_2$, $ITE_3 = T_1 \cap \overline{T_2}$, $ITE_4 = T_1 \cap T_2$. For event $ITE_1 = \overline{T_1} \cap \overline{T_2}$ (the occurrence of ITE_1 implies that neither T_1 nor T_2 occurs), set S_{ITE_1} is simply an empty set (no components are affected). For event $ITE_2 = \overline{T_1} \cap T_2$ (the occurrence of ITE_2 implies the occurrence of T_2 only), set S_{ITE_2} is simply FDG_{T_2}. For event $ITE_3 = T_1 \cap \overline{T_2}$ (the occurrence of ITE_3 implies the occurrence of T_1 only), set S_{ITE_3} is simply FDG_{T_1}. For event $ITE_4 = T_1 \cap T_2$ (the occurrence of ITE_4 implies the occurrence of T_1 and T_2), set S_{ITE_4} is the union of FDG_{T_1}, FDG_{T_2}, and $FDG_{T_1 \cap T_2}$. Note that $FDG_{T_1 \cap T_2}$ is applicable and must be included when $T_1 \cap T_2$ serves as a trigger input of some FDEP gate in the system DFT.

5.2.3.1 Expansion Process

In the case of cascading FDEP appearing in the system DFT, the set S_{ITE_i} should be expanded to include all basic events following the *domino* chain of dependent events involved. The expansion process is iterative. During each iteration, if the input trigger event condition of an FDEP gate is satisfied by any combination of components already in S_{ITE_i}, all the dependent basic events of the corresponding FDEP gate should be added to the expanded set. The expansion process ends when no new components are added to the set S_{ITE_i} after the previous iteration. Refer to Section 5.5 for an example of applying this expansion process.

With S_{ITE_i} being identified, a reduced FT model is generated for the reduced problems P(system fails$|ITE_i$) based on the original DFT using the following procedure.

5.2.3.2 Reduced FT Generation Procedure

1. Remove all the ITEs and related FDEP gates from the original system DFT.
2. Replace basic failure events of components appearing in set S_{ITE_i} in the original system DFT by constant logic value 1 (True). After the replacement, all the components affected by ITE_i disappear from the system model. Consequently, they can be successfully "turned off" so that their UFs (if there are any) are not able to contribute to the system failure.
3. Apply Boolean algebra ($0+x=x$, $1+x=1$, $0\cdot x=0$, $1\cdot x=x$) to simplify the DFT after the replacement.

5.2.3.3 Dual Trigger-Basic Event Handling

Note that in the case of an event playing dual roles (particularly, the event being a trigger event of certain FDEP gate and a basic event of other gates at the same time), in step 1 of the above reduced FT generation procedure, only the trigger event identity of a dual event is removed while its basic event identity stays. In step 2, the dual event's basic event identity is replaced by constant logic value 1 (*True*) when the event itself (e.g. T_1) appears in ITE_i; or by constant logic value 0 (*False*) when the complement of the event (e.g. $\overline{T_1}$) appears in ITE_i. If the basic event identity appears as an input of a logic AND gate and it is replaced by 0, and the other input of the AND gate undergoes UFs, Boolean algebra particularly, $0\cdot x=0$ in step 3 cannot be applied so that the UFs can be considered for evaluating P(system fails$|ITE_i$). Refer to Section 5.6 for an illustration of applying this special handling procedure to a system with dual events.

5.2.3.4 Evaluation of P(system fails$|ITE_i$)

Based on the reduced FT generated, if no FDEP remains, the SEA method (Section 3.3.2) is applied to address UFs of the remaining components connected to the system model (i.e. components not affected by ITE_i). Equation (5.5) shows the evaluation of P(system fails$|ITE_i$) by applying (3.4).

$$P(\text{system fails}|ITE_i) = 1 - P_{u,i}(t) + P_{u,i}(t)\cdot Q_i(t), \tag{5.5}$$

where $P_{u,i}(t) = \prod_{\forall cc-k}(1 - s_k \cdot q_k(t))$ is the probability that no connected components (cc) undergo UFs. $Q_i(t)$ in (5.5) is the system unreliability ignoring all the UFs and effects of FDEP; its evaluation requires modifying each cc's failure probability to a conditional probability computed as $\widetilde{q}_{cc-k}(t) = \frac{c_k\cdot q_k(t)}{(1-s_k\cdot q_k(t))}$ (based on (3.6)).

However, the reduced FT for P(system fails$|ITE_i$) may still contain FDEP gate(s) when the cascading FDEP behavior appears in the original DFT. In this case, Tasks 1–3 need to be repeated for evaluating P(system fails$|ITE_i$). Specifically, UFs of the ITEs in the reduced FT (they are actually dependent trigger events in the original DFT) are first separated from the solution combinatorics using the SEA methodology (Task 1). The ITE space is then built based on those ITEs of the reduced FT (Task 2). Each reduced problem is solved using the SEA method in the case of no further FDEP existing in the reduced FT generated or using steps of Tasks 1–3 when any FDEP still remains.

5.2.4 Task 4: Integrating to Obtain Final System Unreliability

After evaluating all of the reduced problems, the final system unreliability considering effects of both IPC and FDEP is obtained by applying (5.4) and (5.1), consecutively.

5.2.5 Algorithm Summary

A systematic algorithm based on the combinatorial approach detailed in the preceding subsections 5.2.1–5.2.4 is summarized as follows:

1. Find all the elementary ITEs in the FT considered and perform the following computations:
   a. Compute $P_{u,IT} = \prod_{\forall trigger-i}(1 - s_i \cdot q_i(t))$.
   b. Compute $\widetilde{q}_{trigger-i}(t) = \frac{c_i \cdot q_i(t)}{(1-s_i \cdot q_i(t))}$ for each independent trigger component i.
2. Construct the ITE space (5.3) with 2^m disjoint ITEs in the case of m elementary ITEs existing in the FT considered.
3. For each ITE_i,
   a. Compute $P(ITE_i)$ using conditional failure probabilities of trigger components computed in step 1.b.
   b. Identify set S_{ITE_i}.
   c. Build the reduced FT model for $P(\text{system fails}|ITE_i)$ using the generation procedure presented in Section 5.2.3.2.
   d. If the reduced FT contains no further FDEP gates, then
      i. Compute $P_{u,i} = \prod_{\forall cc-k}(1 - s_k \cdot q_k(t))$;
      ii. Compute $\widetilde{q}_{cc-k}(t) = \frac{c_k \cdot q_k(t)}{(1-s_k \cdot q_k(t))}$ for each cc (i.e. component appearing in the reduced FT generated in step 3.c;
      iii. Evaluate the reduced FT model generated in step 3.c by applying the BDD-based method (Section 2.4) to obtain Q_i ($\widetilde{q}_{cc-k}(t)$ is used for evaluating Q_i);
      iv. According to (5.5), compute $P(\text{systemfails}|ITE_i) = 1 - P_{u,i} + P_{u,i} \cdot Q_i$.
   e. If the reduced FT contains further FDEP gates, then go back to step 1 and repeat the evaluation procedure based on this reduced FT until $P(\text{system fails}|ITE_i)$ is evaluated.
4. According to (5.4), compute $Q = \sum_{i=1}^{2^m}[P(\text{system fails}|ITE_i) \cdot P(ITE_i)]$ (in the case of cascading FDEPs, this step may also generate Q_i).
5. According to (5.1), compute $UR_{system} = 1 - P_{u,IT} + P_{u,IT} \cdot Q$ (in the case of cascading FDEPs, this step may also generate $P(\text{system fails}|ITE_i)$).

5.2.6 Algorithm Complexity

Step 3 of the algorithm in Section 5.2.5 dominates the complexity of the entire algorithm, involving $O(2^M)$ reduced reliability problems to be solved using any traditional FT reliability analysis method, where M represents the number of trigger components (including independent and dependent) [3]. The BDD-based method adopted was shown to require less memory and be computationally more efficient than other FT reliability analysis methods [3, 8, 9, 23, 24]. Particularly, applying the dynamic programming [25] concept of *memoization*, the computational complexity of the BDD model evaluation is

O(number of non-sink nodes in BDD). Refer to [26] for a detailed storage and computation complexity analysis of an efficient BDD implementation method.

As illustrated through case studies in Sections 5.3–5.6, the reduced reliability problems generated are simplified and independent problems, which could be solved concurrently given available computing resources.

The combinatorial algorithm is applicable to arbitrary types of component *ttf* distributions. It is also computationally more efficient than the CTMC-based method. The number of states (or state equations) in the CTMC model is $w = O(2^n)$ (exponential to the number of components n, usually far greater than the number of trigger components M). The computational time for solving a Markov model with w-states is $O(w^3)$ [27].

Note that in [21] a combinatorial method was suggested for handling FDEP in IPC systems, which can handle noncascading FDEP accurately. However, in the case of cascading FDEP, the trigger components that also simultaneously serve as dependent components (i.e. dependent trigger components) cannot be isolated or turned off appropriately in the algorithm of [21]; their UFs can always contribute to the system failure, leading to overestimated system unreliability results. But the algorithm of [21] is still more accurate than the logic OR replacement method for handling cascading FDEP systems because it can always isolate the purely dependent components (without the trigger role) appropriately. The algorithm presented in this section improves the method of [21] through handling the trigger events involved in cascading FDEP in a hierarchical manner. By doing so, both dependent trigger components and purely dependent components can be turned off appropriately during the evaluation process.

5.3 Case Study 1: Combined Trigger Event

Figure 5.2 illustrates the DFT model of the memory subsystem for a computer system [17, 28]. The memory subsystem has five memory units M_i (i = 1, 2, 3, 4, 5); they are accessible through two memory interface units (MIU_1, MIU_2). In other words, the

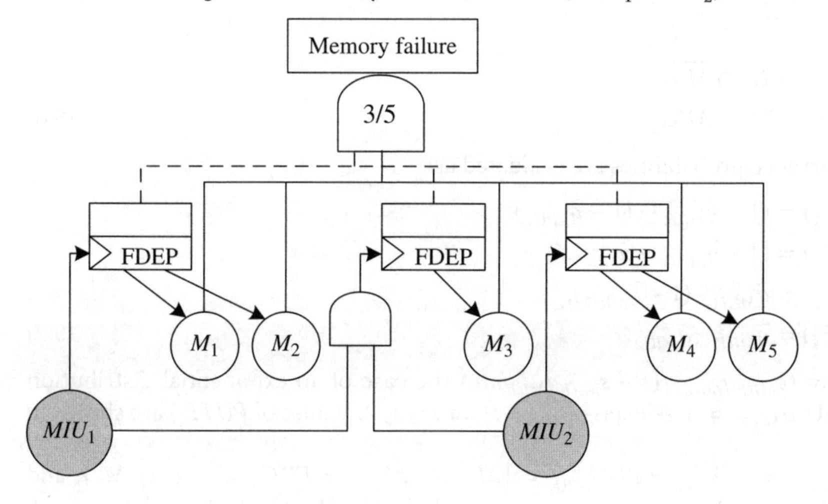

Figure 5.2 DFT of an example memory system [17, 28].

Table 5.1 Input component failure parameters (/hr) and coverage factors.

	Set 1	Set 2	Set 3	Set 4	Set 5
$\lambda_{MIU1}, \lambda_{MIU2}$	1e−6	1e−6	1e−6	1e−6	1e−6
$\lambda_{Mi}, i = 1, 2, \ldots, 5$	1e−4	1e−4	1e−4	1e−4	1e−4
$c_{MIUi}, i = 1, 2$	1	0.9	1	1	1
c_{M1}	0.9	0.9	0.5	1	1
c_{M2}	0.9	0.9	0.5	1	1
c_{M3}	0.9	0.9	0.5	0.5	1
c_{M4}	0.9	0.9	1	0.5	1
c_{M5}	0.9	0.9	1	0.5	1

memory units have FDEP on the MIUs (elementary trigger events). Particularly, as illustrated in the DFT, the following FDEP relationships exist in the system:

- M_1 and M_2 are connected to the system bus via MIU_1; M_1 and M_2 are functionally dependent on MIU_1 (i.e. $FDG_{MIU1} = \{M_1, M_2\}$).
- M_3 is connected to both interfaces implying that M_3 is accessible as long as one of the two interface units is functioning, or when both interface units malfunction M_3 becomes inaccessible. M_3 is functionally dependent on $MIU_1 \cap MIU_2$, which is referred to as a combined trigger event (i.e. $FDG_{MIU1 \cap MIU2} = \{M_3\}$).
- M_4 and M_5 are connected to the system bus via MIU_2; M_4 and M_5 are functionally dependent on MIU_2 (i.e. $FDG_{MIU2} = \{M_4, M_5\}$).

Table 5.1 presents five distinct sets of failure rates and coverage factors for components of the memory subsystem. The r factor is assumed to be ZERO for all the system components. Mission time of $t = 10\,000$ hours is considered.

Applying the combinatorial method, an ITE space with four ITEs is constructed:

$$\begin{aligned} ITE_1 &= \overline{MIU_1} \cap \overline{MIU_2}, \\ ITE_2 &= \overline{MIU_1} \cap MIU_2, \\ ITE_3 &= MIU_1 \cap \overline{MIU_2}, \\ ITE_4 &= MIU_1 \cap MIU_2. \end{aligned} \tag{5.6}$$

Their occurrence probabilities are evaluated as:

$$\begin{aligned} P(ITE_1) &= (1 - \widetilde{q}_{MIU1}) \cdot (1 - \widetilde{q}_{MIU2}), \\ P(ITE_2) &= (1 - \widetilde{q}_{MIU1}) \cdot \widetilde{q}_{MIU2}, \\ P(ITE_3) &= \widetilde{q}_{MIU1} \cdot (1 - \widetilde{q}_{MIU2}), \\ P(ITE_4) &= \widetilde{q}_{MIU1} \cdot \widetilde{q}_{MIU2}, \end{aligned} \tag{5.7}$$

where $\widetilde{q}_{MIUi} = (c_{MIUi} q_{MIUi})/(1 - s_{MIUi} q_{MIUi})$. In the case of an exponential distribution for ttf of MIU, $q_{MIUi} = 1 - \exp(-\lambda_{MIUi} \cdot t)$ for $i = 1, 2$. Values of $P(ITE_i)$ are shown in Table 5.2.

Further, $S_{ITE_1} = \emptyset$, $S_{ITE_2} = FDG_{MIU_2} = \{M_4, M_5\}$, $S_{ITE_3} = FDG_{MIU_1} = \{M_1, M_2\}$, and $S_{ITE_4} = FDG_{MIU_1} \cup FDG_{MIU_2} \cup FDG_{MIU_1 \cap MIU_2} = \{M_1, M_2, M_3, M_4, M_5\}$ are determined.

Table 5.2 Occurrence probabilities of ITEs.

	Set 1	Set 2	Set 3	Set 4	Set 5
$P(ITE_1)$	0.980199	0.982152	0.980199	0.980199	0.980199
$P(ITE_2)$	9.85116e−3	8.883714e−3	9.85116e−3	9.85116e−3	9.85116e−3
$P(ITE_3)$	9.85116e−3	8.883714e−3	9.85116e−3	9.85116e−3	9.85116e−3
$P(ITE_4)$	9.90058e−5	8.035453e−5	9.90058e−5	9.90058e−5	9.90058e−5

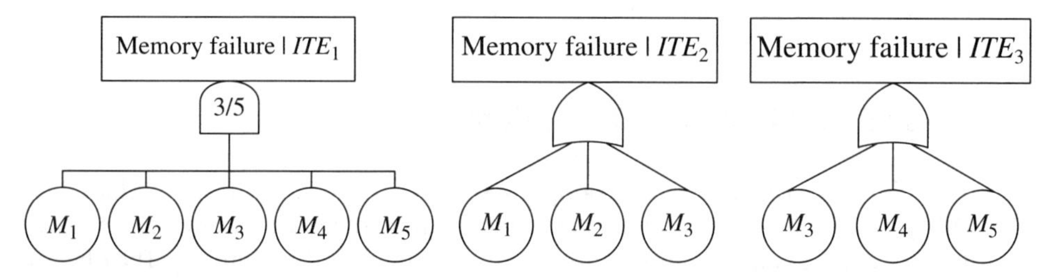

Figure 5.3 Reduced FT models.

Table 5.3 Evaluation results of P(system fails|ITE_i) and Q.

	Set 1	Set 2	Set 3	Set 4	Set 5
P(system fails|ITE_1)	0.780023	0.780023	0.858244	0.858244	0.736436
P(system fails|ITE_2)	0.950212	0.950212	0.950212	0.950212	0.950212
P(system fails|ITE_3)	0.950212	0.950212	0.950212	0.950212	0.950212
P(system fails|ITE_4)	1	1	1	1	1
Q	0.783398	0.783064	0.860070	0.860070	0.740674

Next reduced FT models are generated to evaluate P(system fails|ITE_i). Apparently, P(system fails|ITE_4) = 1 for all the five different parameter settings. Figure 5.3 shows the reduced FTs for the first three reduced problems. Based on the evaluation of these reduced FTs (using the BDD-based method) and (5.5), P(system fails|ITE_i) are obtained as shown in Table 5.3.

According to (5.4), $P(ITE_i)$ and P(system fails|ITE_i) are integrated to obtain Q (last row of Table 5.3). Lastly, according to (5.1) the final system unreliability is obtained. Table 5.4 presents values of the final system unreliability evaluated. It also presents the system unreliability analyzed using the logic OR replacement method.

Table 5.4 Final system unreliability.

Approaches	Set 1	Set 2	Set 3	Set 4	Set 5
Combinatorial	0.783398	0.783496	0.86007	0.86007	0.740674
Logic OR	0.783518	0.783604	0.860331	0.860331	0.740674

For parameters sets 1–4, the logic OR replacement method generates overestimated system unreliability. For parameter set 5, the two methods generate identical results because the dependent memory units experience no UFs (the c factor is ONE).

Since no cascading FDEPs are involved in this case study, the algorithm of [21] can generate results that are identical to those of the combinatorial algorithm presented in this section. If the CTMC-based method is applied to analyze this example memory system, a compact CTMC model (after merging all system failed states and related transitions) containing 87 states and 190 transitions [17] has to be solved. In contrast, the combinatorial method involves only solving three static FTs with five, three, and three components, respectively (Figure 5.3).

5.4 Case Study 2: Shared Dependent Event

Figure 5.4 illustrates the DFT of an example system with dependent event C shared by two FDEP gates. Table 5.5 gives five sets of input component failure parameters and coverage factors. All the system components have exponential ttf distributions. The r factor is assumed to be ZERO for all the system components. Mission time of $t = 10\,000$ hours is considered.

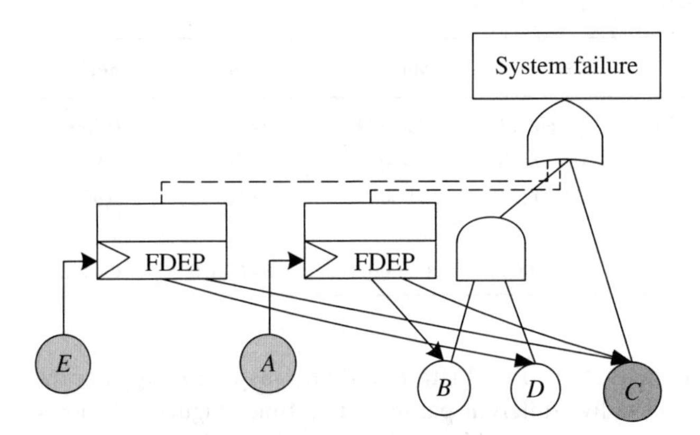

Figure 5.4 Example system with shared dependent events.

Table 5.5 Input component failure parameters and coverage factors.

	Set 1	Set 2	Set 3	Set 4	Set 5
λ_A	1e–5	1e–5	1e–5	1e–5	1e–5
$\lambda_{B,C,D,E}$	1e–5	1e–4	1e–4	1e–4	1e–4
c_A	0.9	0.9	0.9	0.9	1
c_B	0.9	0.9	0.5	0.9	1
c_C	0.9	0.9	0.5	0.9	1
c_D, c_E	1	1	1	0.9	1

The ITE space contains four ITEs:

$$\begin{aligned}
ITE_1 &= \overline{A} \cap \overline{E},\\
ITE_2 &= \overline{A} \cap E,\\
ITE_3 &= A \cap \overline{E},\\
ITE_4 &= A \cap E.
\end{aligned} \tag{5.8}$$

Their occurrence probabilities are:

$$\begin{aligned}
P(ITE_1) &= (1-\widetilde{q}_A)\cdot(1-\widetilde{q}_E),\\
P(ITE_2) &= (1-\widetilde{q}_A)\cdot\widetilde{q}_E,\\
P(ITE_3) &= \widetilde{q}_A\cdot(1-\widetilde{q}_E),\\
P(ITE_4) &= \widetilde{q}_A\cdot\widetilde{q}_E.
\end{aligned} \tag{5.9}$$

Values of $P(ITE_i)$ are shown in Table 5.6. Further, $S_{ITE_1} = \varnothing$, $S_{ITE_2} = \{C, D\}$, $S_{ITE_3} = \{B, C\}$, and $S_{ITE_4} = \{B, C, D\}$ are obtained. Applying the reduced FT generation procedure in Section 5.2.3, the reduced FT model for $P(\text{system fails}|ITE_1)$ is obtained in Figure 5.5. $P(\text{system fails}|ITE_i)$ for $i = 2$, 3, and 4 is simply constant 1, meaning that the system always fails when the corresponding ITE_i occurs.

Based on the evaluation of the reduced FT in Figure 5.5 and (5.5), $P(\text{system fails}|ITE_1)$ is obtained as shown in Table 5.7. According to (5.4), $P(ITE_i)$ in Table 5.6 and $P(\text{system fails}|\ ITE_i)$ in Table 5.7 are integrated to obtain Q (last row of Table 5.7). Lastly, according to (5.1), the final system unreliability is obtained. Table 5.8 presents values of the final system unreliability evaluated. It also presents the system unreliability analyzed using the logic OR replacement method. The final system unreliability results obtained using

Table 5.6 Occurrence probabilities of ITEs.

	Set 1	Set 2	Set 3	Set 4	Set 5
$P(ITE_1)$	0.826597	0.336069	0.336069	0.358746	0.332871
$P(ITE_2)$	0.086934	0.577462	0.577462	0.554785	0.571966
$P(ITE_3)$	0.078241	0.03181	0.03181	0.033957	0.035008
$P(ITE_4)$	0.008229	0.054659	0.054659	0.052512	0.060154

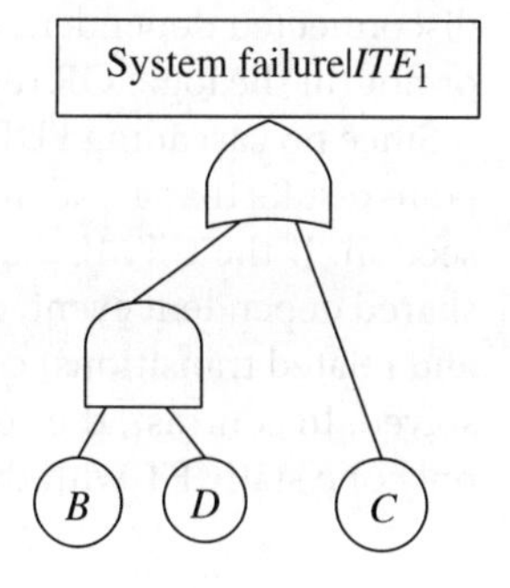

Figure 5.5 Reduced FT model.

Table 5.7 Evaluation results of P(system fails|ITE_i) and Q.

	Set 1	Set 2	Set 3	Set 4	Set 5
P(system fails$\|ITE_1$)	0.111148	0.787671	0.821891	0.796226	0.779117
P(system fails$\|ITE_2$)	1	1	1	1	1
P(system fails$\|ITE_3$)	1	1	1	1	1
P(system fails$\|ITE_4$)	1	1	1	1	1
Q	0.265278	0.928643	0.940143	0.926897	0.926474

Table 5.8 Final system unreliability.

Approaches	Set 1	Set 2	Set 3	Set 4	Set 5
Combinatorial	0.27227	0.929322	0.940713	0.93217	0.926474
Logic OR	0.27227	0.929322	0.940713	0.93217	0.926474

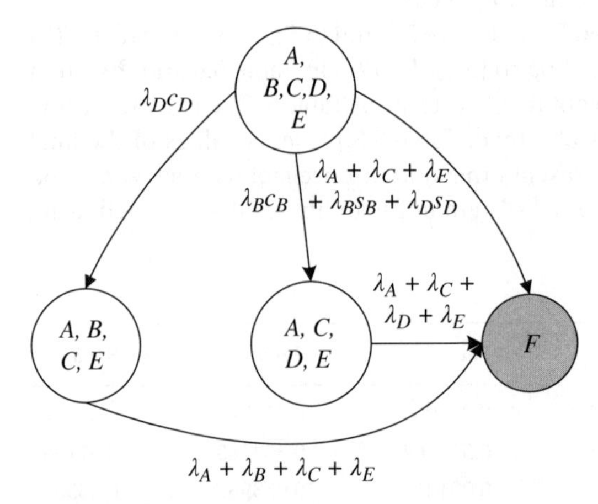

Figure 5.6 CTMC for the example system (F: system failure).

the two methods are identical for this case study. The reason is that when either of the two trigger events (A or E) happens, the entire system fails. In this case, whether the disconnected dependent components can be turned off (in the combinatorial method) or not (in the logic OR replacement method) does not matter.

Since no cascading FDEP is involved in this case study, the algorithm of [21] can generate results that are identical to those of the combinatorial algorithm presented in this section. If the CTMC-based method is applied to analyze the example system with a shared dependent event, a compact CTMC model (after merging all system failed states and related transitions) containing four states and five transitions (Figure 5.6) has to be solved. In contrast, the combinatorial method presented in Section 5.2 involves solving only one static FT with three components (Figure 5.5).

5.5 Case Study 3: Cascading FDEP

The cascading behavior occurs when the failure of one component in the system results in a chain reaction or *domino* effect [29]. Consider a hierarchical hub network where its nodes and hubs are organized into multiple levels. A node at lower levels can be accessible via multiple hubs of different levels. If the top-level hub undergoes a failure, then all its child or grandchild hubs and nodes connected to these hubs become inaccessible in a cascading manner [30]. Cascading effects can be modeled using multiple cascading FDEP gates in the DFT model.

Figure 5.7 illustrates the DFT model of a system with a two-stage domino chain behavior, modeled using two cascading FDEP gates. When event A occurs, both events B and E are forced to occur; consequently, event C also occurs due to the occurrence of event B. In this example system, event A is an ITE while event B is a dependent trigger event. Five sets of input parameters in Table 5.9 are considered. It is assumed that all the system components fail exponentially with identical constant failure rate $\lambda_{A,B,C,D,E,F} = 1e{-}5/\text{hour}$. The r factor is assumed to be ZERO for all the system components. Mission time of $t = 10\,000$ hours is used for the analysis.

Applying the algorithm in Section 5.2.5, since there is a single ITE A in the original system DFT, one obtains $P_{u,IT} = 1 - s_A \cdot q_A(t)$ and $\widetilde{q}_A(t) = c_A \cdot q_A(t)/(1 - s_A \cdot q_A(t))$, where $q_A(t) = 1 - \exp(-\lambda_A \cdot t)$. An ITE space with two ITEs is then constructed: $ITE_1 = \overline{A}$, $ITE_2 = A$. Their occurrence probabilities are: $P(ITE_1) = (1 - \widetilde{q}_A)$, $P(ITE_2) = \widetilde{q}_A$. Values of $P(ITE_i)$ are shown in Table 5.10.

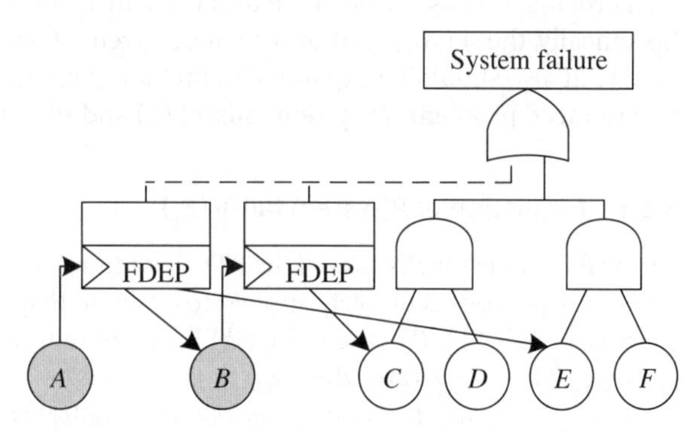

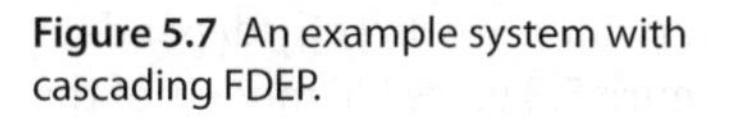
Figure 5.7 An example system with cascading FDEP.

Table 5.9 Input component coverage factors.

	Set 1	Set 2	Set 3	Set 4	Set 5
c_A	0.9	0.9	0.9	1	1
c_B	0.9	0.9	0.9	0.8	1
c_C	0.9	0.9	1	1	1
c_D	1	0.9	1	1	1
c_E	0.9	0.9	1	1	1
c_F	1	0.9	1	1	1

Table 5.10 Occurrence probabilities of ITEs.

	Set 1	Set 2	Set 3	Set 4	Set 5
$P(ITE_1)$	0.913531	0.913531	0.913531	0.904837	0.904837
$P(ITE_2)$	0.086469	0.086469	0.086469	0.095163	0.095163

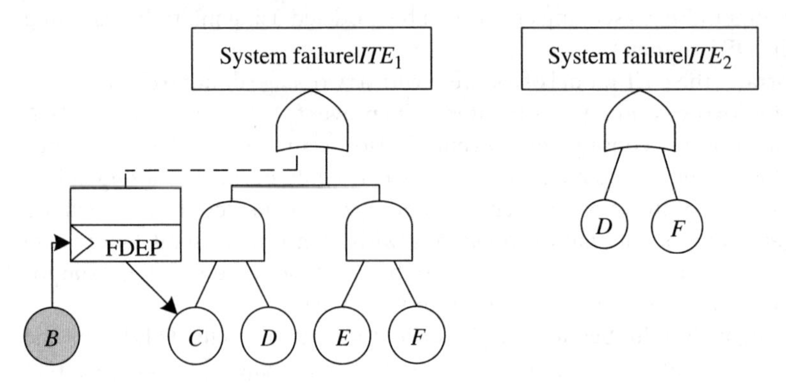

Figure 5.8 Reduced FT models.

Further, $S_{ITE_1} = \emptyset$ and $S_{ITE_2} = FDG_A{}^+ = \{B, C, E\}$ are obtained. Note that the "Expansion Process" for cascading FDEP described in Section 5.2.3.1 is applied to generate S_{ITE_2}. Specifically, the set S_{ITE_2} not only includes events B and E (which are directly dependent on A) but also should be expanded to include event C. Figure 5.8 gives FT models of the two reduced problems $P(\text{system fails}|ITE_1)$ and $P(\text{system fails}|ITE_2)$.

5.5.1 Evaluation of *P*(system fails|ITE_1)

As the FT model for $P(\text{system fails}|ITE_1)$ contains an FDEP gate, the combinatorial algorithm needs to be recursively applied to evaluate $P(\text{system fails}|ITE_1)$. Specifically, since there is a single ITE B in the reduced FT, one obtains $P_{u,IT} = 1 - s_B \cdot q_B(t)$ and $\widetilde{q}_B(t) = c_B \cdot q_B(t)/(1 - s_B \cdot q_B(t))$, where $q_B(t) = 1 - \exp(-\lambda_B \cdot t)$. An ITE space with two ITEs is then constructed for evaluating $P(\text{system fails}|ITE_1)$: $ITE_{1,1} = \overline{B}$, $ITE_{1,2} = B$. Their occurrence probabilities are evaluated as: $P(ITE_{1,1}) = (1 - \widetilde{q}_B)$, $P(ITE_{1,2}) = \widetilde{q}_B$ as shown in Table 5.11.

Further, $S_{ITE_{1,1}} = \emptyset$ and $S_{ITE_{1,2}} = \{C\}$ are obtained. Figure 5.9 gives FT models of the two reduced problems $P(\text{system fails}|ITE_{1,1})$ and $P(\text{system fails}|ITE_{1,2})$. Since both of

Table 5.11 Occurrence probabilities of ITEs for evaluating *P*(system fails|ITE_1).

	Set 1	Set 2	Set 3	Set 4	Set 5
$P(ITE_{1,1})$	0.913531	0.913531	0.913531	0.922393	0.904837
$P(ITE_{1,2})$	0.086469	0.086469	0.086469	0.077607	0.095163

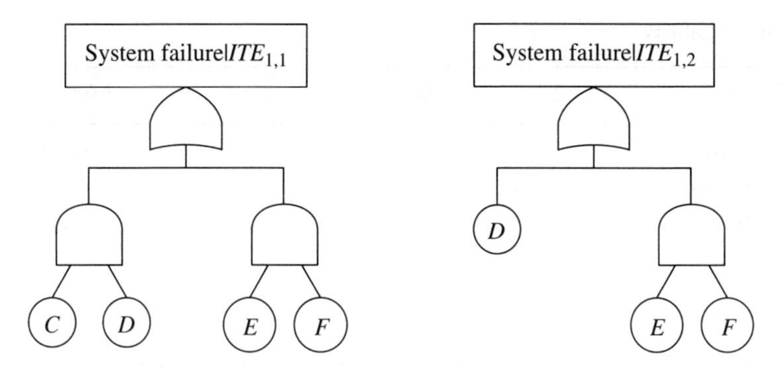

Figure 5.9 Reduced FT models for evaluating P(system fails|ITE_1).

Table 5.12 Evaluation results of P(system fails|$ITE_{1,i}$) and P(system fails|ITE_1).

	Set 1	Set 2	Set 3	Set 4	Set 5
P(system fails$\|ITE_{1,1}$)	0.035021	0.051864	0.018030	0.018030	0.018030
P(system fails$\|ITE_{1,2}$)	0.111148	0.118939	0.103357	0.103357	0.103357
P(system fails$\|ITE_1$)	0.050724	0.066631	0.034682	0.043215	0.026150

the reduced FTs are free of FDEP gates, the SEA method, i.e. Eq. (5.5) is applicable to evaluating the reduced problems. Table 5.12 presents the evaluation results for the five sets of input parameters. Using (5.4), P(system fails$|ITE_{1,1}$) and P(system fails$|ITE_{1,2}$) can then be integrated with $P(ITE_{1,1})$ and $P(ITE_{1,2})$ in Table 5.11 to find $Q(t)$, which is further combined with $P_{u,IT} = 1 - s_B \cdot q_B(t)$ using (5.1) to find P(system fails$|ITE_1$). The last row of Table 5.12 shows the evaluation results for P(system fails$|ITE_1$).

5.5.2 Evaluation of P(system fails|ITE_2)

As the FT model for P(system fails$|ITE_2$) is free of FDEP gates, the SEA method, i.e. Eq. (5.5) is directly applicable to evaluating P(system fails$|ITE_2$). Specifically, $P_{u,2}(t) = (1 - s_D \cdot q_D(t))(1 - s_F \cdot q_F(t))$, $\tilde{q}_D(t) = c_D \cdot q_D(t)/(1 - s_D \cdot q_D(t))$, and $\tilde{q}_F(t) = c_F \cdot q_F(t)/(1 - s_F \cdot q_F(t))$. Further $Q_2(t) = \tilde{q}_D(t) + (1 - \tilde{q}_D(t))\tilde{q}_F(t)$. Based on (5.5), $P_{u,2}(t)$ and $Q_2(t)$ are combined to obtain P(system fails$|ITE_2$). Table 5.13 presents the evaluation results.

Table 5.13 Evaluation results of P(system fails|ITE_2).

	Set 1	Set 2	Set 3	Set 4	Set 5
$P_{u,2}(t)$	1	0.981058	1	1	1
$Q_2(t)$	0.181269	0.165461	0.181269	0.181269	0.181269
P(system fails$\|ITE_2$)	0.181269	0.181269	0.181269	0.181269	0.181269

Table 5.14 Final system unreliability.

	Set 1	Set 2	Set 3	Set 4	Set 5
Combinatorial	0.070938	0.085333	0.056423	0.056353	0.040911
Algorithm [21]	0.071605	0.085999	0.057090	0.057835	0.040911
Logic OR	0.073576	0.087964	0.057090	0.057835	0.040911

5.5.3 Evaluation of UR_{system}

According to (5.4), $P(ITE_i)$ in Table 5.10, P(system fails$|ITE_1$) (last row of Table 5.12) and P(system fails$|ITE_2$) (last row of Table 5.13) are integrated to obtain $Q(t)$, which is further combined with $P_{u,IT} = 1 - s_A \cdot q_A(t)$ using (5.1) to obtain the final system unreliability. Table 5.14 presents values of the final system unreliability evaluated using the combinatorial algorithm. It also presents the system unreliability analyzed using the algorithm of [21] and the logic OR replacement method. For parameter sets 1–4, the algorithm of [21] and the logic OR replacement method both give overestimated system unreliability results. For sets 1 and 2, results obtained using the algorithm of [21] are closer to the accurate results because the purely dependent components C and E can be turned off appropriately in the algorithm of [21], but not in the logic OR replacement method. For sets 3 and 4, the algorithm of [21] and the logic OR replacement method give identical results because the purely dependent components C and E undergo no UFs (the c factor being 1); the dependent trigger component B undergoes UFs but cannot be turned off in both methods. For parameter set 5, the three methods give identical results because none of the dependent trigger component (B) and the purely dependent components (C and E) experience UFs; thus, they do not require the "turn-off" function.

5.6 Case Study 4: Dual Event and Cascading FDEP

Consider an example of systems involving both cascading FDEP and dual trigger-basic events in Figure 5.10. Specifically, T_1 and T_2 are dual events (trigger events for FDEP and basic events contributing to the two AND gates at the same time). All the system components fail exponentially with constant failure rates (/hour) $\lambda_A = 2e-5$, $\lambda_B = 1e-5$,

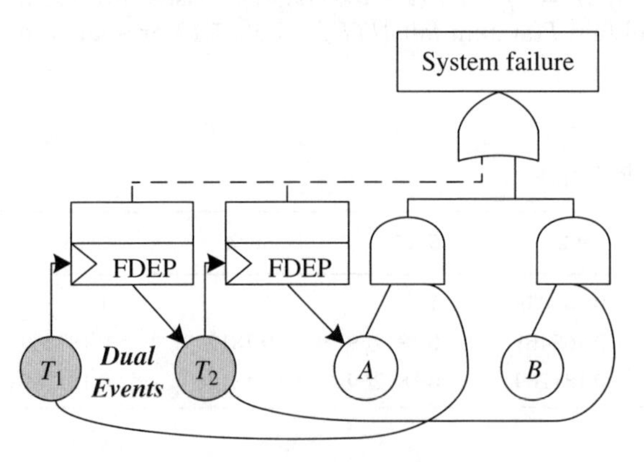

Figure 5.10 An example system containing cascading FDEP and dual events.

Table 5.15 Input component coverage factors.

	Set 1	Set 2	Set 3	Set 4	Set 5
c_A	0.9	0.9	0.9	0.95	1
c_B	0.9	1	1	0.9	1
c_{T1}	0.9	1	1	0.9	1
c_{T2}	0.9	1	0.9	1	1

Table 5.16 Occurrence probabilities of ITEs.

	Set 1	Set 2	Set 3	Set 4	Set 5
$P(ITE_1)$	0.392703	0.367879	0.367879	0.392703	0.367879
$P(ITE_2)$	0.607297	0.632121	0.632121	0.607297	0.632121

$\lambda_{T1} = 1e-4$, and $\lambda_{T2} = 1e-5$. Five sets of coverage factors in Table 5.15 are considered. The r factor is assumed to be ZERO for all the system components. Mission time of $t = 10\,000$ hours is used for the analysis.

Applying the algorithm in Section 5.2.5, since there is a single ITE T_1 in the original system DFT, one obtains $P_{u,IT} = 1 - s_{T_1} \cdot q_{T_1}(t)$ and $\widetilde{q}_{T_1}(t) = c_{T_1} \cdot q_{T_1}(t)/(1 - s_{T_1} \cdot q_{T_1}(t))$, where $q_{T_1}(t) = 1 - \exp(-\lambda_{T_1} \cdot t)$. An ITE space with two ITEs is then constructed: $ITE_1 = \overline{T_1}$, $ITE_2 = T_1$. Their occurrence probabilities are: $P(ITE_1) = (1 - \widetilde{q}_{T_1})$, $P(ITE_2) = \widetilde{q}_{T_1}$. Values of $P(ITE_i)$ are shown in Table 5.16.

Further, $S_{ITE_1} = \varnothing$ and $S_{ITE_2} = FDG_{T_1}{}^{+} = \{T_2, A\}$ are obtained. Note that the "Expansion Process" for cascading FDEP described in Section 5.2.3.1 is applied to generate S_{ITE_2}. Specifically, the set S_{ITE_2} not only includes event T_2 (which is directly dependent on T_1) but also should be expanded to include event A. Figure 5.11 gives FT models of the two reduced problems $P(\text{system fails}|ITE_1)$ and $P(\text{system fails}|ITE_2)$. Since T_1 is a dual event, the special handling procedure in Section 5.2.3.3 is applied to generate the reduced FT for $P(\text{system fails}|ITE_1)$, where only the trigger event identity of T_1 is

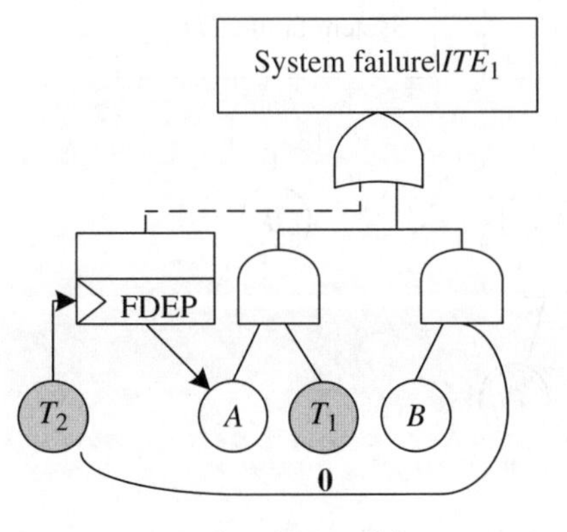

Figure 5.11 Reduced FT models.

Table 5.17 Occurrence probabilities of ITEs for evaluating *P*(system fails|*ITE*$_1$).

	Set 1	Set 2	Set 3	Set 4	Set 5
$P(ITE_{1,1})$	0.913531	0.904837	0.913531	0.904837	0.904837
$P(ITE_{1,2})$	0.086469	0.095163	0.086469	0.095163	0.095163

removed while its basic event identity stays and is replaced by constant logic value 0 (as $\overline{T_1}$ appears in ITE_1). The reduced FT for P(system fails$|ITE_2$) is simply constant 1 after applying the Boolean algebra reduction.

5.6.1 Evaluation of *P*(system fails|*ITE*$_1$)

As the FT model for P(system fails$|ITE_1$) contains an FDEP gate, the combinatorial algorithm needs to be recursively applied to evaluate P(system fails$|ITE_1$). Specifically, since there is a single ITE T_2 in the reduced FT, one obtains $P_{u,IT} = 1 - s_{T_2} \cdot q_{T_2}(t)$ and $\widetilde{q}_{T_2}(t) = c_{T_2} \cdot q_{T_2}(t)/(1 - s_{T_2} \cdot q_{T_2}(t))$, where $q_{T_2}(t) = 1 - \exp(-\lambda_{T_2} \cdot t)$. An ITE space with two ITEs is then constructed for evaluating P(system fails$|ITE_1$): $ITE_{1,1} = \overline{T_2}$, $ITE_{1,2} = T_2$. Their occurrence probabilities are evaluated as: $P(ITE_{1,1}) = (1 - \widetilde{q}_{T_2})$, $P(ITE_{1,2}) = \widetilde{q}_{T_2}$ as shown in Table 5.17.

Further, $S_{ITE_{1,1}} = \varnothing$ and $S_{ITE_{1,2}} = \{A\}$ are obtained. Figure 5.12 gives FT models of the two reduced problems P(system fails$|ITE_{1,1}$) and P(system fails$|ITE_{1,2}$). Since both of the reduced FTs are free of FDEP gates, the SEA method, i.e. Eq. (5.5) is directly applicable to evaluating the reduced problems. Table 5.18 presents the evaluation results for the five sets of input parameters. Note that in the reduced FT for P(system fails$|ITE_{1,1}$), the Boolean algebra $0 \cdot x = 0$ cannot be applied because UFs of A and B can still contribute to the system failure (since they are not isolated). When applying the SEA method, $P_{u,1,1}(t) = (1 - s_A q_A(t)) \cdot (1 - s_B q_B(t))$ and $Q_{1,1}(t) = 0$. Applying (5.5), one obtains P(system fails$|ITE_{1,1}$) $= 1 - P_{u,1,1}(t)$ representing the probability that at least one of the two components A and B undergoes a UF.

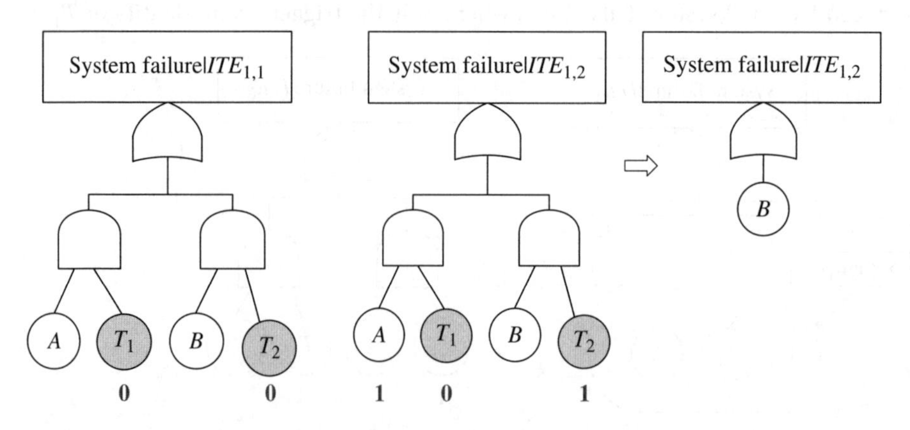

Figure 5.12 Reduced FT models for evaluating *P*(system fails|*ITE*$_1$).

Table 5.18 Evaluation results of $P(\text{system fails}|ITE_{1,i})$ and $P(\text{system fails}|ITE_1)$.

	Set 1	Set 2	Set 3	Set 4	Set 5
$P(\text{system fails}\|ITE_{1,1})$	0.027471	0.018127	0.018127	0.018493	0.000000
$P(\text{system fails}\|ITE_{1,2})$	0.095163	0.095163	0.095163	0.095163	0.095163
$P(\text{system fails}\|ITE_1)$	0.042523	0.025458	0.034069	0.025790	0.009056

Table 5.19 Final system unreliability.

	Set 1	Set 2	Set 3	Set 4	Set 5
Combinatorial	0.647764	0.641486	0.644654	0.641608	0.635452
Logic OR	0.648281	0.642060	0.645170	0.641895	0.635452

Using (5.4), $P(\text{system fails}|ITE_{1,1})$ and $P(\text{system fails}|ITE_{1,2})$ can then be integrated with $P(ITE_{1,1})$ and $P(ITE_{1,2})$ in Table 5.17 to find $Q(t)$, which is further combined with $P_{u,IT} = 1 - s_{T_2} \cdot q_{T_2}(t)$ using (5.1) to find $P(\text{system fails}|ITE_1)$. The last row of Table 5.18 shows the evaluation results for $P(\text{system fails}|ITE_1)$.

5.6.2 Evaluation of UR_{system}

According to (5.4), $P(ITE_i)$ in Table 5.16, $P(\text{system fails}|ITE_1)$ (last row of Table 5.18) and $P(\text{system fails}|ITE_2) = 1$ are integrated to obtain $Q(t)$, which is further combined with $P_{u,IT} = 1 - s_{T_1} \cdot q_{T_1}(t)$ using (5.1) to obtain the final system unreliability. Table 5.19 presents values of the final system unreliability evaluated using the combinatorial algorithm. It also presents the system unreliability analyzed using the logic OR replacement method.

For parameter sets 1–4, the logic OR replacement method generates overestimated system unreliability results; for parameter set 5 (all the system components experience no UFs), the two methods give identical results. In the case of all the system components having exponential *ttf* distributions, this example system can be analyzed using the CTMC-based method. A compact Markov chain as shown in Figure 5.13 must be solved, which contains 6 states and 11 transitions. In contrast, the combinatorial method presented in this chapter involves only analyzing the two reduced FTs in Figure 5.12.

5.7 Summary

This chapter discusses methods for reliability analysis of systems subject to the FDEP behavior. The logic OR replacement method is only applicable to systems with perfect fault coverage, but generates overestimated system unreliability results for systems with imperfect fault coverage. The algorithm presented in [21] works perfectly for IPC systems subject to noncascading FDEPs, but generates overestimated system unreliability results in the case of the occurrence of cascading FDEPs in the system.

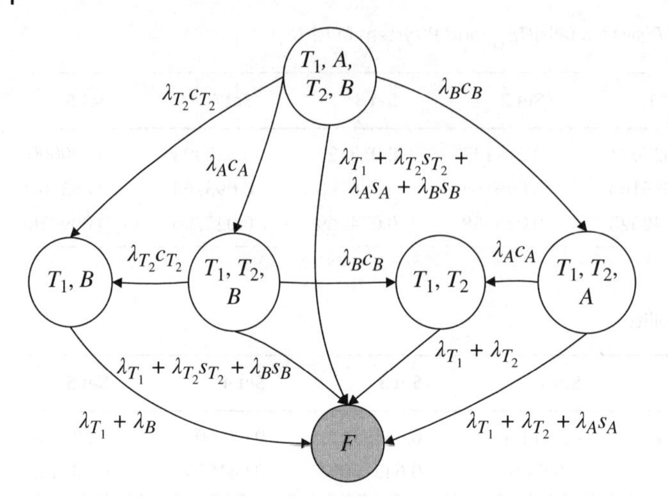

Figure 5.13 CTMC for the example system with dual events.

The combinatorial algorithm described in Section 5.2 can handle various complicated cases including cascading FDEPs, combined trigger events, shared dependent events, and dual-role events. The algorithm, which practices the divide and conquer principle based on the total probability law, is computationally efficient. The reduced reliability problems generated are simplified and independent problems without FDEP, which can be solved in parallel given available computing resources using any traditional FT analysis method.

The methods explained in this chapter do not address possible competitions existing between a covered failure of a trigger component and UFs of the corresponding dependent components. Refer to Chapters 8 and 9 for models and methods of addressing deterministic and probabilistic competing effects, respectively.

References

1 Stallings, W. (2009). *Computer Organization and Architecture*, 8e. Prentice Hall.
2 Xing, L., Levitin, G., Wang, C., and Dai, Y. (2013). Reliability of systems subject to failures with dependent propagation effect. *IEEE Transactions Systems, Man, and Cybernetics: Systems* 43 (2): 277–290.
3 Dugan, J.B. and Doyle, S.A. (1996). New results in fault-tree analysis. In: *Tutorial Notes of Annual Reliability and Maintainability Symposium.* Las Vegas, Nevada, USA.
4 Dugan, J.B., Bavuso, S.J., and Boyd, M.A. (1992). Dynamic fault-tree models for fault-tolerant computer systems. *IEEE Transactions on Reliability* 41 (3): 363–377.
5 Merle, G., Roussel, J-M., and Lesage, J-J. (2010). Improving the efficiency of dynamic fault tree analysis by considering gates FDEP as static. In: *Proceeding of European Safety and Reliability Conference.* Rhodes, Greece.
6 Misra, K.B. (2008). *Handbook of Performability Engineering.* London: Springer-Verlag.

7 Bryant, R.E. (1986). Graph-based algorithms for Boolean function manipulation. *IEEE Transactions on Computers* 35 (8): 677–691.
8 Rauzy, A. (1993). New algorithms for fault tree analysis. *Reliability Engineering and System Safety* 40: 203–211.
9 Zang, X., Sun, H., and Trivedi, K.S. (1999). A BDD-based algorithm for reliability analysis of phased-mission systems. *IEEE Transactions on Reliability* 48 (1): 50–60.
10 Bouricius, W.G., Carter, W.C., and Schneider, P.R. (1969). Reliability modeling techniques for self-repairing computer systems. In: *Proceedings of 24th Ann. ACM., National Conference*, 295–309, New York, NY, USA.
11 Arnold, T.F. (1973). The concept of coverage and its effect on the reliability model of a repairable system. *IEEE Transactions on Computers* C-22: 325–339.
12 Levitin, G. and Amari, S.V. (2007). Reliability analysis of fault tolerant systems with multi-fault coverage. *International Journal of Performability Engineering* 3 (4): 441–451.
13 Myers, A.F. (2007). *k*-out-of-*n*: G system reliability with imperfect fault coverage. *IEEE Transactions on Reliability* 56 (3): 464–473.
14 Myers, A.F. (2008). Achievable limits on the reliability of *k*-out-of-*n*: G systems subject to imperfect fault coverage. *IEEE Transactions on Reliability* 57: 349–354.
15 Dugan, J.B. (1991). Automated analysis of phased-mission reliability. *IEEE Transactions on Reliability* 40 (1): 45–52,55.
16 Gulati, R. and Dugan, J.B. (1997). A modular approach for analyzing static and dynamic fault trees. In: *Proceedings of the Annual Reliability and Maintainability Symposium*. Philadelphia, PA, USA.
17 Manian, R., Dugan, J.B., Coppit, D., and Sullivan, K.J. (1998). Combining various solution techniques for dynamic fault tree analysis of computer systems. In: *Proceedings of the 3rd IEEE International High-Assurance Systems Engineering Symposium*, 21–28. Washington, DC, USA.
18 Meshkat, L., Xing, L., Donohue, S., and Ou, Y. (2003). An overview of the phase-modular fault tree approach to phased-mission system analysis. In: *Proceedings of The International Conference on Space Mission Challenges for Information Technology*. Pasadena, CA, USA.
19 Sune, V. and Carrasco, J.A. (1997). A method for the computation of reliability bounds for non-repairable fault-tolerant systems In: *Proceedings of the 5th IEEE International Symposium on Modeling, Analysis, and Simulation of Computers and Telecommunication System*, 221–228. Haifa, Israel.
20 Sune, V. and Carrasco, J.A. (2001). A failure-distance based method to bound the reliability of non-repairable fault-tolerant systems without the knowledge of minimal cutsets. *IEEE Transactions on Reliability* 50 (1): 60–74.
21 Xing, L., Morrissette, B.A., and Dugan, J.B. (2014). Combinatorial reliability analysis of imperfect coverage systems subject to functional dependence. *IEEE Transactions on Reliability* 63 (1): 367–382.
22 Xing, L. (2008). Handling functional dependence without using Markov models. *International Journal of Performability Engineering, Short Communications* 4 (1): 95–97.

23 Xing, L. and Dugan, J.B. (2002). Analysis of generalized phased mission system reliability, performance and sensitivity. *IEEE Transactions on Reliability* 51 (2): 199–211.
24 Chang, Y., Amari, S.V., and Kuo, S. (2004). Computing system failure frequencies and reliability importance measures using OBDD. *IEEE Transactions on Computers* 53 (1): 54–68.
25 Cormen, T.H., Leiserson, C.E., Rivest, R.L., and Stein, C. (2001). *Introduction to Algorithms*, 2e. MIT Press.
26 Brace, K.S., Rudell, R.L., and Bryant, R.E. (1990). Efficient implementation of a BDD package In: *Proceedings of the 27th ACM/IEEE Design Automation Conference*, 40–45. Orlando, FL, USA.
27 Reibman, A., Smith, R., and Trivedi, K.S. (1989). Markov and Markov reward model transient analysis: an overview of numerical approaches. *European Journal of Operational Research* 40 (2): 257–267.
28 Dugan, J.B., Venkataraman, B., and Gulati, R. (1997). DIFtree: A software package for the analysis of dynamic fault tree models. In: *Proceedings of the Annual Reliability and Maintainability Symposium*, 64–70. Philadelphia, PA, USA.
29 Rausand, M. and Hoyland, A. (2003). *System Reliability Theory: Models and Statistical Methods, Wiley Series in Probability and Mathematical Statistics*. Wiley.
30 Davari, S. and Zarandi, M.H.F. (2013). The single-allocation hierarchical hub-median problem with fuzzy flows. In: *Proceedings of the 5th International Workshop Soft Computing Applications*, 165–181.

6

Deterministic Common-Cause Failure

According to [1], common-cause failures (CCFs) are "A subset of dependent events in which two or more component fault states exist at the same time, or in a short time interval, and are direct results of a shared cause." There are two types of shared root causes or common causes (CCs): *external causes* and *internal causes*. Examples of external causes include floods, lightning strikes, earthquakes, sudden changes in environments, malicious attacks, design mistakes, power-supply disturbances, human errors, radiations, computer viruses, etc. Internal causes are mainly propagated failures (PFs) or destructive effects like fire, overheating, short circuit, blackout, or explosions originating from some component within the system, which may destroy or incapacitate other system components.

CCFs typically happen in systems that are designed with redundancy techniques based on the use of s-identical components [2]. It has been shown by numerous studies that CCFs tend to increase joint component failure probabilities, thus contributing significantly to the overall system unreliability [3–14]. For example, according to studies performed by Fleming et al. [14], the unreliability of a system with CCFs can be increased by more than a factor of 10 in varying CCs contribution up to 1%. In other words, failure to consider CCFs can lead to overestimated system reliability metrics [12, 13]. Therefore, it is crucial to incorporate effects of CCFs for the accurate reliability analysis of systems subject to CCFs.

Dependent on the effect from a CCF, deterministic and probabilistic CCFs can be differentiated. For deterministic CCFs, a CC, upon occurring, results in guaranteed failures of all components affected by this CC; for probabilistic CCFs, the occurrence of a CC results in failures of different components with different probabilities. This chapter focuses on the deterministic CCFs while Chapter 7 addresses the probabilistic CCFs.

For incorporating deterministic CCFs into the system reliability analysis, an explicit method based on expanding the system fault tree (FT) model is first presented in Section 6.1. An implicit method called the efficient decomposition and aggregation (EDA) approach is then presented in Section 6.2. In Section 6.3, decision diagrams-based methods are discussed. In Section 6.4, a universal generating function-based method is discussed for series-parallel systems with random failure propagation time.

Dynamic System Reliability: Modeling and Analysis of Dynamic and Dependent Behaviors, First Edition. Liudong Xing, Gregory Levitin and Chaonan Wang.

6.1 Explicit Method

The basic idea of the explicit method is to evaluate an expanded system model that is constructed by modeling the occurrence of each CC as a basic event shared by all the components affected by this CC in the original system model [8, 9].

6.1.1 Two-Step Method

The explicit method involves a two-step process given below.

Step 1: Build an expanded FT. To consider effects of CCFs, a logic OR gate is introduced for each component affected by at least one CC, and each CC is modeled as a repeated input event to all component OR-gates failed by this CC.

Step 2: Evaluate the expanded FT. The expanded FT can be evaluated using any traditional FT reliability analysis method, e.g. the binary decision diagram (BDD)-based method (Section 2.4) to obtain the system unreliability, considering effects of CCFs.

6.1.2 Illustrative Example

Example 6.1 Consider an example FT shown in Figure 6.1. The failure probability of each component is given as $q_A = 0.05$, $q_B = 0.10$, $q_C = 0.15$, $q_D = 0.20$, $q_E = 0.25$. The system is subject to CCFs from two independent CCs: CC_1 would cause B and C to fail simultaneously with occurrence probability $P(CC_1) = 0.001$; and CC_2 would cause A and E to fail simultaneously with occurrence probability $P(CC_2) = 0.002$.

In step 1, the FT in Figure 6.1 is expanded by adding four component OR gates for components A, B, C, and E. CC_1 is connected to the OR-gates of components B and C; CC_2 is connected to the OR-gates of components A and E.

In step 2, the expanded FT in Figure 6.2 is analyzed by applying the BDD-based method. Figure 6.3 shows the BDD model generated using the ordering of $CC_1 < CC_2 < A < B < C < D < E$. There are eight paths from the root node CC_1 to sink node 1.

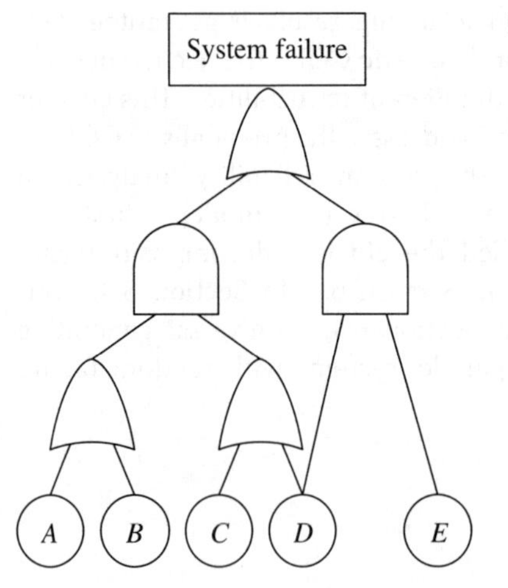

Figure 6.1 An example FT.

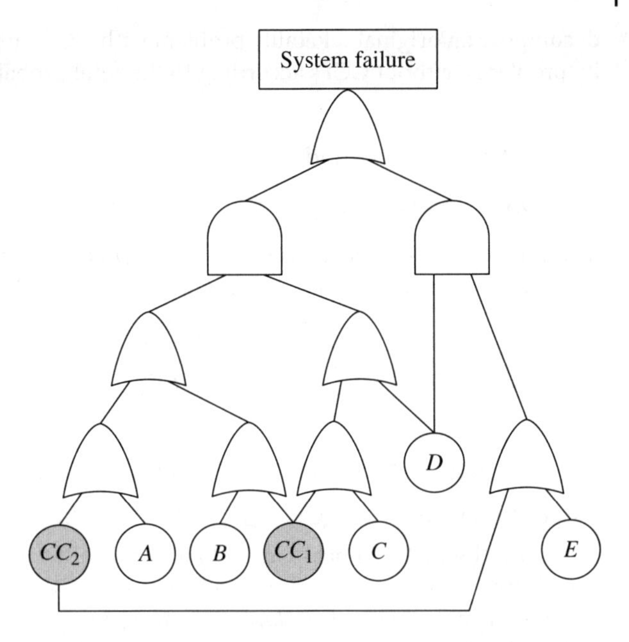

Figure 6.2 Expanded FT considering CCFs.

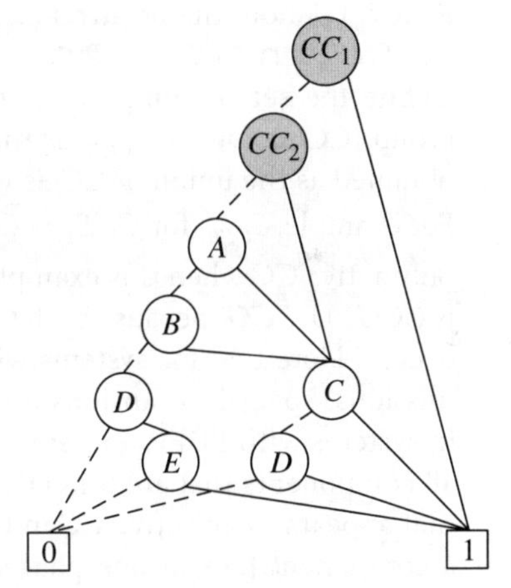

Figure 6.3 BDD for the expanded FT.

The summation of probabilities of these paths gives the system unreliability, which is 0.090 522.

6.2 Efficient Decomposition and Aggregation Approach

Different from the explicit method where the system model is developed with the consideration of contributions from CCs, in the implicit method, the system model and reliability expression are developed without considering CCFs. However, special treatments must be performed to include the contributions of CCs in the expression evaluation [2, 13, 15]. This section introduces such an implicit method called the EDA approach. Based on the "divide and conquer" principle, the basic idea of the EDA method is to decompose an original reliability problem with CCFs into a number of reduced reliability problems without CCFs according to the total probability theorem [16, 17].

6.2.1 Three-Step Method

The EDA method can be applied in three steps:

Step 1: Construct common-cause event (CCE) space. Assume m CCs exist in the system. The m CCs form 2^m disjoint combinations, each called a CCE:

$$
\begin{aligned}
CCE_0 &= \overline{CC_1} \cap \overline{CC_2} \cap \ldots \cap \overline{CC_m}, \\
CCE_1 &= \overline{CC_1} \cap \overline{CC_2} \cap \ldots \cap CC_m, \\
&\ldots\ldots \\
CCE_{2^m-1} &= CC_1 \cap CC_2 \cap \ldots \cap CC_m.
\end{aligned}
\tag{6.1}
$$

Specifically, for CCE_j $(0 \le j \le 2^m-1)$, if the binary representation of j is $a_1 a_2 \ldots a_{m-1} a_m$, then CC_{a_i} appears in CCE_j if $a_i = 1$, otherwise $\overline{CC}_{a_i}$ appears in CCE_j. These collectively exhaustive and mutually exclusive CCEs constitute an event space called "CCE space" (denoted by Ω_{CCE}). That is, $\Omega_{CCE} = \{CCE_0, CCE_1, \ldots, CCE_{2^m-1}\}$. Let $P(CCE_j)$ denote the occurrence probability of each CCE_j. Thus, $\sum_{j=0}^{2^m-1} P(CCE_j) = 1$ and $P(CCE_i \cap CCE_j) = P(\emptyset) = 0$ for any $i \neq j$.

Define the set of components that fail due to the same CC as a common-cause group (CCG). Define S_{CCE_i} as the set of components affected by CCE_i. S_{CCE_i} can be obtained as the union of CCGs whose corresponding CC occurs when CCE_i occurs. For example, S_{CCE_1} for $CCE_1 = \overline{CC_1} \cap \overline{CC_2} \cap CC_3$ is simply CCG_3 because CC_3 is the only active CC when this example CCE_1 occurs; S_{CCE_3} for $CCE_3 = \overline{CC_1} \cap CC_2 \cap CC_3$ is $CCG_2 \cup CCG_3$ because both CC_2 and CC_3 are active CCs when this example CCE_3 occurs. Note that for systems with functional dependence (FDEP) or multi-phased missions, special expansions are necessary for generating S_{CCE_i} [15]. In particular, in systems with FDEP, an iterative expansion process must be performed to include all components that are (directly or indirectly) dependent on the trigger component that appears as an active CC in CCE_i. In a nonrepairable multi-phased system, once a component fails in one phase, it remains failed in all of the subsequent phases. Therefore, the set S_{CCE_i} must be expanded to incorporate the affected components in all of the subsequent mission phases. Refer to [15] for more details and illustrative examples.

Step 2: Generate and solve reduced reliability problems. Based on the CCE space and the total probability theorem, the system unreliability can be evaluated as:

$$
\begin{aligned}
UR_{system} &= \sum_{i=0}^{2^m-1} [P(\text{system fails} \mid CCE_i) \cdot P(CCE_i)] \\
&= \sum_{i=0}^{2^m-1} [UR_i \cdot P(CCE_i)].
\end{aligned}
\tag{6.2}
$$

UR_i in (6.2) is a conditional system failure probability given that CCE_i occurs. The evaluation of UR_i is a reduced problem where components of S_{CCE_i} do not appear in the reliability model. Specifically, each basic event corresponding to the failure of a component in S_{CCE_i} in the original system FT is replaced by a constant logic value 1 (*True*). Boolean algebra rules are then applied to generate a reduced FT for evaluating UR_i. More importantly, the evaluation of UR_i can proceed without the consideration of CCFs. Thus, the BDD-based approach (Section 2.4) can be applied to solve each reduced problem. The studies in [15, 17] showed that most of the FTs after reduction are trivial to solve. In addition, the 2^m reduced problems involved in the EDA approach are independent, enabling parallel evaluations given available computing resources.

Step 3: Integrate for the final *system unreliability*. After all the reduced problems UR_i are evaluated, based on (6.2), results of UR_i are integrated with the occurrence probabilities of CCE, i.e. $P(CCE_i)$ to obtain the final system unreliability considering effects of CCFs.

The EDA approach is applicable to systems subject to multiple CCs that can affect different subsets of system components and can have diverse *s*-relationship (independent, dependent, mutually exclusive). The EDA approach allows utilization of any reliability analysis software package without considering CCFs for computing reliability. It also allows the inputs and outputs of the program to be adjusted to generate the system reliability considering effects of CCFs.

6.2.2 Illustrative Example

Example 6.2 Consider the example system in Example 6.1, where $CCG_1 = \{B, C\}$, $CCG_2 = \{A, E\}$. The EDA approach can be applied as follows.

Step 1: Construct CCE space. Since two CCs exist in the example system, the CCE space contains four CCEs: $\Omega_{CCE} = \{CCE_0, CCE_1, CCE_2, CCE_3\}$ as defined in the first column of Table 6.1. Table 6.1 also gives the occurrence probability of each CCE and the set of components affected by each CCE.

Step 2: Generate and solve reduced reliability problems. Based on the CCE space constructed in Table 6.1 and (6.2), the example system unreliability is evaluated as:

$$UR_{system} = \sum_{i=0}^{3}[P(\text{system fails} \mid CCE_i) \cdot P(CCE_i)] = \sum_{i=0}^{3}[UR_i \cdot P(CCE_i)]. \quad (6.3)$$

Since no components are affected by CCE_0, UR_0 can be evaluated by the BDD-based method using the original system FT model of Figure 6.1. Figure 6.4 shows the BDD

Table 6.1 CCE space for the example system.

i	CCE_i	$P(CCE_i)$	S_{CCE_i}
0	$\overline{CC_1} \cap \overline{CC_2}$	$(1 - 0.001) \times (1 - 0.002)$	$\varnothing$
1	$\overline{CC_1} \cap CC_2$	$(1 - 0.001) \times 0.002$	$\{A, E\}$
2	$CC_1 \cap \overline{CC_2}$	$0.001 \times (1 - 0.002)$	$\{B, C\}$
3	$CC_1 \cap CC_2$	0.001×0.002	$\{A, B, C, E\}$

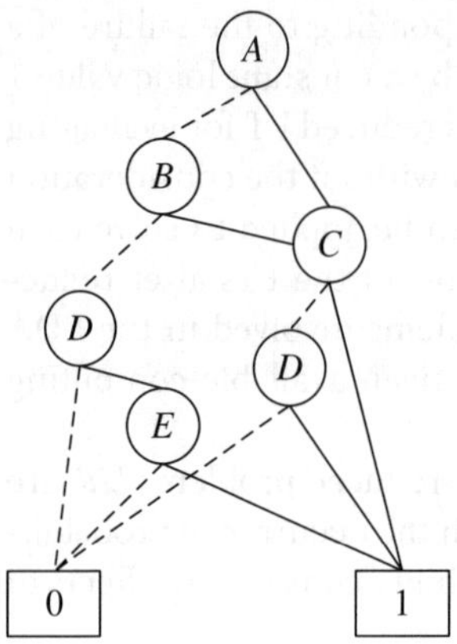

Figure 6.4 BDD for evaluating UR_0.

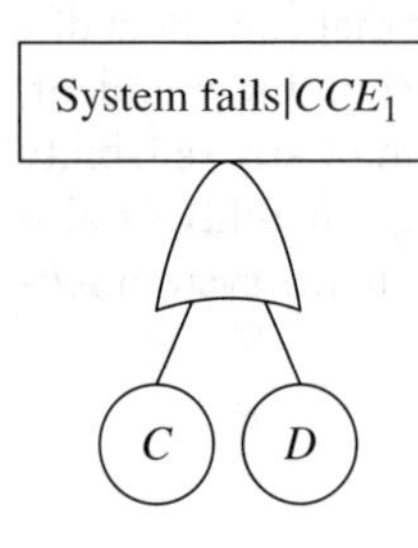

Figure 6.5 Reduced FT for evaluating UR_1.

generated from the system FT using the BDD generation algorithm of Section 2.4. The evaluation of the BDD model in Figure 6.4 gives

$$\begin{aligned} UR_0 &= q_A \cdot q_C + q_A \cdot (1 - q_C) \cdot q_D + (1 - q_A) \cdot q_B \cdot q_C \\ &\quad + (1 - q_A) \cdot q_B \cdot (1 - q_C) \cdot q_D + (1 - q_A) \cdot (1 - q_B) \cdot q_D \cdot q_E \\ &= 0.05 \times 0.15 + 0.05 \times 0.85 \times 0.2 + 0.95 \times 0.1 \times 0.15 \\ &\quad + 0.95 \times 0.1 \times 0.85 \times 0.2 + 0.95 \times 0.9 \times 0.2 \times 0.25 \\ &= 0.08915. \end{aligned} \tag{6.4}$$

To evaluate UR_1, events A and E in the FT of Figure 6.1 are replaced with constant 1 and then the Boolean algebra rules are applied, leading to a reduced FT model shown in Figure 6.5. The evaluation of the reduced FT using the BDD-based method gives

$$UR_1 = 1 - (1 - q_C) \cdot (1 - q_D) = 1 - (1 - 0.15) \cdot (1 - 0.2) = 1 - 0.85 \times 0.8 = 0.32. \tag{6.5}$$

To evaluate UR_2, events B and C in the FT of Figure 6.1 are replaced with constant 1 and then the Boolean algebra rules are applied, leading to constant logic value 1 for the reduced FT model. That is UR_2 is simply 1, meaning that when CCE_2 occurs, the system fails.

Similarly, to evaluate UR_3, events A, B, C, and E in the FT of Figure 6.1 are replaced with constant 1 and then the Boolean algebra rules are applied, leading to constant logic value 1 for the reduced FT model. That is UR_3 is simply 1, meaning that when CCE_3 occurs, the entire system fails.

Step 3: Integrate for the final system reliability result. After all the four reduced problems in (6.3) are evaluated in step 2, their results are integrated with $P(CCE_i)$ listed in Table 6.1 to obtain the final system unreliability considering effects of CCFs as

$$\begin{aligned} UR_{system} &= \sum_{i=0}^{3}[UR_i \cdot P(CCE_i)] \\ &= 0.08915 \times 0.999 \times 0.998 + 0.001 \times 0.998 \\ &\quad + 0.999 \times 0.002 \times 0.32 + 0.001 \times 0.002 \\ &= 0.090522. \end{aligned} \tag{6.6}$$

This result matches the system unreliability obtained using the explicit method in Section 6.1.

6.3 Decision Diagram–Based Aggregation Method

The EDA approach discussed in Section 6.2 involves generating and solving a number of reduced reliability problems separately (and possibly in parallel if adequate computing resources are available). Very often, reliability models of the reduced problems can share common submodels, which are stored and evaluated multiple times. To improve computational efficiency and reduce storage requirements, an enhanced decision diagram (DD)-based analysis method is discussed in this section, where a single compact system DD is generated to model all the reduced problems sharing their isomorphic sub-DDs, implementing an automatic aggregation process of the original EDA approach [18].

6.3.1 Three-Step Method

The DD-based aggregation approach can be conducted in the following three steps:

Step 1: Conduct CCF modeling. Given that m elementary causes CC_i exist in the system, a CCE space is first built based on the statistical relationship of CCs. CCF is then modeled using a multi-valued variable CCE.

Specifically, when CCs are s-dependent (occurrence of one CC influences the likelihood of occurrence of the other CCs) or s-independent (occurrence of one CC does not have any influence on occurrence of the other CCs), similar to the EDA approach (Section 6.2), the CCE space contains 2^m CCEs. Each CCE is a distinct, disjoint combination of m elementary CCs. The CCE space is modeled by a 2^m-valued variable CCE, which is associated with the root node of the system DD model to be built in step 2. This root node has 2^m outgoing edges: 0-edge corresponds to CCE_0 (all the m CCs do not occur during the mission); j-edge ($1 \leq j \leq 2^m-1$, and the binary representation of j is $a_1a_2 \ldots a_{m-1}a_m$) corresponds to CCE_j (CC_{a_i} occurs *iff* $a_i = 1$).

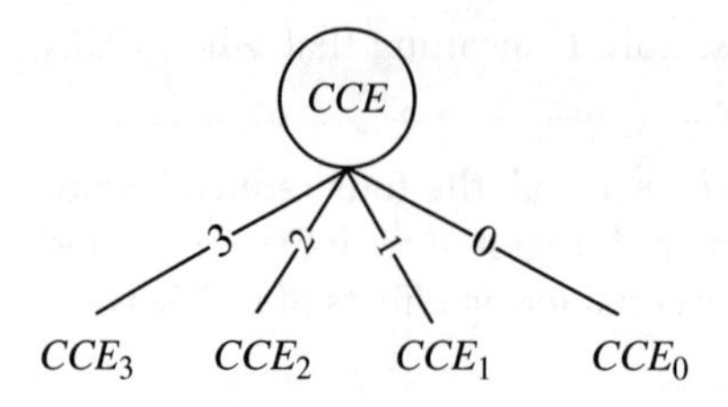

Figure 6.6 Root node for two s-independent or s-dependent CCs.

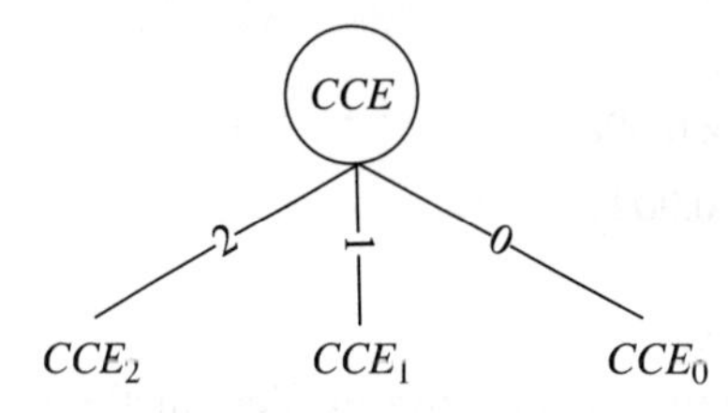

Figure 6.7 Root node for two mutually exclusive CCs.

Example 6.3 Consider a system subject to two *s*-independent/dependent CCs. The CCE space contains four CCEs:

$$\begin{aligned} CCE_0 &= \overline{CC_1} \cap \overline{CC_2}, \\ CCE_1 &= \overline{CC_1} \cap CC_2, \\ CCE_2 &= CC_1 \cap \overline{CC_2}, \\ CCE_3 &= CC_1 \cap CC_2. \end{aligned} \tag{6.7}$$

The root node is illustrated in Figure 6.6.

When CCs are mutually exclusive (they cannot occur at the same time), the CCE space contains $m + 1$ CCEs: $CCE_0 = \overline{CC_1} \cap \overline{CC_2} \cap \ldots \cap \overline{CC_m}$ (none of the CCs occurs); $CCE_i = \overline{CC_1} \cap \ldots \cap \overline{CC_{i-1}} \cap CC_i \cap \overline{CC_{i+1}} \cap \ldots \cap \overline{CC_m}$ (only CC_i occurs, $1 \leq i \leq m$). The CCE space is modeled by a $(m + 1)$-valued variable *CCE*. In this case, the node *CCE* has $m + 1$ outgoing edges: 0-edge corresponds to CCE_0 (none of the m CCs occurs during the mission); j-edge ($1 \leq j \leq m$) corresponds to CCE_j (only CC_j occurs during the mission time).

Example 6.4 Consider a system subject to two disjoint CCs. The CCE space contains three CCEs:

$$\begin{aligned} CCE_0 &= \overline{CC_1} \cap \overline{CC_2}, \\ CCE_1 &= CC_1 \cap \overline{CC_2}, \\ CCE_2 &= \overline{CC_1} \cap CC_2. \end{aligned} \tag{6.8}$$

The root node is illustrated in Figure 6.7.

Step 2: Construct system DD model. The system DD generation can start with generating a reduced FT model FT_i for each CCE_i defined in step 1. Similar to the EDA approach, the reduced FT generation is based on the original system FT and S_{CCE_i} associated with each CCE_i. A BDD model is then generated for each reduced FT based on the BDD generation algorithm of Section 2.4. The entire system DD model considering CCFs is then constructed by attaching the BDD of each reduced FT_i to the corresponding branch of the root *CCE* node and then merging isomorphic sub-BDDs from those BDDs.

The system DD model can also be generated without using reduced FTs. Specifically, the BDD is first generated from the original system FT without CCFs. This BDD is also the BDD model of the reduced problem for CCE_0. For BDD generation given the occurrence of CCE_j ($j > 0$), a recursive removal process is applied to remove all non-sink nodes modeling components affected/failed by CCE_j (i.e. components in S_{CCE_i}). Each removed node is replaced with its child node connected by its 1-edge. The child node connected by the 0-edge of the removed node is simply removed from the BDD structure. During the removal, useless nodes that have identical left and right child nodes can appear and should be removed too. After BDDs for all CCEs are generated using the above removal process, the final system DD model is obtained by attaching them to the corresponding branch of the root *CCE* node and then merging isomorphic sub-BDDs from those BDDs.

Step 3: Evaluate system DD model. Similar to the traditional BDD evaluation in Section 2.4, each path from root node *CCE* to sink node 1 (0) represents a disjoint combination of events leading to the system failure (success). Thus, the system unreliability (reliability) is given by adding probabilities of all the paths from root node *CCE* to sink node 1 (0). The probability of each path is given by multiplying the probability of each edge appearing on the path. For root node *CCE*, the *j*-edge is associated with the probability of CCE_j, i.e. $P(CCE_j)$. For a non-root node *x*, its 1-edge (0-edge) is associated with the unreliability (reliability) of the corresponding component.

6.3.2 Illustrative Example

Example 6.5 Figure 6.8 illustrates the FT model of a computer system subject to CCFs. There are three subsystems composing this system: three processors (P_1, P_2, and P_3), two buses (B_1 and B_2), and three memory modules (M_1, M_2, and M_3). The system functions correctly when at least two processors, one bus, and one memory module are functioning.

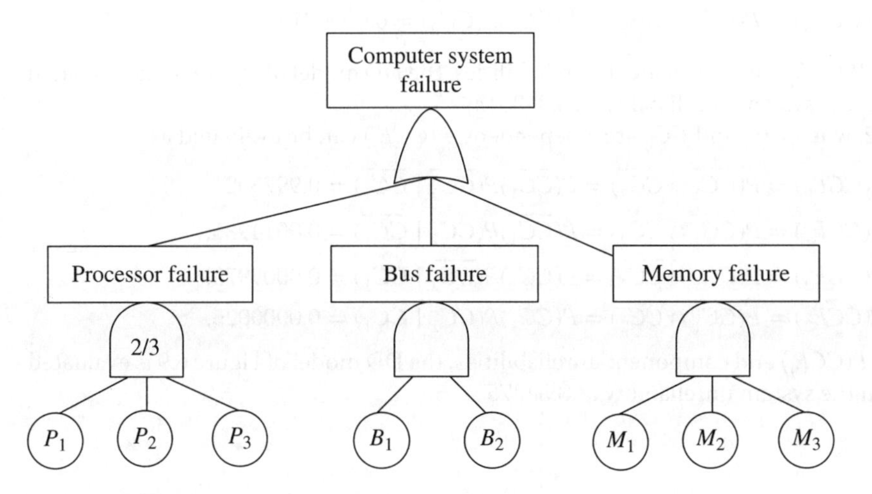

Figure 6.8 FT model of an example computer system [18].

Input Parameters. Unreliability of each processor is assumed to be 0.002, each bus has a failure probability of 0.001, and each memory module has a failure probability of 0.003. The system is subject to CCFs from CC_1, and CC_2 and $CCG_1 = \{P_1, M_1\}$, $CCG_2 = \{P_1, B_2, M_2\}$.

The following three cases are considered for CCs:

Case 1 (CC_1 and CC_2 are s-independent). $P(CC_1) = 0.001$, $P(CC_2) = 0.0015$.
Case 2 (CC_1 and CC_2 are s-dependent). $P(CC_1) = 0.001$, $P(CC_2|CC_1) = 0.0025$, $P(CC_2|\overline{CC_1}) = 0.0015$. Thus, $P(CC_2) = P(CC_2|CC_1) \cdot P(CC_1) + P(CC_2|\overline{CC_1}) \cdot P(\overline{CC_1}) = 0.0015$.
Case 3 (CC_1 and CC_2 are mutually exclusive). $P(CC_1) = 0.001$, $P(CC_2) = 0.0015$.

Example Analysis. Applying the DD-based method, the unreliability of the example computer system is analyzed as follows:

Step 1: Conduct CCF modeling. In the case of CC_1 and CC_2 being *s*-independent or *s*-dependent, the CCE space consists of four CCEs as defined in (6.7). Figure 6.6 illustrates the root node *CCE* encoding the CCE space in this case.
In the case of CC_1 and CC_2 being mutually exclusive, the CCE space consists of three CCEs as defined in (6.8). Figure 6.7 illustrates the root node *CCE* encoding the CCE space in the disjoint case.
Step 2: Construct system DD model. Figure 6.9 illustrates the final DD generated for CCs being *s*-independent or *s*-dependent. Figure 6.10 illustrates the final DD generated for CCs being disjoint.
Step 3: Evaluate system DD model. The evaluation of $P(CCE_j)$ and evaluation of the system unreliability based on the system DD model generated in step 2 are given for the three different CC cases.

Case 1. when CC_1 and CC_2 are *s*-independent, $P(CCE_j)$ can be evaluated as

$$\begin{aligned}
P(CCE_0) &= P(\overline{CC_1} \cap \overline{CC_2}) = P(\overline{CC_1})P(\overline{CC_2}) = 0.997502,\\
P(CCE_1) &= P(\overline{CC_1} \cap CC_2) = P(\overline{CC_1})P(CC_2) = 0.0014985,\\
P(CCE_2) &= P(CC_1 \cap \overline{CC_2}) = P(CC_1)P(\overline{CC_2}) = 0.0009985,\\
P(CCE_3) &= P(CC_1 \cap CC_2) = P(CC_1)P(CC_2) = 0.0000015.
\end{aligned}$$

Using $P(CCE_j)$ and component unreliabilities, the DD model of Figure 6.9 is evaluated to obtain the system unreliability as $3.34351e-5$.

Case 2. when CC_1 and CC_2 are *s*-dependent, $P(CCE_j)$ can be evaluated as

$$\begin{aligned}
P(CCE_0) &= P(\overline{CC_1} \cap \overline{CC_2}) = P(\overline{CC_1})P(\overline{CC_2} \mid \overline{CC_1}) = 0.997502,\\
P(CCE_1) &= P(\overline{CC_1} \cap CC_2) = P(\overline{CC_1})P(CC_2 \mid \overline{CC_1}) = 0.0014985,\\
P(CCE_2) &= P(CC_1 \cap \overline{CC_2}) = P(CC_1)P(\overline{CC_2} \mid CC_1) = 0.0009975,\\
P(CCE_3) &= P(CC_1 \cap CC_2) = P(CC_1)P(CC_2 \mid CC_1) = 0.0000025.
\end{aligned}$$

Using $P(CCE_j)$ and component unreliabilities, the DD model of Figure 6.9 is evaluated to obtain the system unreliability as $3.34375e-5$.

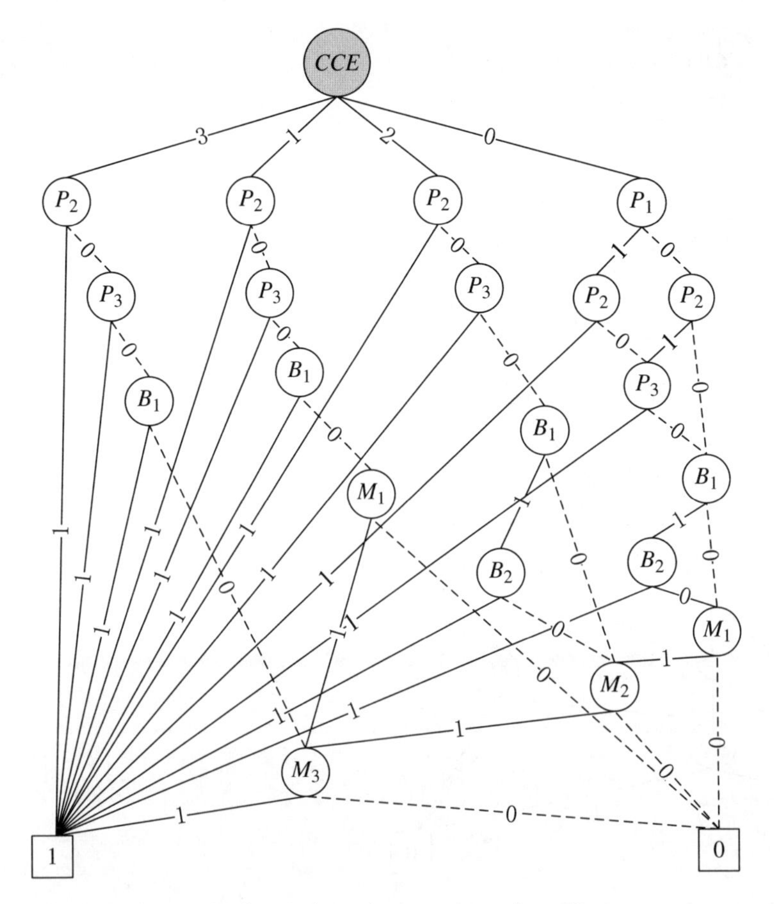

Figure 6.9 System DD for *s*-independent or *s*-dependent CCs.

Case 3. when CC_1 and CC_2 are mutually exclusive, $P(CCE_j)$ can be evaluated as

$$P(CCE_0) = P(\overline{CC_1} \cap \overline{CC_2}) = 1 - P(CC_1) - P(CC_2) = 0.9975,$$

$$P(CCE_1) = P(CC_1 \cap \overline{CC_2}) = P(CC_1) = 0.001,$$

$$P(CCE_2) = P(\overline{CC_1} \cap CC_2) = P(CC_2) = 0.0015.$$

Using $P(CCE_j)$ and component unreliabilities, the DD model of Figure 6.10 is evaluated to obtain the system unreliability as $3.34475e - 5$.

Note that the DD-based method described in this section is applicable to systems involving only single-phased missions. Refer to [19] for a BDD method based on the system FT expansion for reliability analysis of nonrepairable phased-mission systems (PMSs) subject to CCFs. Also refer to [20] for a recursive method developed for PMSs with CCFs.

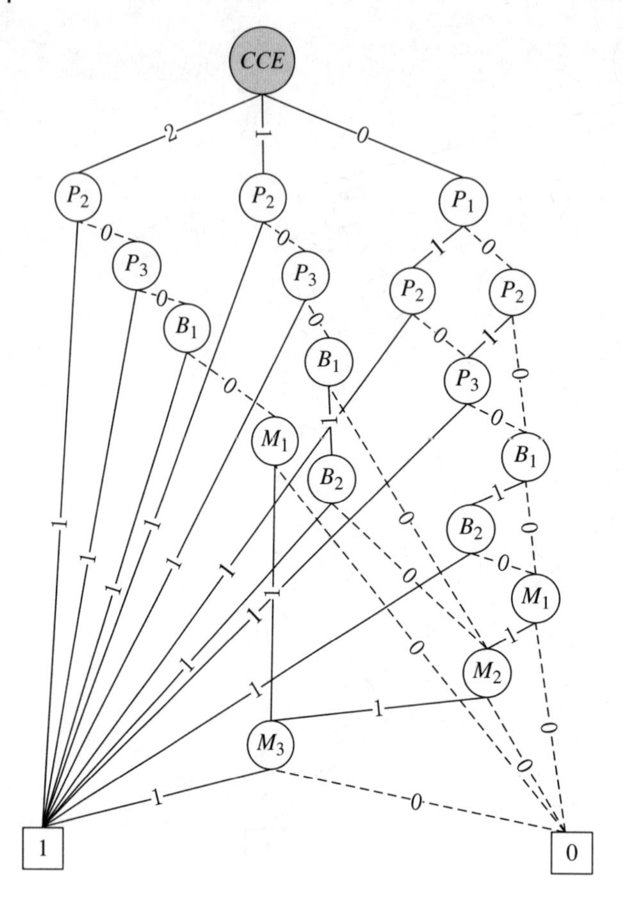

Figure 6.10 System DD for two disjoint CCs.

6.4 Universal Generating Function–Based Method

This section presents a universal generating function (u-function) based approach for the reliability analysis of nonrepairable series-parallel systems with CCFs caused by propagation of failures of system components. The failure propagation may have selective effects, meaning that CCFs originating from different components can cause failures of different subsets of system components [21]. Moreover, the failure propagation can take time to take effect. For example, infection, fire, or corrosion takes time to propagate; some component failures cause factors (e.g. overheating or humidity) that affect other components, but not immediately [22]. In this section, the failure propagation time is assumed to be a random value following a certain given distribution. The set of components that can generate CCFs and the set of components affected by the CCFs are disjoint. Also, any system component can be affected by at most one CCF.

6.4.1 System Model

There are n system components. Any system component j either functions with nominal performance g_j, or is in the failed state with performance 0. The performance rate of component j at any time instant is a random variable $G_j \in \{0, g_j\}$. Given the *pdf* $f_j(t)$ of *ttf* of component *j*, Eq. (6.9) gives the probability $p_j(T)$ that component *j* remains in the functioning state during mission time T. Thus, $P(G_j(T) = 0) = 1 - p_j(T)$.

$$p_j(T) = P(G_j(T) = g_j) = 1 - \int_0^T f_j(\tau)d\tau. \tag{6.9}$$

Let $\varphi(G_1, \ldots, G_n)$ be the system structure function, which maps the space of the components' performance rates to the space of the system's performance rate V.

Equations (6.10) and (6.11) constitute the generic model of systems with capacitated binary components.

$$g_j \ (1 \le j \le n),$$

$$p_j(T) \ (1 \le j \le n),$$

$$V = \varphi(G_1, \ldots, G_n). \tag{6.10}$$

Equation (6.11) gives the probability mass function (*pmf*) of the entire system random performance V.

$$q_i\,(T) = P(V = v_i), 0 \le i \le K. \tag{6.11}$$

Thus, the expected system performance during the mission time T is

$$W(T) = E[V] = \sum_{i=0}^{K} q_i(T)v_i. \tag{6.12}$$

Define $\pi(V, D)$ as the system acceptability function: $\pi(V, D) = 1$ if the system performance V is acceptable for a desired system demand D, and $\pi(V, D) = 0$ otherwise. The system reliability is the expected acceptability given by (6.13) [23].

$$R(D, T) = E[\pi(V, D)] = \sum_{i=0}^{K} (q_i(T)\pi(v_i, D)). \tag{6.13}$$

As an example, in applications where the system performance is indicated by productivity or capacity, and D is the minimum allowed productivity, the system reliability can be in the form of (6.14).

$$R(D, T) = \sum_{i=0}^{K} (q_i(T)1(v_i \ge D)). \tag{6.14}$$

To evaluate $W(T)$ and R(D,T), the *pmf* of the system random performance in the form of (6.11) has to be obtained using the u-function technique [23] discussed in Section 6.4.2.

6.4.2 *u*-Function Method for Series-Parallel Systems

Equation (6.15) gives the u-function representing the *pmf* of random performance of component *j* at time T.

$$u_j(z) = \sum_{m_j=0}^{1} a_{j,m_j} z^{b_{j,m_j}}, \tag{6.15}$$

where $a_{j,0} = 1 - p_j(T)$, $b_{j,0} = 0$, $a_{j,1} = p_j(T)$, $b_{j,1} = g_j$.

Equation (6.16) describes the composition operator for obtaining the u-function representing the *pmf* of the system random performance $V = \varphi(G_1, \dots, G_n)$.

$$U(z) = \otimes_\varphi(u_1(z), \dots, u_n(z)) = \otimes_\varphi\left(\sum_{m_1=0}^{1} a_{1,m_1} z^{b_{1,m_1}}, \dots, \sum_{m_n=0}^{1} a_{n,m_n} z^{b_{n,m_n}}\right)$$

$$= \sum_{m_1=0}^{1} \dots \sum_{m_n=0}^{1} \left(\prod_{i=1}^{n} a_{i,m_i} z^{\varphi(b_{1,m_1},\dots,b_{n,m_n})}\right). \tag{6.16}$$

The polynomial $U(z)$ in (6.16) represents all possible disjoint combinations of realizations of the s-independent variables $G_1, \dots, G_n$ by relating the probability of each combination to the value of function $\varphi(G_1, \dots, G_n)$ for this particular combination. Eventually, this polynomial can take the form of (6.17), representing the entire system performance distribution at time T.

$$U(z) = \sum_{i=0}^{K} q_i z^{v_i}. \tag{6.17}$$

The system reliability can thus be obtained using operator δ_D over $U(z)$, as shown in (6.18).

$$R(D, T) = \delta_D(U(z)) = \delta_D\left(\sum_{i=0}^{K} q_i z^{v_i}\right) = \sum_{i=0}^{K} q_i 1(v_i \geq D). \tag{6.18}$$

The expected system performance can be obtained using operator ε over $U(z)$, as shown in (6.19).

$$W(T) = \varepsilon(U(z)) = \sum_{i=0}^{K} q_i v_i. \tag{6.19}$$

In summary, with (6.10), the system reliability and expected performance can be obtained through the following steps.

1. Represent the *pmf* of random performance of each system component j in the u-function (6.15).
2. Obtain the u-function of the entire system $U(z)$ by applying the composition operator $\otimes_\varphi$ (6.16).
3. Compute the system reliability and expected performance by applying (6.18) and (6.19) over system performance *pmf* (6.11) represented by $U(z)$.

Among the three steps, step 2 often involves complicated computations. The structure function of a complex series-parallel system can typically be represented as composition of structure functions of s-independent subsystems containing only components connected in parallel or in series. Hence, to obtain the u-function of a series-parallel system, the composition operators can be recursively applied to obtain u-functions of intermediate purely series or purely parallel subsystem structures using the following procedure.

1. Identify any pair of system components (i and j) connected in parallel or in series.

2. Obtain the u-function of this pair using the corresponding composition operator $\otimes_\sigma$ over u-functions of the two components:

$$U_{\{i,j\}}(z) = u_i(z)\otimes_\sigma u_j(z) = \sum_{m_i=0}^{1}\sum_{m_j=0}^{1} a_{i,m_i} a_{j,m_j} z^{\sigma(b_{i,m_i}, b_{j,m_j})}. \tag{6.20}$$

   The function σ of (6.20) is decided by the nature of interaction between the components' performances. For example, in a production system with performance defined as its throughput, if two components i and j work in parallel, the total throughput is the sum of throughputs of the two components. The performance of the pair of components is determined by (6.21) in this case.

$$\sigma(G_i, G_j) = G_i + G_j. \tag{6.21}$$

   If two components consecutively process some material, then the total throughput is determined by the bottleneck (i.e. the component with the minimal performance). In this case, the function σ takes the form of (6.22).

$$\sigma(G_i, G_j) = \min(G_i, G_j). \tag{6.22}$$

   Refer to [24] for other types of function σ.
3. Replace the pair with a single component having the u-function obtained in step 2.
4. If the system contains more than one component, go back to step 1.

6.4.3 *u*-Function Method for CCFs

Let c_j denote the probability that given component j fails, this failure can propagate to other system components. Thus, $c_j(1-p_j(T))$ gives the probability that a CCF or PF originating from component j occurs during the mission time T, and $(1-c_j)(1-p_j(T))$ gives the probability that a local failure occurs in this component. Equation (6.23) gives the u-function representing the conditional performance distribution of component j at time T given that no PF originates from this component.

$$\tilde{u}_j(z) = \sum_{m_j=0}^{1} \tilde{a}_{j,m_j} z^{b_{j,m_j}}, \tag{6.23}$$

where

$$\tilde{a}_{j,0} = \frac{(1-c_j)(1-p_j(T))}{1-c_j(1-p_j(T))}, b_{j,0} = 0,$$

$$\tilde{a}_{j,1} = \frac{p_j(T)}{1-c_j(1-p_j(T))}, b_{j,1} = g_j.$$

If $c_j = 0$ (i.e. component j cannot cause PFs), then (6.23) is reduced to the form of (6.15).

Assume that the set of $w \le n$ components $\Theta = \{j(1), j(2), \ldots, j(w)\}$ can cause independent PFs. Thus, there are 2^w possible combinations of PFs. For any combination s $(0 \le s \le 2^w-1)$ the PF of component $j(k)$ happens if

$$\alpha(k, s) = \text{mod}_2 \lfloor s/2^{k-1} \rfloor = 1. \tag{6.24}$$

Equation (6.24) defines the subset of components θ_s such that $j(k) \in \theta_s$ *iff* $\alpha(k, s) = 1$. All possible PF combinations (subsets θ_s) can be obtained by running from $s = 0$ to $s = 2^w - 1$. Equation (6.25) gives the probability Q_s of combination s of the PFs during the mission time T.

$$Q_s = \prod_{k=1}^{w} (c_{j(k)}(1 - p_{j(k)}(T)))^{\alpha(k,s)}(1 - c_{j(k)}(1 - p_{j(k)}(T)))^{1-\alpha(k,s)}$$

$$= \prod_{k=1}^{w} \left(c_{j(k)} \int_0^T f_{j(k)}(\tau)d\tau \right)^{\alpha(k,s)} \left(1 - c_{j(k)} \int_0^T f_{j(k)}(\tau)d\tau \right)^{1-\alpha(k,s)}. \tag{6.25}$$

Let $A_{j(k)}$ be a subset of system components affected by a PF originating from component $j(k)$. According to the assumption that any system component can be affected by at most one PF, subsets $A_{j(k)}$ for different $j(k)$ are disjoint, i.e. $A_{j(k)} \cap A_{j(e)} = \emptyset$ for any $k \neq e$. If the PF of component $j(k)$ occurs, it can propagate to different combinations of components from $A_{j(k)}$ before mission time T. There are $2^{|A_{j(k)}|}$ such combinations. In analogy with (6.24), $\beta(x, d_k) = \text{mod}_2 \lfloor d_k/2^{x-1} \rfloor = 1$ means that the PF of component $j(k)$ propagates to component x from set $A_{j(k)}$ before T. Thus, for any d_k $(0 \le d_k \le 2^{|A_{j(k)}|} - 1)$, $\beta(x, d_k)$ defines the subset $A_{j(k),d_k}$ of components affected by the PF originating from component $j(k)$ before time T. Equation (6.26) gives the conditional probability that the PF originating from component $j(k)$ propagates to the combination d_k of components from set $A_{j(k)}$ before T, given that this PF happens.

$$\phi_{k,d_k} = \frac{\int_0^T f_{j(k)}(\tau) \left(\prod_{x \in A_{j(k)}} \left(\int_\tau^T h_{j(k),x}(t-\tau)dt \right)^{\beta(x,d_k)} \left(1 - \int_\tau^T h_{j(k),x}(t-\tau)dt \right)^{1-\beta(x,d_k)} \right) d\tau}{\int_0^T f_{j(k)}(\tau)d\tau}. \tag{6.26}$$

For any combination s of PFs, there exist $\prod_{k=1}^{w} 2^{|A_{j(k)}|\alpha(k,s)}$ combinations of components affected by the PFs. When the combination s of PFs takes place, and these PFs cause failures of subsets of components $A_{j(k),d_k}$ in each $A_{j(k)}$, $\Lambda_{s,d_1,\dots,d_w} = (\cup_{j(k)\in\theta_s} A_{j(k),d_k}) \cup \theta_s$ gives the entire set of components failed as the result of this combination of PFs. In this case, the system performance distribution can be obtained using the u-function method presented in Section 6.4.2 after replacing all the u-functions $\tilde{u}_j(z)$ corresponding to components belonging to $\Lambda_{s,d_1,\dots,d_w}$ with u-function z^0 corresponding to the component failure. The resulting u-function $U^*(\Lambda_{s,d_1,\dots,d_w})$ denotes the entire system conditional performance distribution given that all of the components from the set $\Lambda_{s,d_1,\dots,d_w}$ have failed. The occurrence probability of this event is $Q_s \prod_{k=1}^{w} (\phi_{k,d_k})^{\alpha(k,s)}$.

By combining all the possible PF outcome events, the u-function representing the unconditional system performance distribution is obtained as

$$U_{system}(z) = \sum_{s=0}^{2^w-1} Q_s \sum_{d_1=0}^{(2^{A_{j(1)}}-1)\alpha(1,s)} \dots \sum_{d_w=0}^{(2^{A_{j(w)}}-1)\alpha(w,s)} \dots \prod_{k=1}^{w} (\phi_{k,d_k})^{\alpha(k,s)} U^*(\Lambda_{s,d_1,\dots,d_w}). \tag{6.27}$$

Based on (6.27), the system reliability and its conditional expected performance can be obtained based on (6.18) and (6.19), respectively. Note that

$$\delta_D(U_{system}(z)) = \delta_D\left(\sum_{s=0}^{2^w-1} Q_s \sum_{d_1=0}^{(2^{A_{j(1)}}-1)\alpha(1,s)} \cdots \sum_{d_w=0}^{(2^{A_{j(w)}}-1)\alpha(w,s)} \cdots \prod_{k=1}^{w}(\phi_{k,d_k})^{\alpha(k,s)} U^*(\Lambda_{s,d_1,\ldots,d_w})\right)$$

$$= \sum_{s=0}^{2^w-1} Q_s \sum_{d_1=0}^{(2^{A_{j(1)}}-1)\alpha(1,s)} \cdots \sum_{d_w=0}^{(2^{A_{j(w)}}-1)\alpha(w,s)} \cdots \prod_{k=1}^{w}(\phi_{k,d_k})^{\alpha(k,s)} \delta_D(U^*(\Lambda_{s,d_1,\ldots,d_w})), \quad (6.28)$$

which can avoid the summation of polynomials in (6.27).

In summary, the u-function based algorithm for evaluating the system reliability and expected performance can be described as the following procedure [22].

1. Assign $U_{\text{system}}(z) = z^0$.
2. Determine u-functions $u_j(z)$ and $\tilde{u}_j(z)$ for all of the system components using (6.15) and (6.23), respectively.
3. For any combination s of PFs ($s = 0, \ldots, 2^w - 1$):
   3.1 Determine Q_s using (6.25), and define set θ_s of components that can cause the PFs;
   3.2 For any k such that $j(k)\in\theta_s$, generate all combinations $d_k = 0, \ldots, 2^{|A_{j(k)}|} - 1$.
   3.3 For any combination d_k, define set $A_{j(k),d_k}$ of components affected by the PFs, and compute probability ϕ_{k,d_k} using (6.26).
   3.4 For any group of combinations $d_1, \ldots, d_w$:
      3.4.1 Compute $\rho = Q_s \prod_{k=1}^{w} (\phi_{k,d_k})^{\alpha(k,s)}$.
      3.4.2 Determine set $\Lambda_{s,d_1,\ldots,d_w}$.
      3.4.3 Replace u-functions $\tilde{u}_j(z)$ of all the components belonging to set $\Lambda_{s,d_1,\ldots,d_w}$ with u-function z^0.
      3.4.4 Obtain u-function $U^*(\Lambda_{s,d_1,\ldots,d_w})$ using the method described in Section 6.4.2.
      3.4.5 Update u-function $U_{\text{system}}(z) = U_{\text{system}}(z) + \rho U^*(\Lambda_{s,d_1,\ldots,d_w})$.
4. Find the system reliability and expected performance by applying operators (6.18) and (6.19) over the obtained $U_{\text{system}}(z)$, respectively.

6.4.4 Illustrative Example

Example 6.6 Figure 6.11 illustrates an electronic device with two s-identical processor units (components 1 and 2) and three interface units (components 3–5) configured in a series-parallel structure. Components 1 and 2 have nominal performances of 4, *ttf pdf* $f(t)$, and failure propagation probability c. Components 3, 4, and 5 have the same *ttf pdf* $g(t)$, but different nominal performances 2, 1, and 2, respectively. The entire device is reliable if the system performance is at least $D = 4$.

Due to the particular components layout, the failure originating from component 1 (e.g. overheating) can propagate independently to components 3 and 4, and the failure originating from component 2 can propagate to component 5, illustrated by the dashed

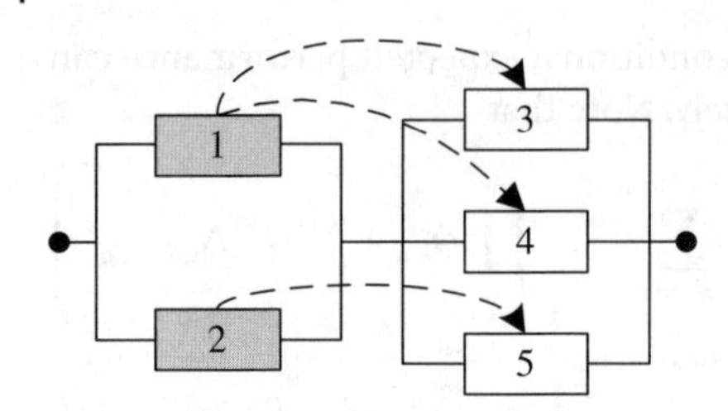

Figure 6.11 Structure of an example series-parallel system.

lines with arrows in Figure 6.11. Components 3, 4, and 5 do not have PFs. Thus, $\Theta = \{1, 2\}$, $A_1 = \{3, 4\}$, $A_2 = \{5\}$.

Define $p = \left(1 - \int_0^T f(\tau)d\tau\right)$ and $y = \left(1 - \int_0^T g(\tau)d\tau\right)$. Thus, u-functions representing the *pmf* of the component performance are

$$\begin{aligned} u_1(z) &= u_2(z) = pz^4 + (1-p)(1-c)z^0 + c(1-p)z^0, \\ u_3(z) &= u_5(z) = yz^2 + (1-y)z^0, \\ u_4(z) &= yz^1 + (1-y)z^0. \end{aligned} \tag{6.29}$$

According to (6.23), define $x = \frac{p}{1-c(1-p)}$. The u-functions representing the conditional performance *pmf* of the system components, given that no PF originates from the component are

$$\begin{aligned} \widetilde{u}_1(z) &= \widetilde{u}_2(z) = xz^4 + (1-x)z^0, \\ \widetilde{u}_3(z) &= u_3(z), \\ \widetilde{u}_4(z) &= u_4(z), \\ \widetilde{u}_5(z) &= u_5(z). \end{aligned} \tag{6.30}$$

In this example system, performance of a group of components connected in parallel is the sum of individual component performances, and performance of a group of components connected in series is the minimum of individual component performances. Thus, functions (6.21) and (6.22) are used for generating the system performance u-function as follows:

s varies from 0 to 3: $\theta_0 = \emptyset$; $\theta_1 = \{1\}$; $\theta_2 = \{2\}$; $\theta_3 = \{1,2\}$.
d_1 varies from 0 to 3: $A_{1,0} = \emptyset$; $A_{1,1} = \{3\}$; $A_{1,2} = \{4\}$; $A_{1,3} = \{3,4\}$.
d_2 varies from 0 to 1: $A_{2,0} = \emptyset$; $A_{2,1} = \{5\}$.

$$\begin{aligned} R = {} & Q_0\delta_4(U^*(\emptyset)) \\ & + Q_1[\phi_{1,0}\delta_4(U^*(\{1\}) + \phi_{1,1}\delta_4(U^*(\{1,3\}) + \phi_{1,2}\delta_4(U^*(\{1,4\}) \\ & + \phi_{1,3}\delta_4(U^*(\{1,3,4\})] \\ & + Q_2[\phi_{2,0}\delta_4(U^*(\{2\}) + \phi_{2,1}\delta_4(U^*(\{2,5\})] \\ & + Q_3[\phi_{1,0}\phi_{2,0}\delta_4(U^*(\{1,2\}) + \phi_{1,1}\phi_{2,0}\delta_4(U^*(\{1,2,3\}) \\ & + \phi_{1,2}\phi_{2,0}\delta_4(U^*(\{1,2,4\}) \\ & + \phi_{1,3}\phi_{2,0}\delta_4(U^*(\{1,2,3,4\}) + \phi_{1,0}\phi_{2,1}\delta_4(U^*(\{1,2,5\}) \\ & + \phi_{1,1}\phi_{2,1}\delta_4(U^*(\{1,2,3,5\}) \\ & + \phi_{1,2}\phi_{2,1}\delta_4(U^*(\{1,2,4,5\}) + \phi_{1,3}\phi_{2,1}\delta_4(U^*(\{1,2,3,4,5\})], \end{aligned} \tag{6.31}$$

where

$$Q_0 = (1 - c(1-p))^2,$$

$$Q_1 = Q_2 = c(1-p)(1 - c(1-p)),$$

$$Q_3 = (c(1-p))^2,$$

$$\phi_{1,0} = \int_0^T f(\tau)\left(1 - \int_\tau^T h_{1,3}(t-\tau)dt\right)\left(1 - \int_\tau^T h_{1,4}(t-\tau)dt\right)d\tau/(1-p),$$

$$\phi_{1,1} = \int_0^T f(\tau)\left(\int_\tau^T h_{1,3}(t-\tau)dt\right)\left(1 - \int_\tau^T h_{1,4}(t-\tau)dt\right)d\tau/(1-p),$$

$$\phi_{1,2} = \int_0^T f(\tau)\left(1 - \int_\tau^T h_{1,3}(t-\tau)dt\right)\left(\int_\tau^T h_{1,4}(t-\tau)dt\right)d\tau/(1-p),$$

$$\phi_{1,3} = \int_0^T f(\tau)\left(\int_\tau^T h_{1,3}(t-\tau)dt\right)\left(\int_\tau^T h_{1,4}(t-\tau)dt\right)d\tau/(1-p),$$

$$\phi_{2,0} = \int_0^T f(\tau)\left(1 - \int_\tau^T h_{1,5}(t-\tau)dt\right)d\tau/(1-p),$$

$$\phi_{2,1} = \int_0^T f(\tau)\left(\int_\tau^T h_{1,5}(t-\tau)dt\right)d\tau/(1-p),$$

$$\begin{aligned}
U^*(\varnothing) &= (\widetilde{u}_1(z)\otimes_+\widetilde{u}_2(z))\otimes_{\min}(u_3(z)\otimes_+u_4(z)\otimes_+u_5(z)) \\
&= ((xz^4 + (1-x)z^0)\otimes_+(xz^4 + (1-x)z^0)) \\
&\quad \otimes_{\min}((yz^2 + (1-y)z^0)\otimes_+(yz^1 + (1-y)z^0)\otimes_+(yz^2 + (1-y)z^0)) \\
&= (x^2z^8 + 2x(1-x)z^4 + (1-x)^2z^0) \\
&\quad \otimes_{\min}(y^3z^5 + y^2(1-y)z^4 + 2y^2(1-y)z^3 + 2y(1-y)^2z^2 \\
&\qquad + y(1-y)^2z^1 + (1-y)^3z^0) \\
&= x^2y^3z^5 + x^2y^2(1-y)z^4 + x^2 2y^2(1-y)z^3 + x^2 2y(1-y)^2z^2 + x^2y(1-y)^2z^1 \\
&\quad + x^2(1-y)^3z^0 + 2x(1-x)y^2z^4 + 4x(1-x)y^2(1-y)z^3 \\
&\quad + 4x(1-x)y(1-y)^2z^2 + 2x(1-x)y(1-y)^2z^1 \\
&\quad + 2x(1-x)(1-y)^3z^0 + (1-x)^2z^0,
\end{aligned}$$

$$\delta_4(U^*(\varnothing)) = x^2y^3 + x^2y^2(1-y) + 2x(1-x)y^2 = 2xy^2 - x^2y^2.$$

Replacing $\widetilde{u}_1(z)$ with z^0, one obtains

$$\begin{aligned}
U^*(\{1\}) &= (z^0\otimes_+(xz^4 + (1-x)z^0))\otimes_{\min}((yz^2 + (1-y)z^0)\otimes_+(yz^1 + (1-y)z^0) \\
&\quad \otimes_+(yz^2 + (1-y)z^0)) \\
&= (xz^4 + (1-x)z^0)\otimes_{\min}(y^3z^5 + y^2(1-y)z^4 + 2y^2(1-y)z^3 + 2y(1-y)^2z^2 \\
&\quad + y(1-y)^2z^1 + (1-y)^3z^0) \\
&= xy^3z^4 + xy^2(1-y)z^4 + 2xy^2(1-y)z^3 + 2xy(1-y)^2z^2 + xy(1-y)^2z^1 \\
&\quad + x(1-y)^3z^0 + (1-x)z^0.
\end{aligned}$$

Replacing $\widetilde{u}_1(z)$ and $\widetilde{u}_3(z)$ with z^0, one obtains

$$\begin{aligned}
&U^*(\{1,3\})\\
&\quad = (z^0\otimes_+(xz^4+(1-x)z^0))\otimes_{\min}(z^0\otimes_+(yz^1+(1-y)z^0)\otimes_+(yz^2+(1-y)z^0))\\
&\quad = (xz^4+(1-x)z^0)\otimes_{\min}(y^2z^3+y(1-y)z^2+y(1-y)z^1+(1-y)^2z^0)\\
&\quad = xy^2z^3+xy(1-y)z^2+xy(1-y)z^1+x(1-y)^2z^0+(1-x)z^0.
\end{aligned}$$

$$\delta_4(U^*(\{1,3\}))=0.$$

Replacing $\widetilde{u}_1(z)$ and $\widetilde{u}_4(z)$ with z^0, one obtains

$$\begin{aligned}
&U^*(\{1,4\})\\
&\quad = (z^0\otimes_+(xz^4+(1-x)z^0))\otimes_{\min}(z^0\otimes_+(yz^2+(1-y)z^0)\otimes_+(yz^2+(1-y)z^0))\\
&\quad = (xz^4+(1-x)z^0)\otimes_{\min}(y^2z^4+2y(1-y)z^2+(1-y)^2z^0)\\
&\quad = xy^2z^4+2xy(1-y)z^2+x(1-y)^2z^0+(1-x)z^0.
\end{aligned}$$

$$\delta_4(U^*(\{1,4\}))=xy^2.$$

Replacing $\widetilde{u}_1(z)$, $\widetilde{u}_3(z)$, and $\widetilde{u}_4(z)$ with z^0, one obtains

$$\begin{aligned}
&U^*(\{1,3,4\})\\
&\quad = (z^0\otimes_+(xz^4+(1-x)z^0))\otimes_{\min}(z^0\otimes_+z^0\otimes_+(yz^2+(1-y)z^0))\\
&\quad = (xz^4+(1-x)z^0)\otimes_{\min}(yz^2+(1-y)z^0)=xyz^2+(1-xy)z^0.
\end{aligned}$$

$$\delta_4(U^*(\{1,3,4\}))=0.$$

Replacing $\widetilde{u}_2(z)$ with z^0, one obtains

$$\begin{aligned}
&U^*(\{2\})\\
&\quad = ((xz^4+(1-x)z^0)\otimes_+z^0)\otimes_{\min}\\
&\qquad ((yz^2+(1-y)z^0)\otimes_+(yz^1+(1-y)z^0)\otimes_+(yz^2+(1-y)z^0))\\
&\quad = (xz^4+(1-x)z^0)\otimes_{\min}(y^3z^5+y^2(1-y)z^4+2y^2(1-y)z^3\\
&\qquad +2y(1-y)^2z^2+y(1-y)^2z^1+(1-y)^3z^0)\\
&\quad = xy^3z^4+xy^2(1-y)z^4+2xy^2(1-y)z^3+2xy(1-y)^2z^2+xy(1-y)^2z^1\\
&\qquad +x(1-y)^3z^0+(1-x)z^0.
\end{aligned}$$

$$\delta_4(U^*(\{2\}))=xy^3+xy^2(1-y)=xy^2.$$

Replacing $\widetilde{u}_2(z)$ and $\widetilde{u}_5(z)$ with z^0, one obtains

$$\begin{aligned}
&U^*(\{2,5\})\\
&\quad = (xz^4+(1-x)z^0)\otimes_{\min}(y^2z^3+y(1-y)z^2+y(1-y)z^1+(1-y)^2z^0)\\
&\quad = xy^2z^3+xy(1-y)z^2+xy(1-y)z^1+x(1-y)^2z^0+(1-x)z^0.
\end{aligned}$$

$$\delta_4(U^*(\{2,5\}))=0.$$

After replacing $\widetilde{u}_1(z)$ and $\widetilde{u}_2(z)$ with z^0, the resulting system u-functions take the form

$$\begin{aligned} U^*(\{1,2\}) &= U^*(\{1,2,3\}) = U^*(\{1,2,4\}) = U^*(\{1,2,5\}) = U^*(\{1,2,3,4\}) \\ &= U^*(\{1,2,3,5\}) = U^*(\{1,2,4,5\}) = U^*(\{1,2,3,4,5\}) = z^0. \end{aligned}$$

Finally,

$$\begin{aligned} R &= Q_0\,(2xy^2 - x^2y^2) + Q_1[\phi_{1,0}xy^2 + \phi_{1,3}xy^2] + Q_2\phi_{2,0}xy^2 \\ &= [(2-x)Q_0 + Q_1(\phi_{1,0} + \phi_{1,2}) + Q_2\phi_{2,0}]xy^2 \\ &= \Bigg\{\left(1 + (1-2c)\int_0^T f(\tau)d\tau\right) \\ &\quad + c\Bigg[\int_0^T f(\tau)\left(1 - \int_\tau^T h_{1,3}(t-\tau)dt\right)\left(1 - \int_\tau^T h_{1,4}(t-\tau)dt\right)d\tau \\ &\quad + \int_0^T f(\tau)\left(1 - \int_\tau^T h_{1,3}(t-\tau)dt\right)\left(\int_\tau^T h_{1,4}(t-\tau)dt\right)d\tau \\ &\quad + \int_0^T f(\tau)\left(1 - \int_\tau^T h_{1,5}(t-\tau)dt\right)d\tau\Bigg]\Bigg\} \\ &\quad \left(1 - \int_0^T f(\tau)d\tau\right)\left(1 - \int_0^T g(\tau)d\tau\right)^2. \end{aligned} \tag{6.32}$$

For the special case where $c = 0$ (i.e. no PFs can happen), the reliability evaluation equation becomes

$$\begin{aligned} R &= \left(1 + \int_0^T f(\tau)d\tau\right)\left(1 - \int_0^T f(\tau)d\tau\right)\left(1 - \int_0^T g(\tau)d\tau\right)^2 \\ &= \left(1 - \left(\int_0^T f(\tau)d\tau\right)^2\right)\left(1 - \int_0^T g(\tau)d\tau\right)^2 = (1-(1-p)^2)y^2. \end{aligned} \tag{6.33}$$

Equation (6.33) can be easily verified by the fact that the system can satisfy the desired demand $D = 4$ when at least one of the two processor components is working (occurrence probability is $(1-(1-p)^2)$), and both components 3 and 5 are working (occurrence probability is y^2).

For another special case where the PFs take zero time to propagate, i.e. $h_{k,j}(t) = 1$, and $\int_\tau^T h_{k,j}(t-\tau)dt = 1$, the reliability evaluation equation becomes

$$\begin{aligned} R &= \left(1 + (1-2c)\int_0^T f(\tau)d\tau\right)\left(1 - \int_0^T f(\tau)d\tau\right)\left(1 - \int_0^T g(\tau)d\tau\right)^2 \\ &= (1 + (1-2c)(1-p))py^2. \end{aligned} \tag{6.34}$$

In this case, the system cannot satisfy the desired demand if at least one of the PFs occurs, which causes component 3 or 5 or both to fail. In other words, the system is reliable only when at least one processor is functioning and no PFs occur (occurrence probability is $(p^2 + 2p\,(1-p)(1-c) = (1+(1-2c)(1-p))p)$), and both components 3 and 5 are working (occurrence probability is y^2).

6.5 Summary

This chapter discusses the reliability modeling of deterministic CCFs, where the occurrence of a root cause results in deterministic failures of multiple system components simultaneously. Several different combinatorial approaches are presented. The explicit approach based on expanding the system reliability model is straightforward. However, it can become computationally inefficient for large-scale systems. In addition, the explicit approach is only applicable to systems subject to *s*-independent CCs. The implicit approaches (the EDA and DD-based aggregation methods) require no expansion on the system reliability model and are flexible in handling the *s*-relationship among CCs (which can be *s*-independent, *s*-dependent, or mutually exclusive). The UGF method presented provides a reliability evaluation solution for series-parallel systems subject to CCFs that take random time to take effect. All of the approaches discussed in this chapter are applicable to arbitrary types of component *ttf* distributions.

References

1 NUREG/CR-4780 (1988). *Procedure for Treating Common-Cause Failures in Safety and Reliability Studies*, vol. I and II. Washington DC, USA: U.S. Nuclear Regulatory Commission.
2 Tang, Z. and Dugan, J.B. (2004). An integrated method for incorporating common cause failures in system analysis. In: *Proceedings of the 50th Annual Reliability and Maintainability Symposium*, 610–614. Los Angeles, CA, USA.
3 Bai, D.S., Yun, W.Y., and Chung, S.W. (1991). Redundancy optimization of k-out-of-n systems with common-cause failures. *IEEE Transactions on Reliability* 40 (1): 56–59.
4 Pham, H. (1993). Optimal cost-effective design of triple-modular-redundancy-with-spares systems. *IEEE Transactions on Reliability* 42 (3): 369–374.
5 Anderson, P.M. and Agarwal, S.K. (1992). An improved model for protective-system reliability. *IEEE Transactions on Reliability* 41 (3): 422–426.
6 Chae, K.C. and Clark, G.M. (1986). System reliability in the presence of common-cause failures. *IEEE Transactions on Reliability* R-35: 32–35.
7 Fleming, K.N., Mosleh, N., and Deremer, R.K. (1986). A systematic procedure for incorporation of common cause events into risk and reliability models. *Nuclear Engineering and Design* 93: 245–273.
8 Dai, Y.S., Xie, M., Poh, K.L., and Ng, S.H. (2004). A model for correlated failures in N-version programming. *IIE Transactions* 36 (12): 1183–1192.
9 Fleming, K.N. and Mosleh, A. (1985). Common-cause data analysis and implications in system modeling. In: *Proceedings of the International Topical Meeting on Probabilistic Safety Methods and Applications*, USA.
10 Amari, S.V., Dugan, J.B., and Misra, R.B. (1999). Optimal reliability of systems subject to imperfect fault-coverage. *IEEE Transactions on Reliability* 48 (3): 275–284.
11 Vaurio, J.K. (2003). Common cause failure probabilities in standby safety system fault tree analysis with testing – scheme and timing dependencies. *Reliability Engineering & System Safety* 79 (1): 43–57.
12 Mitra, S., Saxena, N.R., and McCluskey, E.J. (2000). Common-mode failures in redundant VLSI systems: a survey. *IEEE Transactions on Reliability* 49 (3): 285–295.

13 Vaurio, J.K. (1998). An implicit method for incorporating common-cause failures in system analysis. *IEEE Transactions on Reliability* 47 (2): 173–180.
14 Fleming, K.N., Mosleh, A., and Kelly, A.P. (1983). On the analysis of dependent failures in risk assessment and reliability evaluation. *Nuclear Safety* 24: 637–657.
15 Xing, L., Shrestha, A., Meshkat, L., and Wang, W. (2009). Incorporating common-cause failures into the modular hierarchical systems analysis. *IEEE Transactions on Reliability* 58 (1): 10–19.
16 Xing, L. (2004). Fault-tolerant network reliability and importance analysis using binary decision diagrams. In: *Proceedings of the 50th Annual Reliability and Maintainability Symposium*, Los Angeles, CA, USA.
17 Xing, L. (2005). Reliability modeling and analysis of complex hierarchical systems. *International Journal of Reliability, Quality and Safety Engineering* 12 (6): 477–492.
18 Mo, Y. and Xing, L. (2013). An enhanced decision diagram-based method for common-cause failure analysis. *Proc IMechE, Part O, Journal of Risk and Reliability* 227 (5): 557–566.
19 Xing, L. and Levitin, G. (2013). BDD-based reliability evaluation of phased-mission systems with internal/external common-cause failures. *Reliability Engineering & System Safety* 112: 145–153.
20 Levitin, G., Xing, L., Amari, S.V., and Dai, Y. (2013). Reliability of non-repairable phased-mission systems with common-cause failures. *IEEE Transactions on Systems, Man, and Cybernetics: Systems* 43 (4): 967–978.
21 Levitin, G. and Xing, L. (2010). Reliability and performance of multi-state systems with propagated failures having selective effect. *Reliability Engineering & System Safety* 95 (6): 655–661.
22 Levitin, G., Xing, L., Ben-Haim, H., and Dai, Y. (2013). Reliability of series-parallel systems with random failure propagation time. *IEEE Transactions on Reliability* 62 (3): 637–647.
23 Lisnianski, A. and Levitin, G. (2003). *Multi-State System Reliability: Assessment, Optimization and Applications*. World Scientific.
24 Levitin, G. (2005). *Universal Generating Function in Reliability Analysis and Optimization*. London: Springer-Verlag.

7

Probabilistic Common-Cause Failure

The occurrence of a probabilistic common-cause failure (PCCF) can result in failures of multiple system components with different probabilities [1, 2]. Consider a real-world example of PCCFs, where multiple gas detectors are installed in a production room [3]. These gas detectors may be purchased from different companies and at different times, and thus can be resistant to different levels of humidity. The increased humidity in the production room may fail the gas detectors installed at different locations of the production room with different occurrence probabilities. The increased humidity serves as a shared root cause of the example PCCFs.

PCCFs can be caused by propagated failures originating from certain components within the system or by external shocks/factors such as malicious attacks and environmental conditions. This chapter presents explicit and implicit methods to analyze reliability of systems subject to internal or external PCCFs. Both single-phase and multi-phase systems are considered.

7.1 Single-Phase System

The system contains components with (different) individual failure probabilities. Some components can fail also as a result of different common causes (CCs). These CCs can be associated with failures of other system components or with some external shocks. The system structure function is given in the form of a fault tree (FT) model [4], which defines the entire system state for any combination of the component states.

The PCCF gate (Figure 7.1) is used to model the PCCF behavior [5]. This gate is designed based on the FDEP gate [2]. The input of the PCCF gate is a trigger event representing the occurrence of a certain CC. The gate also has one or more dependent events representing failures of system components affected by the CC; these dependent components form a probabilistic common-cause group (PCCG). When the trigger event occurs, the dependent events are forced to occur with certain (maybe different) probabilities, represented by switch symbols in Figure 7.1.

Both an explicit method and an implicit method are presented for the reliability analysis of single-phased systems subject to PCCFs. The following assumptions are made:

- There are no failure cascading or loops. In other words, the failure of a dependent component for a PCCF gate cannot trigger another PCCF gate.

Dynamic System Reliability: Modeling and Analysis of Dynamic and Dependent Behaviors,
First Edition. Liudong Xing, Gregory Levitin and Chaonan Wang.

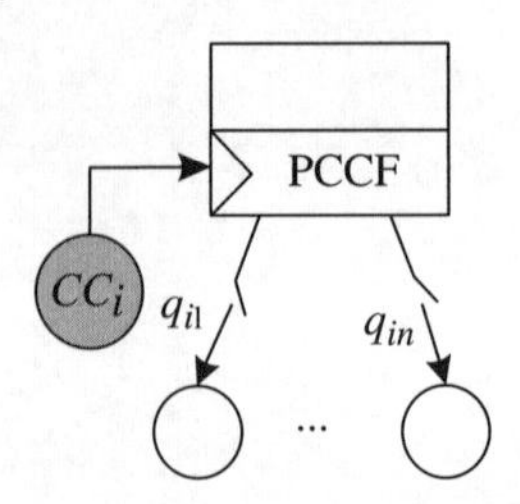

Figure 7.1 General structure of the PCCF gate.

- The individual failure event and failure event caused by a CC for the same component are *s*-independent, and their occurrence probabilities are given input parameters or can be derived from the input parameter.

7.1.1 Explicit Method

In the explicit method, the system model is developed with the consideration of contributions from CCs causing PCCFs. The explicit algorithm involves the following two-step process:

Step 1: Build an expanded FT. To explicitly consider effects of PCCFs, independent pseudo nodes denoting component failure events triggered by CCs are added to the original system FT model (excluding PCCF gates). Specifically, if component X is affected by n CCs (CC_1, CC_2, ... CC_n), that is, it appears in n PCCGs, then n pseudo nodes (X_1, X_2, ... X_n) are added to FT for this component, where X_i represents the failure event of component X caused by CC_i. Since a component fails either locally or due to the occurrence of a CC affecting this component, its total failure behavior can be modeled by logical expression (7.1), where X represents its local failure. The FT model of (7.1) is given in Figure 7.2.

$$X_{TF} = (CC_1 \cdot X_1 + CC_2 \cdot X_2 + \ldots + CC_n \cdot X_n) + X \tag{7.1}$$

The probability of X_i is the conditional failure probability of component X conditioned on the occurrence of CC_i, which is denoted by q_{iX}. Because a PCCG may include more than one component, a CC event CC_i can thus appear as an input in more than one component's FT. For an internal CC, CC_i is the failure event of another system component. The FT in the form of Figure 7.2 is built for each of the dependent components appearing in PCCGs. The expanded FT for the entire system is formed by using these FTs to replace the corresponding component failure events in the original system FT model.

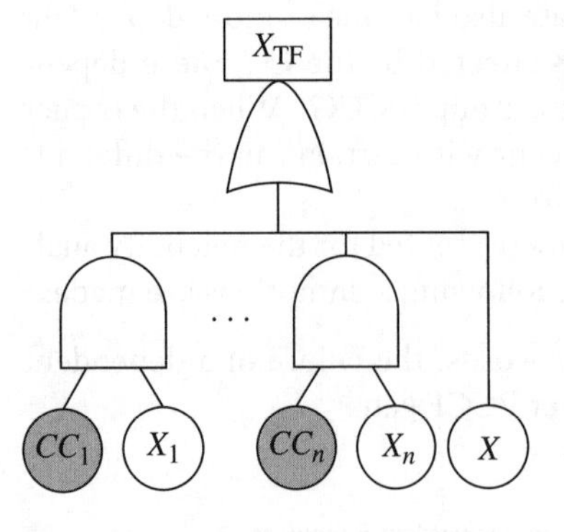

Figure 7.2 FT representing component *X*'s total failure event [5].

Step 2: Evaluate the expanded FT formed in Step 1. The expanded FT can be evaluated using any traditional FT reliability analysis method, e.g. cut/path-sets based methods [6], or binary decision diagrams (BDDs) [7]. In this chapter, the BDD-based method (Section 2.4) is adopted for the example analysis.

Example 7.1 A computer system (Figure 7.3) has two processors (P_1 and P_2), two buses (B_1 and B_2), three memory units (M_1, M_2, and M_3), and input/output (I/O). The system operation requires at least one processor, at least one bus, at least two memory units, and the I/O be functioning correctly.

Figure 7.4 illustrates the FT of the system. The system undergoes two *s*-independent external CCs: CC_1 and CC_2. CC_1 affects processor P_1 and memory unit M_1, thus $PCCG_1 = \{P_1, M_1\}$. CC_2 affects processor P_1, bus B_1 and memory unit M_2, thus, $PCCG_2 = \{P_1, B_1, M_2\}$.

Input Parameters: The following parameter values are used for the illustrative analysis.

- Component local/individual failure probabilities $q_X = 0.01$ for $X \in \{P_1, P_2, B_1, B_2$, I/O, $M_1, M_2, M_3\}$. Note that while the fixed probabilities are assumed in this example system, the method is applicable to any arbitrary types of component time-to-failure (*ttf*) distributions; based on the probability density function (*pdf*) or cumulative distribution function (*cdf*) of the component *ttf*, the component failure probability can be derived.
- Probabilities of CCs $P(CC_i) = 0.001$, $i = 1, 2$.

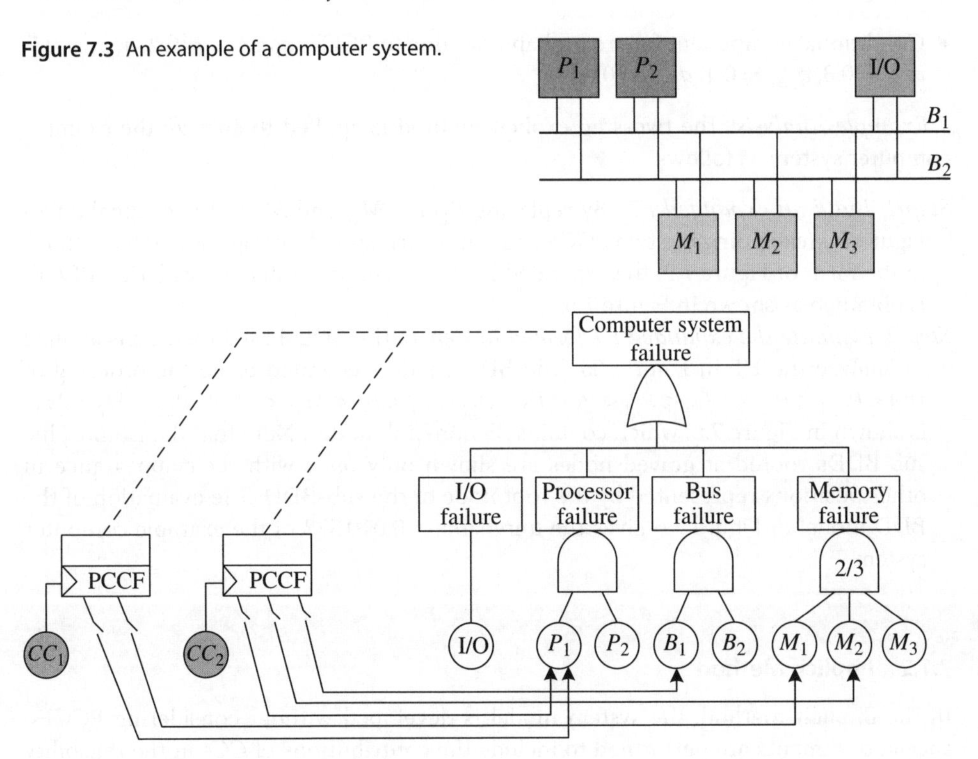

Figure 7.3 An example of a computer system.

Figure 7.4 FT of the example computer system.

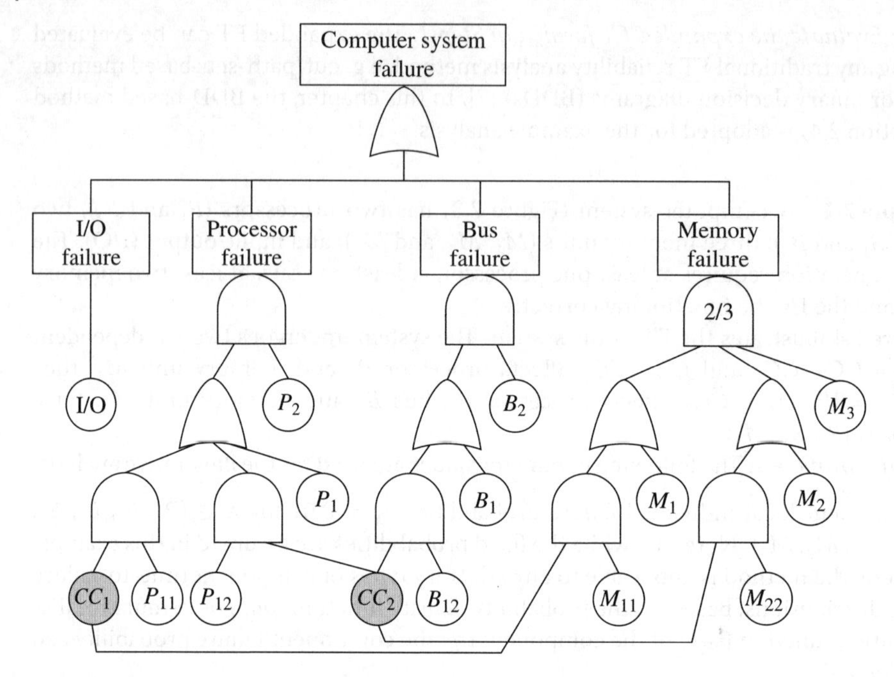

Figure 7.5 Expanded FT for the example computer system.

- Conditional component failure probabilities due to PCCFs q_{iX}: $q_{1P1} = 0.2$, $q_{1M1} = 0.5$, $q_{2P1} = 0.3$, $q_{2B1} = 0.4$, $q_{2M2} = 0.6$.

Example Analysis: The two-step explicit method is applied to analyze the example computer system as follows.

Step 1: Build an expanded FT. By replacing P_1, B_1, M_1, and M_2 in the original FT of Figure 7.4 (excluding the two PCCF gates and corresponding trigger events) with FT in the form of Figure 7.2, the expanded FT for the entire system considering PCCFs is obtained as shown in Figure 7.5.

Step 2: Evaluate the expanded FT formed in step 1. The BDD-based method is applied to analyze the FT in Figure 7.5. The BDD model generated using the ordering of $I/O < CC_1 < P_{11} < CC_2 < P_{12} < P_1 < P_2 < B_{12} < B_1 < B_2 < M_{11} < M_1 < M_{22} < M_2 < M_3$ is shown in Figure 7.6, which contains 35 non-sink nodes. Note that the isomorphic sub-BDDs rooted at grayed nodes are shown only once with their appearance in other positions represented by the root node of the sub-BDD. The evaluation of the BDD model in Figure 7.6 gives the unreliability 0.010 523 of the example computer system.

7.1.2 Implicit Method

In the implicit method, the system model is developed without considering PCCFs; special treatments are performed to include the contributions of CCs in the reliability evaluation. The implicit algorithm presented in this section extends the efficient

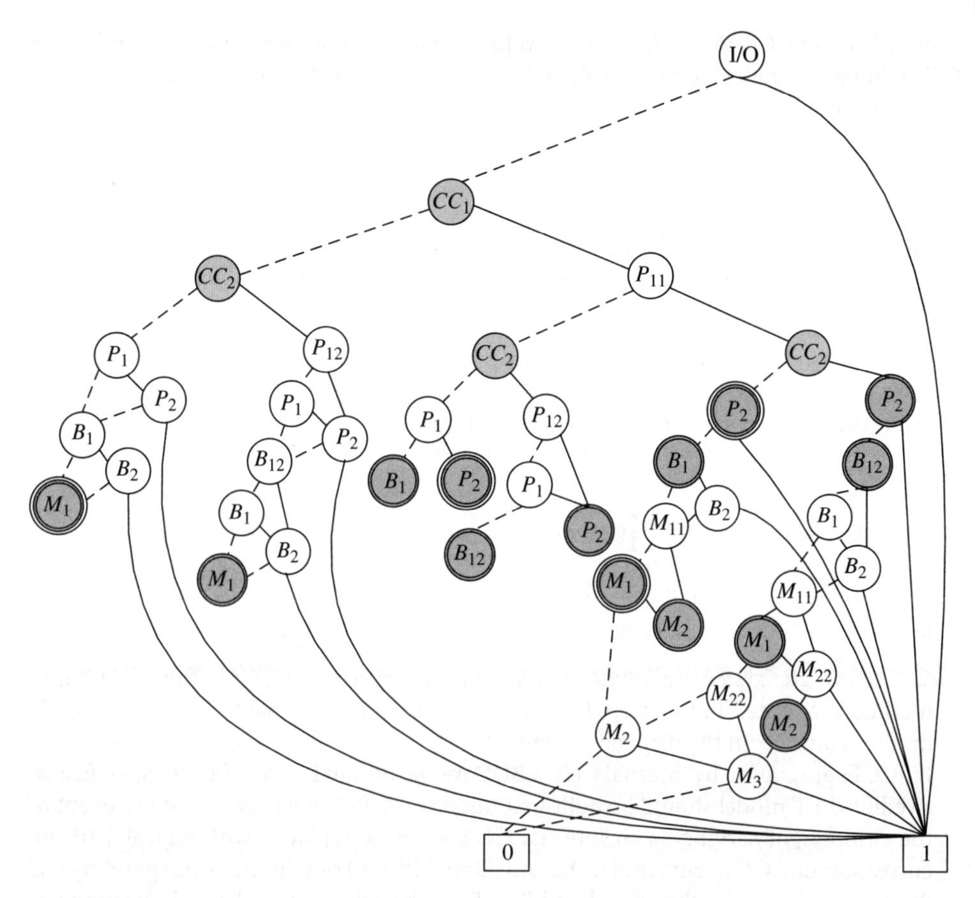

Figure 7.6 BDD model of the expanded FT.

decomposition and aggregation (EDA) method presented in Chapter 6.2 to address PCCFs. It can be described as the following five-step process.

Step 1: Construct a probabilistic common-cause event (PCCE) space. The space contains all possible combinations of occurrence or nonoccurrence of CCs. Specifically, given that m CCs occur in the system, the PCCE space consists of 2^m disjoint events, each called a PCCE, illustrated in (7.2).

$$PCCE_0 = \overline{CC_1} \cap \overline{CC_2} \cap \ldots \cap \overline{CC_m},$$
$$PCCE_1 = \overline{CC_1} \cap \overline{CC_2} \cap \ldots \cap CC_m,$$
$$\ldots\ldots$$
$$PCCE_{2^m-1} = CC_1 \cap CC_2 \cap \ldots \cap CC_m. \quad (7.2)$$

In (7.2), for $PCCE_j$ $(0 \leq j \leq 2^m - 1)$, if the binary representation of j is $a_1a_2\ldots a_{m-1}a_m$, then CC_{a_i} appears in $PCCE_j$ if $a_i = 1$, otherwise $\overline{CC_{a_i}}$ appears in the definition of $PCCE_j$. For example, $PCCE_0$ is an event that all of the m CCs do not occur, $PCCE_1$ is an event that only CC_m occurs while other CCs do not occur. The occurrence probability of each $PCCE_i$, denoted by $P(PCCE_i)$, is also evaluated in this step. Given the occurrence probabilities of CCs, $P(PCCE_i)$ can be evaluated based on the s-relationship among the CCs (similar to the evaluation of $P(CCE_i)$ as illustrated in Section 6.3.2). In addition, Eq. (7.3) should be satisfied.

$$\sum_{i=0}^{2^m-1} P(PCCE_i) = 1 \tag{7.3}$$

Note that the implicit algorithm can handle diverse s-relationships among the CCs including mutually exclusive, s-independent, or s-dependent CCs, while the explicit algorithm presented in Section 7.1.1 can handle only s-independent CCs.

Step 2: Evaluate the component total failure probability. For each component affected by at least one CC, its total failure probability is evaluated under each PCCE. Specifically, given the occurrence of $PCCE_j$, if component X is affected by k CCs (CC_1, CC_2, … CC_k), then its total failure probability under $PCCE_j$, denoted by Q_{jX}, is calculated as

$$Q_{jX} = 1 - (1 - q_X)\prod_{i=1}^{k}(1 - q_{iX}), \tag{7.4}$$

where q_X is the local failure probability of component X and q_{iX} is the conditional failure probability of component X given that CC_i takes place.

Step 3: Build the system reliability model ignoring effects of PCCFs. The BDD-based method is adopted in this step. If PCCFs are caused by external CCs, only one BDD model is built from the original system FT.
If PCCFs are caused by internal CCs, a BDD model is built for each PCCE. Specifically, a reduced FT model should be built first under each PCCE where the failure event of the component serving as an internal CC should be replaced with logical 1 (if the corresponding CC occurs under the considered PCCE) or 0 (if the corresponding CC does not occur under the considered PCCE) in the original FT. Then a BDD model is built for each reduced FT.

Step 4: Evaluate P(system fails|$PCCE_i$). Using the total component failure probabilities for components affected by PCCFs obtained in step 2, the conditional system failure probability given that $PCCE_i$ occurs, P(system fails|$PCCE_i$), can be obtained by evaluating the BDD model generated in step 3. Note that for components that are not affected by any CC, their local failure probabilities are used in the BDD evaluation.

Step 5: Integrate for the system unreliability. In this last step, $P(PCCE_i)$ evaluated in step 1 and P(system fails|$PCCE_i$) evaluated in step 4 are integrated using the total probability law to obtain the final system unreliability.

$$UR_{system} = \sum_{i=0}^{2^m-1} [P(\text{system fails} \mid PCCE_i) \cdot P(PCCE_i)] \tag{7.5}$$

Example 7.2 The same example computer system in Example 7.1 is analyzed using the implicit method as follows.

Step 1: Construct a PCCE space. Based on the two external, s-independent CCs, the PCCE space consists of four PCCEs:

$$PCCE_0 = \overline{CC_1} \cap \overline{CC_2},$$
$$PCCE_1 = \overline{CC_1} \cap CC_2,$$
$$PCCE_2 = CC_1 \cap \overline{CC_2},$$
$$PCCE_3 = CC_1 \cap CC_2. \quad (7.6)$$

Their occurrence probabilities are

$$P(PCCE_0) = (1 - P(CC_1))(1 - P(CC_2)) = 0.998001,$$
$$P(PCCE_1) = (1 - P(CC_1))P(CC_2) = 0.000999,$$
$$P(PCCE_2) = P(CC_1)(1 - P(CC_2)) = 0.000999,$$
$$P(PCCE_3) = P(CC_1)P(CC_2) = 0.000001. \quad (7.7)$$

In the case of CCs being mutually exclusive or *s*-dependent, $P(PCCE_i)$ can be evaluated in the similar way to that illustrated in Section 6.3.2.

Step 2: Evaluate the component total failure probability. In this step, the total failure probability for each component affected by at least one CC, i.e. P_1, M_1, B_1, and M_2, is evaluated under each PCCE.

Specifically, under $PCCE_0$, no CC takes place. The total component failure probability is simply the local failure probability of the component (i.e. 0.01).

Under $PCCE_1$, only CC_2 happens and components P_1, B_1, and M_2 fail due to the occurrence of CC_2. According to (7.4), the total failure probabilities for P_1, B_1, and M_2 are

$$Q_{1P_1} = 1 - (1 - q_{P_1})(1 - q_{2P_1}) = 0.307,$$
$$Q_{1B_1} = 1 - (1 - q_{B_1})(1 - q_{2B_1}) = 0.406,$$
$$Q_{1M_2} = 1 - (1 - q_{M_2})(1 - q_{2M_2}) = 0.604.$$

Since M_1 is not affected, its total component failure probability is its local failure probability (i.e. 0.01).

Under $PCCE_2$, only CC_1 happens and components P_1 and M_1 can fail due to the occurrence of CC_1. According to (7.4), the total failure probabilities for component P_1 and M_1 are

$$Q_{2P_1} = 1 - (1 - q_{P_1})(1 - q_{1P_1}) = 0.208,$$
$$Q_{2M_1} = 1 - (1 - q_{M_1})(1 - q_{1M_1}) = 0.505.$$

Since B_1 and M_2 are not affected, their total component failure probabilities are their local failure probability (i.e. 0.01).

Under $PCCE_3$, both CC_1 and CC_2 happen, and components P_1, B_1, M_1, and M_2 fail. According to (7.4), the total failure probabilities for P_1, B_1, M_1, and M_2 are

$$Q_{3P_1} = 1 - (1 - q_{P_1})(1 - q_{1P_1})(1 - q_{2P_1}) = 0.4456,$$
$$Q_{3B_1} = 1 - (1 - q_{B_1})(1 - q_{2B_1}) = 0.406,$$
$$Q_{3M_1} = 1 - (1 - q_{M_1})(1 - q_{1M_1}) = 0.505,$$
$$Q_{3M_2} = 1 - (1 - q_{M_2})(1 - q_{2M_2}) = 0.604.$$

Table 7.1 Total component failure probability evaluation.

Component	$PCCE_0$	$PCCE_1$	$PCCE_2$	$PCCE_3$
P_1	0.01	0.307	0.208	0.4456
B_1	0.01	0.406	0.01	0.406
M_1	0.01	0.01	0.505	0.505
M_2	0.01	0.604	0.01	0.604

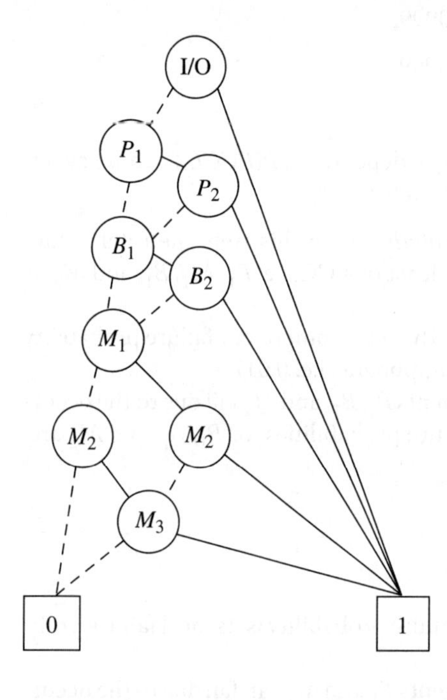

Figure 7.7 BDD model for the example computer system.

Table 7.1 summarizes the total failure probabilities for P_1, M_1, B_1, and M_2 under each PCCE.

Step 3: Build the system reliability model ignoring effects of PCCFs. Based on the original FT (Figure 7.4 excluding the two PCCF gates and their trigger events), the BDD model is generated for the example computer system, as shown in Figure 7.7.

Step 4: Evaluate P(system fails|$PCCE_i$). P(system fails|$PCCE_i$) is computed by evaluating the BDD model in Figure 7.7 using the total failure probabilities in Table 7.1 for components affected by PCCFs and the local failure probabilities given in Section 7.1.1 for components not affected by any CCs.

$$\begin{aligned}
&P(\text{system fails} \mid PCCE_0) = 0.010493,\\
&P(\text{system fails} \mid PCCE_1) = 0.028900,\\
&P(\text{system fails} \mid PCCE_2) = 0.022134,\\
&P(\text{system fails} \mid PCCE_3) = 0.322714.
\end{aligned} \tag{7.8}$$

Step 5: Integrate for the system unreliability. According to (7.5), the system unreliability considering effects of PCCFs is obtained by integrating results of (7.7) and (7.8) as

$$UR_{system} = \sum_{i=0}^{3} [P(\text{system fails} \mid PCCE_i) \cdot P(PCCE_i)] = 0.010523. \quad (7.9)$$

This unreliability result matches that obtained using the explicit algorithm in Section 7.1.1.

7.1.3 Comparisons and Discussions

The explicit algorithm involves only two steps, where external and internal CCs are handled in the same manner. This methodology can become computationally inefficient for large-scale systems due to the large number of pseudo nodes introduced into the expanded FT and the dependencies among sub-FTs sharing the CC nodes. In addition, the explicit method is only applicable to addressing *s*-independent CCs.

The implicit algorithm involves five steps; where external and internal CCs are handled differently particularly in step 3. The implicit method requires no expansion on the system reliability model. Thus, the size of the system model (BDD) constructed is much smaller than that built in the explicit algorithm (9 non-sink nodes in Figure 7.7 vs. 35 non-sink nodes in Figure 7.6). In addition, the implicit method is applicable to handling diverse *s*-relationship among CCs (including *s*-independent, *s*-dependent, and mutually exclusive).

Both the explicit and implicit methods are combinatorial and applicable to arbitrary types of component *ttf* distributions.

7.2 Multi-Phase System

The section presents explicit and implicit methods for reliability analysis of multi-phase systems (also known as phased-mission systems [PMSs]) subject to PCCFs caused by external shocks or factors. The PMS considered can be subject to more than one external CC happening in one phase or multiple different phases. Different CCs are *s*-independent. The local failure event of a component and its failure event(s) caused by external CC(s) within each phase are also *s*-independent.

Assume there are m phases and L_i CCs (denoted by $CC_{i1}, \ldots, CC_{iL_i}$) occurring in phase i. Thus, the total number of CCs occurring during the mission is $L = \sum_{i=1}^{m} L_i$. All components affected by CC_{ij} constitute PCCG$_{ij}$ $(i \leq m, j \leq L_i)$.

7.2.1 Explicit Method

Similar to the explicit method for single-phase systems (Section 7.1.1), the explicit method for PMSs involves constructing and evaluating an expanded system model where each CC is modeled as a basic event shared by all the components in its PCCG. The explicit method involves the following two-step process:

Step 1: Build an expanded PMS FT. Independent pseudo-node(s) representing component failure event(s) caused by CC(s) in a phase are created and added to the original phase FT to generate an expanded PMS FT. Specifically, if component X belongs to h different PCCGs in phase i (implying that this component is affected by h s-independent CCs in this phase, denoted by $CC_{iX(1)}$, $CC_{iX(2)}$, …, $CC_{iX(h)}$), then h pseudo-nodes (X_{i1}, X_{i2}, … X_{ih}) are created for component X and added to the original FT of phase i. The probability of X_{ij} is the conditional probability that component X fails given that $CC_{iX(j)}$ takes place.

The logical expression in (7.10) represents the total failure behavior of component X in phase i (the component fails when it either suffers a local failure denoted by X_i or is affected by a CC):

$$X_{iTF} = (CC_{iX(1)} \cap X_{i1}) \cup (CC_{iX(2)} \cap X_{i2}) \cup \ldots (CC_{iX(h)} \cap X_{ih}) \cup X_i. \quad (7.10)$$

Figure 7.8 illustrates the FT representation of (7.10). Note that it may happen that the local failure event of a component X does not appear in the phase i FT, but the component can still fail if it is affected by certain $CC_{iX(j)}$ occurring in phase i. In this case, a logical AND gate connecting pseudo-node X_{ij} and $CC_{iX(j)}$ (modeling effect of $CC_{iX(j)}$ on X) needs to be added to the later phase FT where the local failure event of component X first appears. If the local failure event of component X does not appear at all after phase i, then the effect of $CC_{iX(j)}$ on X is simply ignored after phase i since this component's failure makes no contribution to the PMS status after phase i. The expanded FT for the entire PMS is constructed by replacing each component failure event in the original PMS FT with the component total failure event FT in the form of Figure 7.8 for all components affected by at least one CC.

Step 2: Evaluate the expanded PMS FT formed in step 1. In this step, the traditional PMS evaluation methods are applicable, for example, the PMS BDD method [9], the multi-valued decision diagrams-based method [10], and the recursive algorithm [11, 12]. The four-step PMS BDD method described in Section 3.5.3 is adopted in Example 7.3 below to evaluate the expanded PMS FT to obtain the reliability of a PMS subject to PCCFs.

Example 7.3 Consider an example of a wireless sensor network (WSN) system in Figure 7.9. All the nodes are perfectly reliable, only links are subject to random failures. The explicit method is applied to analyze the communication reliability between the base station s and sensor node t.

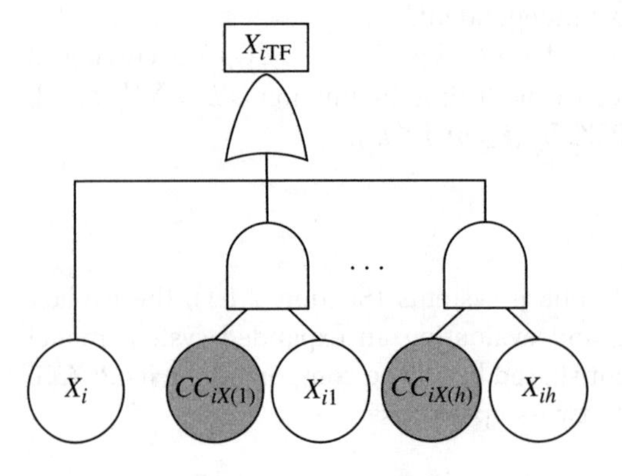

Figure 7.8 FT of component total failure event in phase i [8].

Figure 7.9 An illustrative example of a WSN [8].

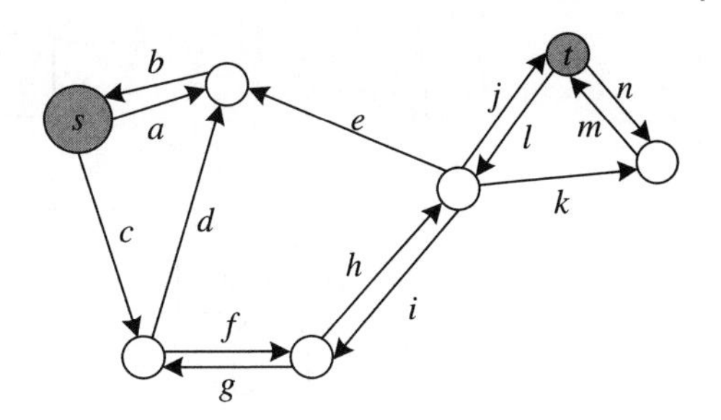

Communications within a WSN have two types [13–15]: infrastructure communication involves the delivery of configuration and maintenance information from the base station to sensor nodes; application communication involves the delivery of sensed data from sensor nodes to the base station. A two-phase communication between the base station s and sensor node t is considered. Phase 1 is an infrastructure communication phase (ICP) from node s to node t; there are two paths: path$_{11}$ ($c \rightarrow f \rightarrow h \rightarrow j$) and path$_{12}$ ($c \rightarrow f \rightarrow h \rightarrow k \rightarrow m$). Phase 2 is an application communication phase (ACP) from t to s; there are also two paths: path$_{21}$ ($l \rightarrow e \rightarrow b$) and path$_{22}$ ($l \rightarrow i \rightarrow g \rightarrow d \rightarrow b$). The communication reliability between s and t is the probability that the communication succeeds in both ICP and ACP. Figure 7.10 gives the FT model of the two-phased communication in the example WSN system.

The system is subject to two s-independent external CCs: CC_{11} in phase 1 affecting links e, h, i, i.e. PCCG$_{11}$ = {e, h, i}; and CC_{21} in phase 2 affecting links e, j, l, i.e. PCCG$_{21}$ = {e, j, l}. Figure 7.11 illustrates the FT modeling PCCFs.

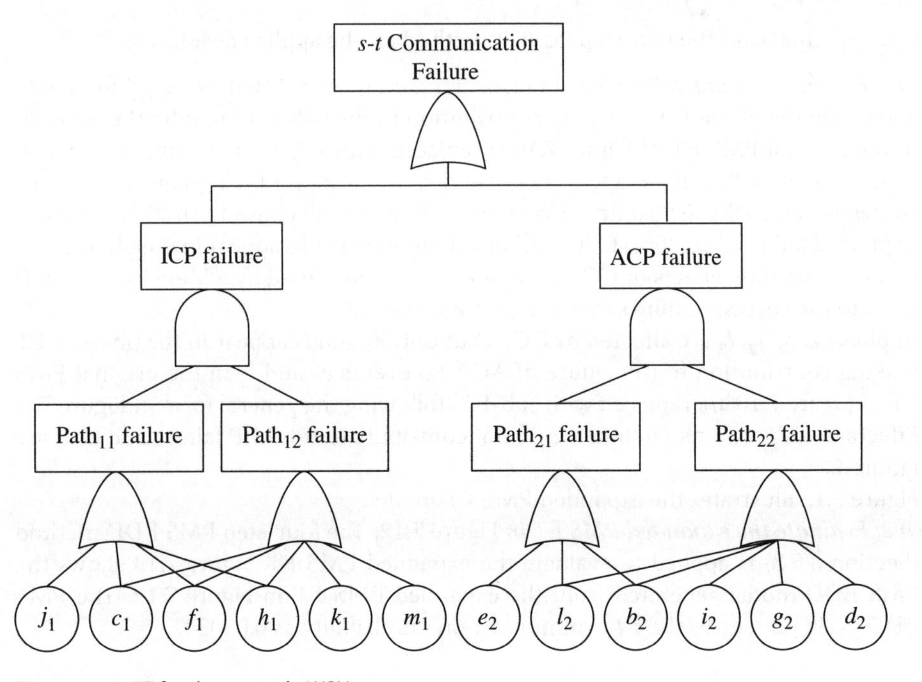

Figure 7.10 FT for the example WSN.

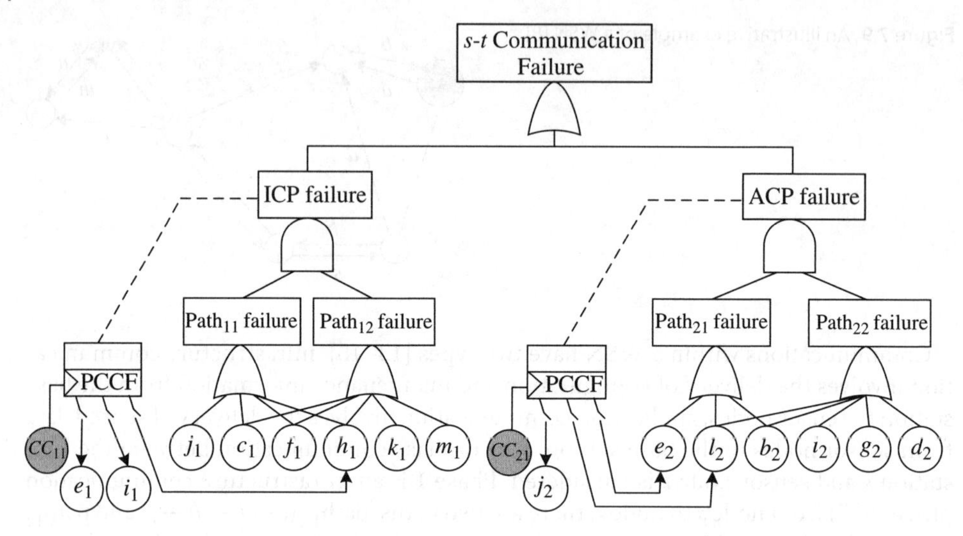

Figure 7.11 FT for the example WSN with PCCFs.

Input Parameters: The following parameter values are assumed for the analysis:

- Link local failure probabilities in phase 1 ($q_{1X} = 0.01$) and phase 2 ($q_{2X} = 0.02$) for $X \in \{a, b, \dots, n\}$.
- Probabilities of CCs: $P(CC_{11}) = P(CC_{21}) = 0.001$.
- Link conditional failure probabilities given a CC occurs: $q_{11e} = 0.2$, $q_{11h} = 0.5$, $q_{11i} = 0.3$, $q_{21e} = 0.3$, $q_{21j} = 0.5$, $q_{21l} = 0.7$.

Example Analysis: The two-step explicit method can be applied as follows:

Step 1: Build an expanded PMS FT. In phase 1, e_1, h_1, i_1 are affected by CC_{11} but only h_1 appears in the phase 1 FT making contribution to the failure of ICP. So the event h_1 in the original PMS FT of Figure 7.10 is replaced with sub-FT following the general form in Figure 7.8. Links e_1 and i_1 do not appear in the phase 1 FT (they do not make contributions to the ICP failure). However, their failures in phase 1 can still contribute to phase 2 failure since local failure events of these two links appear in the phase 2 FT model. Therefore, effects of CC_{11} on e_1 and i_1 are considered by adding logical AND gates to the corresponding sub-FT in phase 2 or ACP.
In phase 2, e_2, j_2, l_2 are affected by CC_{21}, but only e_2 and l_2 appear in the phase 2 FT making contribution to the failure of ACP. So events e_2 and l_2 in the original PMS FT of Figure 7.10 are replaced with sub-FTs following the general form in Figure 7.8. Effects of CC_{21} on link j in phase 2 do not contribute to the ACP failure and thus are ignored.
Figure 7.12 illustrates the expanded PMS FT model.

Step 2: Evaluate the expanded PMS FT in Figure 7.12. The four-step PMS BDD method (Section 3.5.3) is applied to evaluate the expanded PMS FT. Figure 7.13 shows the PMS BDD model generated from the expanded PMS FT in Figure 7.12. The PMS BDD evaluation gives the *s-t* communication unreliability as 0.090347.

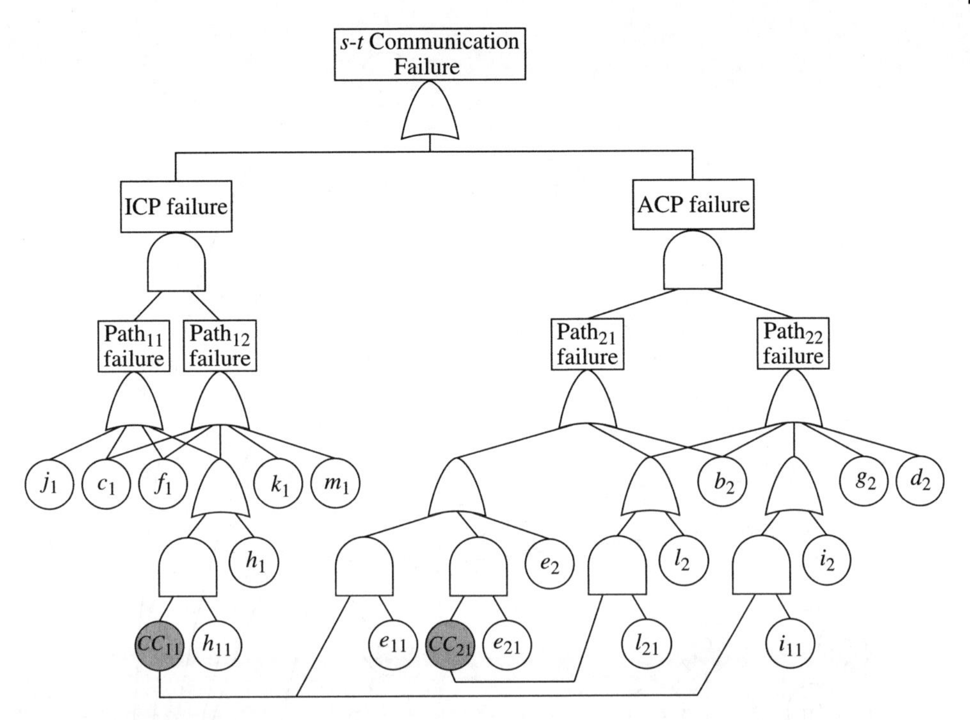

Figure 7.12 Expanded PMS FT.

7.2.2 Implicit Method

Similar to the implicit method for single-phase systems (Section 7.1.2), the implicit method for PMSs involves constructing a system model ignoring effects of PCCFs first and then evaluating the model in a way that incorporates contributions from PCCFs. The implicit method is described as the following five-step process.

Step 1: Construct a PCCE space. Similar to step 1 in Section 7.1.2, the space contains all possible combinations of occurrence or non-occurrence of CCs occurring during the mission. Specifically, given L elementary CCs occur during m phases, the PCCE space contains 2^L disjoint events, as illustrated in (7.11).

$$\begin{aligned}
PCCE_0 &= \overline{CC_{11}} \cap \ldots \cap \overline{CC_{1L_1}} \cap \ldots\ldots \cap \overline{CC_{m1}} \cap \ldots \cap \overline{CC_{mL_m}},\\
PCCE_1 &= \overline{CC_{11}} \cap \ldots \cap \overline{CC_{1L_1}} \cap \ldots\ldots \cap \overline{CC_{m1}} \cap \ldots \cap CC_{mL_m},\\
&\ldots\ldots\\
PCCE_{2^L-1} &= CC_{11} \cap \ldots \cap CC_{1L_1} \cap \ldots\ldots \cap CC_{m1} \cap \ldots \cap CC_{mL_m}.
\end{aligned} \tag{7.11}$$

The occurrence probability of each $PCCE_k$, denoted by $P(PCCE_k)$, is also evaluated in this step based on the occurrence probabilities of CCs. In addition, Eq. (7.12) is satisfied.

$$\sum_{k=0}^{2^L-1} P(PCCE_k) = 1. \tag{7.12}$$

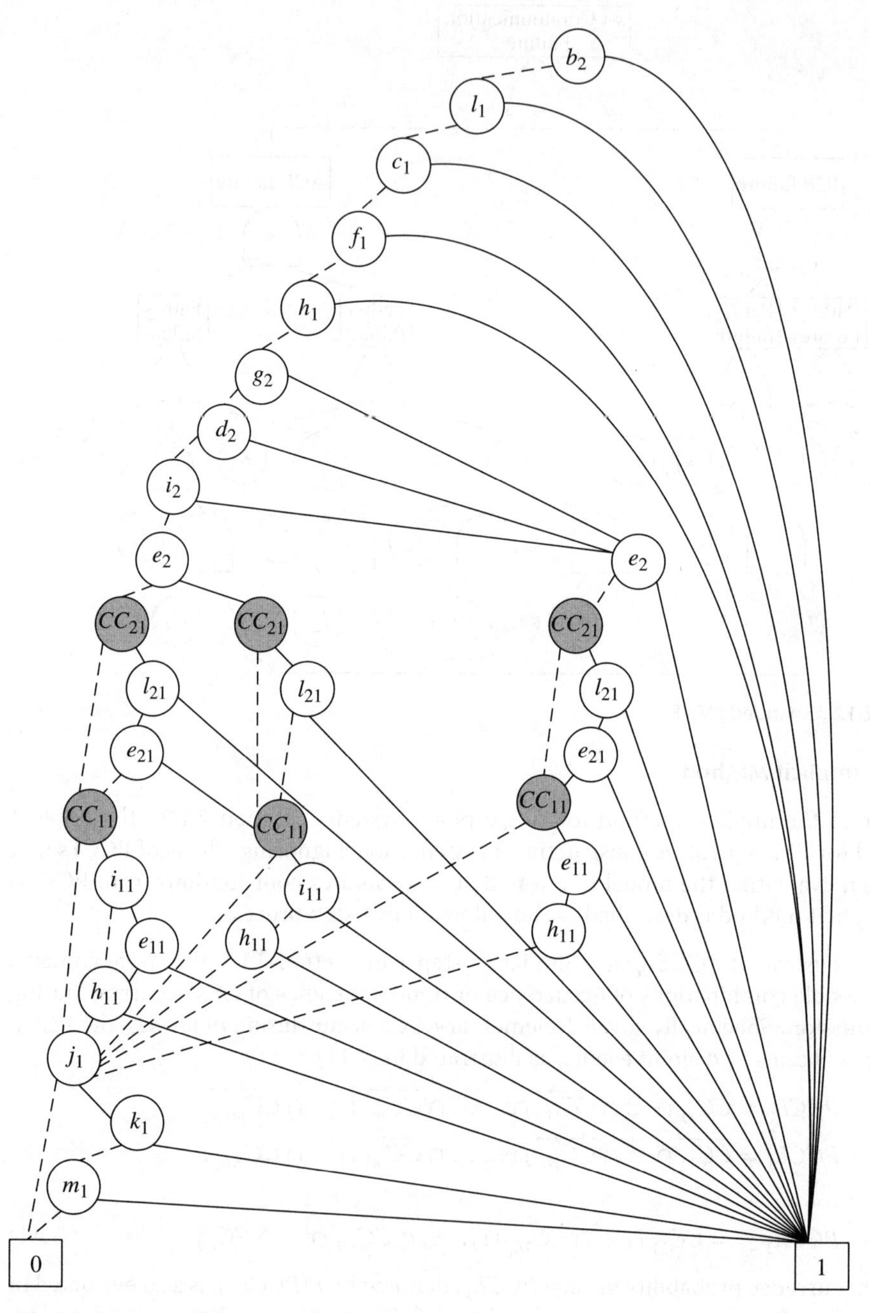

Figure 7.13 PMS BDD for the example WSN system.

Step 2: Evaluate the component total conditional failure probability. In this step, the total conditional failure probability is evaluated for each component affected by at least one CC under each PCCE in each phase.

Consider component X that is affected by h_k CCs (denoted by $CC_{iX(1)}$, $CC_{iX(2)}$, …, $CC_{iX(h_k)}$) in phase i under $PCCE_k$. Its total conditional failure probability in phase i given that component X has survived phase $i-1$ is

$$Q_{ikX} = 1-(1-q_{iX})\prod_{j=1}^{h_k}(1-q_{ijX}), \tag{7.13}$$

where q_{iX} represents the conditional local failure probability given that component X has survived phase $i-1$ and q_{ijX} represents the conditional failure probability of component X given that $CC_{iX(j)}$ occurs in phase i.

Step 3: Build the PMS reliability model ignoring effects of PCCFs. In this step, according to the PMS BDD method described in Section 3.5.3, a PMS BDD model is constructed from the original PMS FT model without considering effects of PCCFs.

Step 4: Evaluate P(PMS fails|PCCE$_k$). The conditional system failure probability given that $PCCE_k$ occurs, $P(\text{PMS fails}|PCCE_k)$, is computed by evaluating the PMS BDD model built in Step 3 using the component total conditional failure probabilities calculated in Step 2.

Step 5: Integrate for the PMS unreliability. According to the total probability law, the final PMS unreliability considering effects of PCCFs is

$$UR_{PMS} = \sum_{k=0}^{2^L-1}[P(\text{PMS fails} \mid PCCE_k)\cdot P(PCCE_k)]. \tag{7.14}$$

Example 7.4 A detailed analysis of the example WSN PMS (Example 7.3) using the implicit method is as follows:

Step 1: Construct a PCCE space. Based on the two external, *s*-independent CCs, the PCCE space consists of four PCCEs:

$$\begin{aligned}
PCCE_0 &= \overline{CC_{11}} \cap \overline{CC_{21}},\\
PCCE_1 &= \overline{CC_{11}} \cap CC_{21},\\
PCCE_2 &= CC_{11} \cap \overline{CC_{21}},\\
PCCE_3 &= CC_{11} \cap CC_{21}.
\end{aligned} \tag{7.15}$$

Their occurrence probabilities are

$$\begin{aligned}
P(PCCE_0) &= (1-P(CC_{11}))(1-P(CC_{21})) = 0.998001,\\
P(PCCE_1) &= (1-P(CC_{11}))P(CC_{21}) = 0.000999,\\
P(PCCE_2) &= P(CC_{11})(1-P(CC_{21})) = 0.000999,\\
P(PCCE_3) &= P(CC_{11})P(CC_{21}) = 0.000001.
\end{aligned} \tag{7.16}$$

Step 2: Evaluate the component total conditional failure probability. Under $PCCE_0$, no CCs take place. Hence, no links are subject to PCCFs under $PCCE_0$. The conditional local failure probability of a link is its total conditional failure probability.

Table 7.2 Component total conditional failure probabilities.

Link	$PCCE_0$	$PCCE_1$	$PCCE_2$	$PCCE_3$
e_1	0.01	0.01	**0.208**	**0.208**
h_1	0.01	0.01	**0.505**	**0.505**
i_1	0.01	0.01	**0.307**	**0.307**
e_2	0.02	**0.314**	0.02	**0.314**
l_2	0.02	**0.706**	0.02	**0.706**

Under $PCCE_1$, only CC_{21} occurs, affecting links e, j, and l in phase 2. The total conditional failure probability for links e, j, and l can be evaluated using (7.13). For example, the total conditional failure probability for link e in phase 2 is evaluated as $Q_{21e} = 1 - (1 - q_{2e})(1 - q_{21e}) = 0.314$.

Under $PCCE_2$, only CC_{11} occurs, affecting links e, h, and i in phase 1. The total conditional failure probability for links e, h, and i can be evaluated using (7.13). For example, the total conditional failure probability for link e in phase 1 is evaluated as $Q_{12e} = 1 - (1 - q_{1e})(1 - q_{11e}) = 0.208$.

Under $PCCE_3$, both CC_{11} and CC_{21} occur, affecting links e, h, and i in phase 1 and links e, j, and l in phase 2. The total conditional failure probability for these links can be evaluated using (7.13). Table 7.2 summarizes the total conditional failure probabilities for links affected by at least one CC.

Note that link j is affected by CC_{21} in phase 2, but does not contribute to phase 2 failure. Thus, it is not necessary to evaluate its total failure probability.

Step 3: Build the PMS reliability model ignoring effects of PCCFs. The PMS BDD model built from the original PMS FT (Figure 7.10) without considering effects of PCCFs is shown in Figure 7.14. The backward ordering is used for the PMS BDD generation.

Step 4: Evaluate P(PMS fails|$PCCE_k$). Using the component total conditional failure probabilities in Table 7.2, P(PMS fails$|PCCE_k$) is computed by evaluating the PMS BDD model in Figure 7.14. The results are:

$$\begin{aligned}
&P(\text{PMS fails} \mid PCCE_0) = 0.08921187,\\
&P(\text{PMS fails} \mid PCCE_1) = 0.7336815,\\
&P(\text{PMS fails} \mid PCCE_2) = 0.5802923,\\
&P(\text{PMS fails} \mid PCCE_3) = 0.88559284.
\end{aligned} \tag{7.17}$$

Step 5: Integrate for the PMS unreliability. According to (7.14), the final PMS unreliability is obtained by integrating results of (7.16) and (7.17) as

$$UR_{PMS} = \sum_{i=0}^{3} [P(\text{PMS fails} \mid PCCE_i) \cdot P(PCCE_i)] = 0.090347.$$

This result matches exactly the unreliability result obtained using the explicit method in Section 7.2.1.

Figure 7.14 PMS BDD model for the example WSN system.

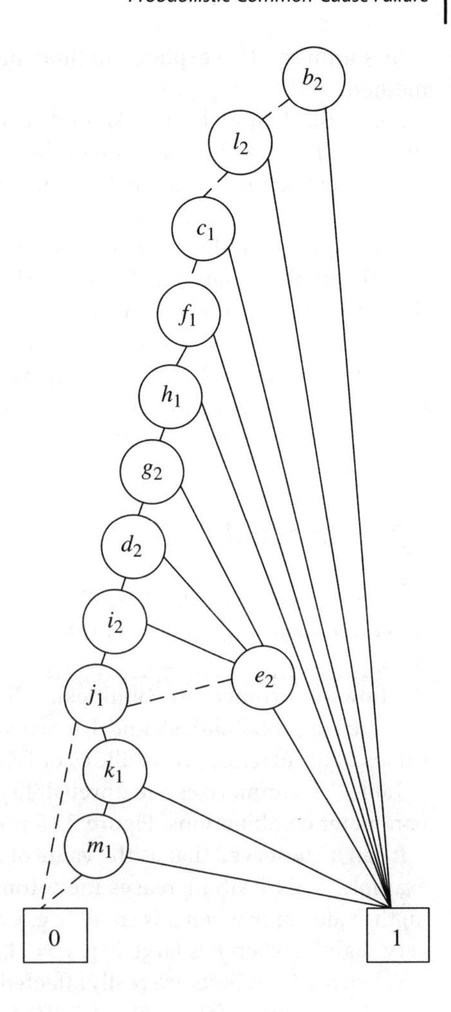

7.2.3 Comparisons and Discussions

The space and computation complexity of the explicit method (Section 7.2.1) and the implicit method (Section 7.2.2) are analyzed in what follows.

Consider a PMS with m phases and n components. In the explicit method, if there exist w CCs, each affecting all of the n components at the worst case, then $n{\cdot}w$ pseudo nodes representing component failures due to PCCFs and w nodes representing occurrences of CCs are added to the expanded FT, which is then used for generating the system BDD model. Thus, at the worst case, $m{\cdot}n+(n+1)\ w$ input variables are used for the system BDD generation. Since the worst-case size of BDD is $O(2^N/N)$ for a system with N variables [16], the worst-case size of BDD is $O(2^{m\cdot n+(n+1)\cdot w}/(m{\cdot}n+(n+1){\cdot}w))$ in the explicit method.

In the implicit method, the PMS BDD is constructed without considering effects of PCCFs. At the worst case, there are $m{\cdot}n$ input variables. Therefore, the worst-case size of BDD is $O(2^{m\cdot n}/(m{\cdot}n))$ in the implicit method.

In summary, the explicit method has greater space complexity than the implicit method.

For a BDD model with M nodes, its evaluation has complexity of $O(M)$ if the *memoization*-based bottom-up evaluation approach is adopted [17].

In the explicit method, the BDD generated from the expanded FT has the worst case size of $O(2^{m\cdot n+(n+1)\cdot w}/(m\cdot n+(n+1)\cdot w))$ and is evaluated only once. Thus, the computation complexity of the explicit method is also $O(2^{m\cdot n+(n+1)\cdot w}/(m\cdot n+(n+1)\cdot w))$.

In the implicit method, given that there are w CCs, the BDD model generated from the original FT without considering PCCFs needs to be evaluated 2^w times with different failure parameters, the computation complexity of the implicit method is thus $2^w\cdot O(2^{m\cdot n}/(m\cdot n)) = O(2^{m\cdot n+w}/(m\cdot n))$. Moreover, the 2^w evaluations are independent and can be performed in parallel given available computing resource. In general, the implicit method is computationally more efficient than the explicit method.

7.3 Impact of PCCF

The example WSN PMS in Example 7.3 is analyzed using five different values of conditional link failure probability conditioned on the occurrence of a CC ($q = q_{11e} = q_{11h} = q_{11i} = q_{21e} = q_{21j} = q_{21l}$), which are 0 (implying no PCCFs occur), 0.2 (low occurrence probabilities), 0.5 (medium occurrence probabilities), 0.8 (high occurrence probabilities), and 1 (corresponding to deterministic CCFs). Four different values of occurrence probabilities of CCs ($p = P(CC_{11}) = P(CC_{21})$) are also considered.

Table 7.3 summarizes the unreliability of the example WSN PMS under the different parameter combinations. Figure 7.15 presents the unreliability results graphically.

It can be observed that as the value of q increases (from 0 to 1), the unreliability of the example WSN PMS increases monotonically. The degree of the increase is dependent on the value of p: when p is small (e.g. $p = 0.001$), the PMS unreliability increases with q very slightly; when p is large (e.g. $p = 1$), the PMS unreliability increases with q rapidly.

When $q = 0$, no links are really affected by CCs. Thus, the system unreliability stays the same irrespective of the value of p. When $p = q = 1$, all links affected suffer deterministic failures, which crashes the $s-t$ communication because no redundant or alternative paths are available. Therefore, the system unreliability is ONE.

Table 7.3 Unreliability of the example WSN PMS.

p	Set 1 ($q = 0$)	Set 2 ($q = 0.2$)	Set 3 ($q = 0.5$)	Set 4 ($q = 0.8$)	Set 5 ($q = 1$)
0.001	0.089212	0.089631	0.090268	0.090801	0.091033
0.1	0.089212	0.130763	0.192068	0.241287	0.262262
0.5	0.089212	0.290421	0.547469	0.712300	0.772303
1	0.089212	0.475268	0.865672	0.992170	1

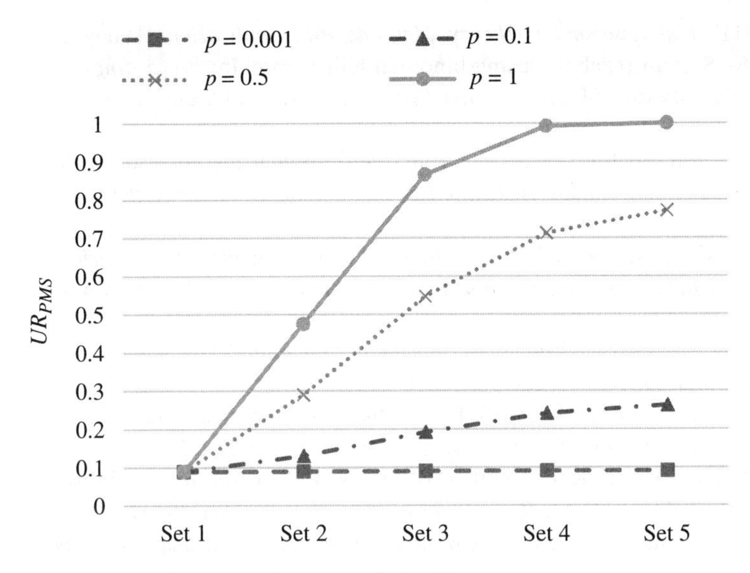

Figure 7.15 Unreliability of the example WSN PMS.

7.4 Summary

PCCFs may contribute significantly to the overall system failure and thus the system unreliability. This chapter discusses the reliability modeling of PCCFs in single-phase systems and multi-phase systems, where the occurrence of a root cause results in failures of multiple system components with different probabilities. Explicit and implicit methodologies of evaluating system unreliability are presented. The explicit method requires expanding the system reliability model to explicitly consider effects of PCCFs. It involves a straightforward two-step procedure. However, the explicit method can become computationally expensive for large-scale systems. The implicit method requires no expansion on the system reliability model and allows parallel evaluation of independent subproblems. The implicit method has lower requirements in both space and time than the explicit method. Both methods are applicable to arbitrary types of component *ttf* distributions.

References

1 Xing, L. and Wang, W. (2008). Probabilistic common-cause failures analysis. In: *Proceedings of the Annual Reliability and Maintainability Symposium*, 354–358. Las Vagas, Nevada, USA.

2 Xing, L., Boddu, P., Sun, Y., and Wang, W. (2010). Reliability analysis of static and dynamic fault-tolerant systems subject to probabilistic common-cause failures. *Proc. IMechE, Part O: Journal of Risk and Reliability* 224 (1): 43–53.

3 Rausand, M. (2011). *Risk Assessment: Theory, Methods, and Applications.* Wiley.
4 Chae, K.C. (1988). System reliability using binomial failure rate. In: *Proceedings of the Annual Reliability and Maintainability Symposium*, 136–138. Los Angeles, California, USA.
5 Wang, C., Xing, L., and Levitin, G. (2014). Explicit and implicit methods for probabilistic common-cause failure analysis. *Reliability Engineering & System Safety* 131: 175–184.
6 Dugan, J.B. and Doyle, S.A. (1996). New results in fault-tree analysis. In: *Tutorial Notes of the Annual Reliability and Maintainability Symposium*, Las Vegas, Nevada, USA.
7 Xing, L. and Amari, S.V. (2015). *Binary Decision Diagrams and Extensions for System Reliability Analysis.* MA: Wiley-Scrivener.
8 Wang, C., Xing, L., and Levitin, G. (2015). Probabilistic common cause failures in phased-mission systems. *Reliability Engineering & System Safety* 144: 53–60.
9 Zang, X., Sun, H., and Trivedi, K.S. (1999). A BDD-based algorithm for reliability analysis of phased-mission systems. *IEEE Transactions on Reliability* 48 (1): 50–60.
10 Mo, Y., Xing, L., and Amari, S.V. (2014). A multiple-valued decision diagram based method for efficient reliability analysis of non-repairable phased-mission systems. *IEEE Transaction on Reliability* 63 (1): 320–330.
11 Levitin, G., Xing, L., Amari, S.V., and Dai, Y. (2013). Reliability of non-repairable phased- mission systems with propagated failures. *Reliability Engineering & System Safety* 119: 218–228.
12 Levitin, G., Xing, L., and Amari, S.V. (2012). Recursive algorithm for reliability evaluation of non-repairable phased mission systems with binary elements. *IEEE Transactions on Reliability* 61 (2): 533–542.
13 Wang, C., Xing, L., Zonouz, A.E. et al. (2017). Communication reliability analysis of wireless sensor networks using phased-mission model. *Quality and Reliability Engineering International* 33 (4): 823–837.
14 Shrestha, A., Xing, L., Sun, Y., and Vokkarane, V.M. (2012). Infrastructure communication reliability of wireless sensor networks considering common-cause failures. *International Journal of Performability Engineering*, special issue on Dependability of Wireless Systems and Networks 8 (2): 141–150.
15 Zonouz, A.E., Xing, L., Vokkarane, V.M., and Sun, Y. (2015). Application communication reliability of wireless sensor networks. *IET Wireless Sensor Systems* 5 (2): 58–67.
16 Liaw, H. and Lin, C. (1992). On the OBDD-representation of general Boolean functions. *IEEE Transactions on Computers* 41 (6): 661–664.
17 Shrestha, A., Xing, L., and Dai, Y. (2010). Decision diagram based methods and complexity analysis for multi-state systems. *IEEE Transactions on Reliability* 59 (1): 145–161.

8

Deterministic Competing Failure

8.1 Overview

A propagated failure with global effect (PFGE) that originates from a system component causes the failure of the entire system [1]. As one type of common-cause failures (CCFs), PFGEs have been investigated intensively in literature (see, e.g. [2–6]). Examples of causes for PFGEs include imperfect coverage (IPC) and destructive effects. Specifically, as discussed in Chapter 3, due to the IPC, a component fault, if not being detected or located successfully by the system recovery mechanism, may propagate and cause an overall system failure even when adequate redundancy remains. Certain types of failures originating from a system component can cause destructive effects on other components, for example, fire, explosion, overheating, blackout, or short circuit may incapacitate or destroy all other system components, causing the failure of the entire system.

However, it is not necessarily always the truth that a PFGE causes the entire system failure, particularly for systems undergoing the Functional DEPendence (FDEP) behavior. As described in Chapter 5, with the FDEP, a trigger event, upon occurring, can isolate the corresponding *dependent components* (making them unusable or inaccessible) deterministically. Due to this *isolation* effect, a PFGE originating from a dependent component can thus be isolated without affecting other portions of the system. For example, in a clustered wireless sensor network (WSN) system, sensor nodes within a cluster are accessed through their cluster head [7]. In other words, these sensor nodes have FDEP on the cluster head. If the cluster head fails, PFGEs originating from any of the sensor nodes within the cluster can be isolated from the rest of the WSN system. Note that the failure isolation effect can take place only when the trigger event occurs before the occurrence of any PFGE originating from the corresponding dependent components. On the other hand, if any PFGE from a dependent component occurs before the trigger event happens, the global failure propagation effect takes place causing an entire system failure.

In summary, competitions exist in the time domain between the *failure isolation* effect and the *failure propagation* effect; different occurrence sequences lead to different system statuses. In this chapter, a separable method for handling PFGEs in system reliability analysis is first discussed. Based on this approach, methods are then presented for addressing the competing effects in the reliability analysis of different types of nonrepairable systems, including single-phase system with single FDEP group, single-phase system with multiple dependent FDEP groups, single-phase system subject to propagated failures (PFs) with global and selective effects, multi-phase system

Dynamic System Reliability: Modeling and Analysis of Dynamic and Dependent Behaviors,
First Edition. Liudong Xing, Gregory Levitin and Chaonan Wang.

(or phased-mission system, PMS) with single FDEP group, and PMS with multiple independent or dependent FDEP groups.

8.2 PFGE Method

T_{il} and T_{ip} are random variables respectively representing the time-to-local-failure and the time-to-PFGE of a system component i. $f_{il}(t)$ and $f_{ip}(t)$ represent the probability density function (*pdf*) of T_{il} and T_{ip}, respectively. $q_{il}(t)$ and $q_{ip}(t)$ are unconditional local and propagated failure probabilities of component i at time t, respectively. Thus, $q_{il}(t) = \int_0^t f_{il}(\tau)d\tau$ and $q_{ip}(t) = \int_0^t f_{ip}(\tau)d\tau$.

According to the simple and efficient algorithm (SEA) in Section 3.3.2 [2, 4], the system unreliability can be evaluated based on the total probability law as:

$$UR_{system}(t) = 1 - P_u(t) + Q(t) \cdot P_u(t), \tag{8.1}$$

with $P_u(t)$ being defined and computed as

$$P_u(t) = P(\text{no PFGEs}) = \prod_{\forall i}[1 - q_{ip}(t)]. \tag{8.2}$$

$Q(t)$ in (8.1) is defined as a conditional system failure probability given that no PFGEs take place during the considered mission time. The evaluation of $Q(t)$ requires no consideration of effects from PFGEs and thus can be performed using any approaches ignoring PFGEs, e.g. the binary decision diagram (BDD)–based methods [4, 7] for single-phase systems (Section 2.4) and PMSs (Section 3.5.3).

As in the SEA method, the evaluation of $Q(t)$ requires the calculation of a conditional component failure probability $q_i(t)$ given that no PFGEs occur to the component. The evaluation method is illustrated for different statistical relationships between the local failure (LF) and PFGE in the following sections.

8.2.1 *s*-Independent LF and PFGE

When the LF and PFGE of the same component are *s*-independent, the conditional component failure probability is evaluated as

$$\begin{aligned} q_i(t) &= P(\text{LF}|\text{no PFGE}) = \frac{P(\text{LF} \cap \text{no PFGE})}{P(\text{no PFGE})} \\ &= \frac{P(\text{LF}) \cdot P(\text{no PFGE})}{P(\text{no PFGE})} = P(\text{LF}) = q_{il}(t) = \int_0^t f_{il}(\tau)d\tau. \end{aligned} \tag{8.3}$$

8.2.2 *s*-Dependent LF and PFGE

When the LF and PFGE of the same component are *s*-dependent, the conditional component failure probability is evaluated as

$$\begin{aligned} q_i(t) &= P(\text{LF}|\text{no PFGE}) = \frac{P(\text{LF} \cap \text{no PFGE})}{P(\text{no PFGE})} \\ &= \frac{P(\text{LF}) \cdot P(\text{no PFGE}|\text{LF})}{P(\text{no PFGE})} \\ &= \frac{q_{il} \cdot q_{\overline{ip}|l}}{1 - q_{ip}} = \frac{q_{il}(1 - q_{ip|l})}{1 - q_{ip}}, \end{aligned} \tag{8.4}$$

where,

$$q_{ip|l}(t) = P(\text{PFGE of component } i|\text{LF of component } i),$$

$$q_{ip|\bar{l}}(t) = P(\text{PFGE of component } i|\text{ no LF of component } i),$$

$$\begin{aligned} q_{ip}(t) &= P(\text{PFGE of component } i) \\ &= P(\text{PFGE of } i|\text{LF of } i)P(\text{LF of } i) \\ &\quad + P(\text{PFGE of } i|\text{no LF of } i)P(\text{no LF of } i) \\ &= q_{il} \cdot q_{ip|l} + (1 - q_{il}) \cdot q_{ip|\bar{l}}. \end{aligned} \tag{8.5}$$

8.2.3 Disjoint LF and PFGE

When the LF and PFGE of the same component are disjoint or mutually exclusive, the conditional component failure probability is evaluated as

$$\begin{aligned} q_i(t) &= P(\text{LF}|\text{no PFGE}) = \frac{P(\text{LF} \cap \text{no PFGE})}{P(\text{no PFGE})} \\ &= \frac{P(\text{LF}) \cdot P(\text{no PFGE}|\text{LF})}{P(\text{no PFGE})} \\ &= \frac{q_{il} \cdot 1}{1 - q_{ip}} = \frac{q_{il}}{1 - q_{ip}}. \end{aligned} \tag{8.6}$$

8.3 Single-Phase System with Single FDEP Group

Based on the PFGE method, a combinatorial methodology is discussed in this section for analyzing reliability of a single-phase system subject to competing failures involved in a single FDEP group or multiple independent (nonoverlapped) FDEP groups. The method is applicable to any arbitrary *ttf* distributions for the system components.

8.3.1 Combinatorial Method

Given that the trigger component(s) can only experience LFs. The method contains the following three steps:

Step 1: Define FDEP-related events and evaluate event occurrence probabilities. Three events representing different occurrence sequences of the trigger event and PFGE events of the corresponding dependent components are defined as follows:

E_1: the *trigger* event does not take place (i.e. the trigger component does not fail locally). Assume that the unconditional LF event of trigger component, e.g. A is Y_{Al}. $P(E_1)$ is calculated as:

$$P(E_1) = P(\overline{Y_{Al}}) = 1 - \int_0^t f_{Al}(\tau)d\tau = 1 - q_{Al}(t). \tag{8.7}$$

Note that in the case of the trigger component being subject to PFGEs in addition to the LF, the PFGE method presented in Section 8.2 should be applied to separate the global failure propagation effect originating from the trigger component before Step 1.

$q_{Al}(t)$ in (8.7) should be, respectively, replaced with $q_A(t)$ evaluated using (8.3), (8.4), or (8.6) when the LF and PFGE of the trigger component are independent, dependent, or disjoint. Accordingly, the *pdf* of time-to-LF of trigger component A involved in (8.11), i.e. $f_{Al}(\tau_2)$ should be evaluated as $dq_A(t)/dt$.

E_2: at least one PFGE of *dependent* components takes place before the trigger LF event occurs. Assume the trigger component A affects n dependent components D_1, D_2, ..., D_n, i.e. the fuctional dependence group (FDG) for component A is $FDG_A = \{D_1, D_2, ..., D_n\}$. The PFGE events of these dependent components are represented by Y_{D1p}, Y_{D2p}, ... , and Y_{Dnp} and are s-independent. Thus, $P(E_2)$ is evaluated as:

$$P(E_2) = P\{(Y_{D1p} \cup Y_{D2p} \cup \ldots \cup Y_{Dnp}) \rightarrow Y_{Al}\}, \tag{8.8}$$

where

$$\begin{aligned} &P\{(Y_{D1p} \cup Y_{D2p} \cup \ldots \cup Y_{Dnp})\} \\ &= P\{\overline{(\overline{Y_{D1p}} \cap \overline{Y_{D2p}} \cap \ldots \cap \overline{Y_{Dnp}})}\} \\ &= 1 - P\{\overline{Y_{D1p}} \cap \overline{Y_{D2p}} \cap \ldots \cap \overline{Y_{Dnp}}\} \\ &= 1 - P(\overline{Y_{D1p}}) \cdot P(\overline{Y_{D2p}}) \cdot \ldots \cdot P(\overline{Y_{Dnp}}). \end{aligned} \tag{8.9}$$

In general, for n components with their *ttf r.v.*s represented by $X_1, \ldots, X_n$, the probability of their sequential failures is evaluated as [8]:

$$P(X_1 \rightarrow X_2 \rightarrow \ldots \rightarrow X_n) = \int_0^t \int_{\tau_1}^t \cdots \int_{\tau_{n-1}}^t \prod_{k=1}^{n} f_{X_k}(\tau_k) d\tau_n \ldots d\tau_2 d\tau_1. \tag{8.10}$$

Thus, (8.8) can be evaluated as

$$\begin{aligned} P(E_2) &= P\{(Y_{D1p} \cup Y_{D2p} \cup \ldots \cup Y_{Dnp}) \rightarrow Y_{Al}\} \\ &= \int_0^t \int_{\tau_1}^t f_{(Y_{D1p} \cup Y_{D2p} \cup \ldots \cup Y_{Dnp})}(\tau_1) f_{Al}(\tau_2) d\tau_2 d\tau_1, \end{aligned} \tag{8.11}$$

where

$$f_{(Y_{D1p} \cup Y_{D2p} \cup \ldots \cup Y_{Dnp})}(t) = \frac{d\{P(Y_{D1p} \cup Y_{D2p} \cup \ldots \cup Y_{Dnp})\}}{dt}. \tag{8.12}$$

By definition, in the case of the dependent components undergoing no PFGEs, $P(E_2) = 0$.

E_3: the *trigger* event takes place before the occurrence of any PFGE originating from the dependent components. Under this event, the failure isolation effect takes place. Since E_1, E_2, and E_3 form a complete event space, one obtain

$$P(E_3) = 1 - P(E_1) - P(E_2). \tag{8.13}$$

Step 2: Evaluate P(system fails|E_i) for i∈{1, 2, 3}.

P(system fails|E_1): based on the system fault tree (FT) after removing the FDEP gate and its trigger component(s), P(system fails|E_1) is evaluated by applying the PFGE method described in Section 8.2.

P(system fails|E_2): because when at least one PFGE takes place, the entire system fails due to the global failure propagation effect, P(system fails|E_2) = 1.

$P(\text{system fails}|E_3)$: a reduced FT that considers the failure isolation effect is generated. Firstly, the FDEP gate and its trigger component(s) are removed from the original system FT. Failure events of the corresponding dependent components are then replaced with constant 1 (TURE). Boolean algebra rules ($1+x=1$ and $1 \cdot x = x$, where x represents a Boolean variable) are finally applied to generate the reduced FT. Based on the reduced FT, if any component appearing in the reduced FT undergoes PFGEs, the PFGE method described in Section 8.2 is applied to evaluate $P(\text{system fails}|E_3)$; otherwise, any traditional approach ignoring PFGEs, e.g. the BDD-based method is applied to solve $P(\text{system fails}|E_3)$.

Step 3: Integrate for final system unreliability. Based on the total probability law [9], the system unreliability is evaluated by integrating $P(E_i)$ and $P(\text{system fails}|E_i)$ as [10]:

$$UR_{system} = \sum_{i=1}^{3} [P(\text{system fails}|E_i) \cdot P(E_i)]. \tag{8.14}$$

8.3.2 Case Study

Example 8.1 Consider a memory subsystem of a computer illustrated in Figure 8.1 [8]. The memory subsystem contains an independent memory module (MM), and two memory chips (MC_1, MC_2) accessible by CPU via a memory interface unit (MIU). Thus, the two memory chips have FDEP on the MIU, so, $FDG_{\text{MIU}} = \{MC_1, MC_2\}$. The memory subsystem functions when both MC_1 and MC_2 function correctly or MM functions correctly. Only MC_1 and MC_2 can undergo PFGEs.

Figure 8.2 shows the FT of the example memory subsystem, where A, B, C, and D, respectively, represent MIU, MC_1, MC_2, and MM for simplifying the representation.

For illustration, the exponential distribution is assumed for both time-to-LF and time-to-PFGE of the system components. The LF rates of components A, B, C, D are respectively represented by λ_{Al}, λ_{Bl}, λ_{Cl}, and λ_{Dl}. The unconditional PFGE rates of components B and C are, respectively, denoted by λ_{Bp} and λ_{Cp}.

The *pdf* of time-to-LF and time-to-PFGE for component $i \in \{A, B, C, D\}$ are, respectively:

$$f_{il}(t) = \lambda_{il} e^{-\lambda_{il} t}, \quad f_{ip}(t) = \lambda_{ip} e^{-\lambda_{ip} t}. \tag{8.15}$$

The LF probability and the PFGE probability are, respectively:

$$q_{il}(t) = \int_0^t f_{il}(\tau) d\tau = 1 - e^{-\lambda_{il} t}, \quad q_{ip}(t) = \int_0^t f_{ip}(\tau) d\tau = 1 - e^{-\lambda_{ip} t}. \tag{8.16}$$

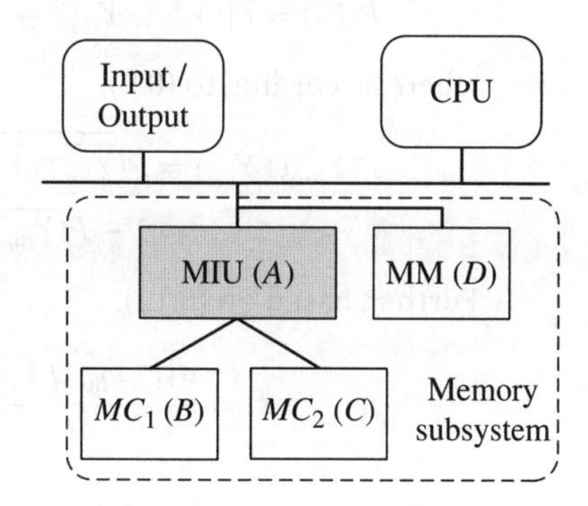

Figure 8.1 An example of a computer system.

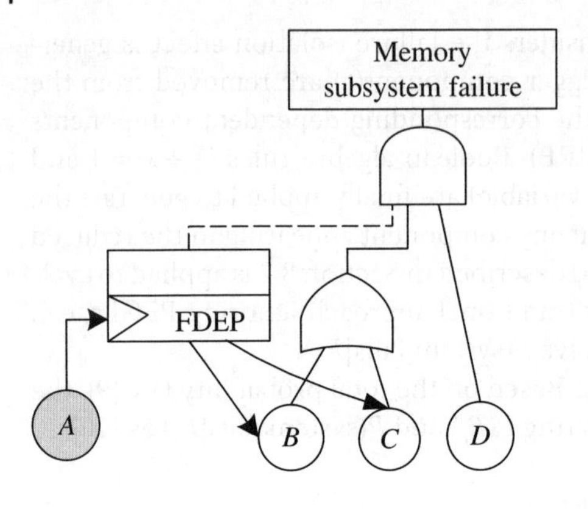

Figure 8.2 FT model of the example memory subsystem.

When the LF and PFGE of component B or C are s-dependent, conditional PGFE failure rates conditioned on occurrence or nonoccurrence of an LF ($\lambda_{Bp|l}$, $\lambda_{Bp|\bar{l}}$, $\lambda_{Cp|l}$, $\lambda_{Cp|\bar{l}}$) are given. Two types of dependencies can be modeled [11]: *positive dependence* takes place if the LF of a component causes an increased tendency of the component's PFGE (thus, e.g. $\lambda_{Bp|l} > \lambda_{Bp|\bar{l}}$); *negative dependence* takes place if the LF of a component causes a reduced tendency of the component's PFGE (thus, e.g. $\lambda_{Bp|l} < \lambda_{Bp|\bar{l}}$).

Input Parameters. The following values are used in the illustrative analysis: $\lambda_{Al} = 0.0001/\text{hr}$, $\lambda_{Bl} = \lambda_{Cl} = \lambda_{Dl} = 0.0002/\text{hr}$. For component B or C, two sets of parameters are considered. If the LF and PFGE are s-independent or disjoint, $\lambda_{Bp} = \lambda_{Cp} = 0.00001/\text{hr}$; if the LF and PFGE are s-dependent, $\lambda_{Bp|l} = \lambda_{Cp|l} = 0.00003/\text{hr}$ and $\lambda_{Bp|\bar{l}} = \lambda_{Cp|\bar{l}} = 0.00001/\text{hr}$ (positive dependence).

Example Analysis. The s-independent case is used to illustrate the combinatorial method in detail.

Step 1: Define FDEP-related events and evaluate their occurrence probabilities, as follows:

E_1: trigger component A does not fail (locally). According to (8.7) and (8.16),

$$P(E_1) = P(\overline{Y_{Al}}) = 1 - (1 - e^{-\lambda_{Al}t}) = e^{-\lambda_{Al}t}. \tag{8.17}$$

E_2: at least one PFGE from B and C takes place before the trigger event occurs. According to (8.8),

$$P(E_2) = P[(Y_{Bp} \cup Y_{Cp}) \to Y_{Al}], \tag{8.18}$$

where, according to (8.9),

$$\begin{aligned} P(Y_{Bp} \cup Y_{Cp}) &= P(\overline{\overline{Y_{Bp}} \cap \overline{Y_{Cp}}}) = 1 - P(\overline{Y_{Bp}} \cap \overline{Y_{Cp}}) \\ &= 1 - P(\overline{Y_{Bp}}) \cdot P(\overline{Y_{Cp}}) = 1 - e^{-(\lambda_{Bp}+\lambda_{Cp})t}. \end{aligned} \tag{8.19}$$

Further, based on (8.12),

$$f_{Y_{Bp}\cup Y_{Cp}}(t) = \frac{d[P(Y_{Bp} \cup Y_{Cp})]}{dt} = \frac{d[1 - e^{-(\lambda_{Bp}+\lambda_{Cp})t}]}{dt} = (\lambda_{Bp} + \lambda_{Cp})e^{-(\lambda_{Bp}+\lambda_{Cp})t}. \tag{8.20}$$

According to (8.11) and (8.19),

$$P(E_2) = P[(Y_{Bp} \cup Y_{Cp}) \rightarrow Y_{Al}] = \int_0^t \int_{\tau_1}^t f_{Y_{Bp}\cup Y_{Cp}}(\tau_1) f_{Al}(\tau_2) d\tau_2 d\tau_1$$

$$= \int_0^t \int_{\tau_1}^t (\lambda_{Bp} + \lambda_{Cp}) e^{-(\lambda_{Bp}+\lambda_{Cp})\tau_1} \lambda_{Al} e^{-\lambda_{Al}\tau_2} d\tau_2 d\tau_1$$

$$= \frac{\lambda_{Al}}{\lambda_{Al} + \lambda_{Bp} + \lambda_{Cp}} e^{-(\lambda_{Al}+\lambda_{Bp}+\lambda_{Cp})t} - e^{-\lambda_{Al}t} + \frac{\lambda_{Bp} + \lambda_{Cp}}{\lambda_{Al} + \lambda_{Bp} + \lambda_{Cp}}. \tag{8.21}$$

E_3: the LF of trigger component A takes place before any PFGE originating from the dependent components happens. According to (8.13),

$$P(E_3) = 1 - P(E_1) - P(E_2)$$
$$= \frac{\lambda_{Al}}{\lambda_{Al} + \lambda_{Bp} + \lambda_{Cp}} - \frac{\lambda_{Al}}{\lambda_{Al} + \lambda_{Bp} + \lambda_{Cp}} e^{-(\lambda_{Al}+\lambda_{Bp}+\lambda_{Cp})t}. \tag{8.22}$$

Step 2: Evaluate P(system fails|E_i) for i∈{1, 2, 3}.

P(system fails|E_1): under E_1, no failure isolation effect takes place. Figure 8.3 shows the FT after removing the FDEP gate and its trigger component A. Based on the FT in Figure 8.3, P(system fails|E_1) is evaluated using the PFGE method (Section 8.2) as follows.

According to (8.1),

$$P(\text{system fails}|E_1) = 1 - P_u(t) + Q(t) \cdot P_u(t), \tag{8.23}$$

where, based on (8.2),

$$P_u(t) = [1 - q_{Bp}(t)] \cdot [1 - q_{Cp}(t)] = e^{-(\lambda_{Bp}+\lambda_{Cp})t}. \tag{8.24}$$

To evaluate $Q(t)$ in (8.23), component conditional failure probabilities are computed. For the s-independent case, (8.3) is adopted for the computation, that is,

$$q_B = q_{Bl} = 1 - e^{-\lambda_{Bl}t},$$
$$q_C = q_{Cl} = 1 - e^{-\lambda_{Cl}t}. \tag{8.25}$$

Figure 8.4 shows the BDD model generated from the FT in Figure 8.3 for $Q(t)$ in (8.23).

Figure 8.3 Reduced FT for P(system fails|E_1).

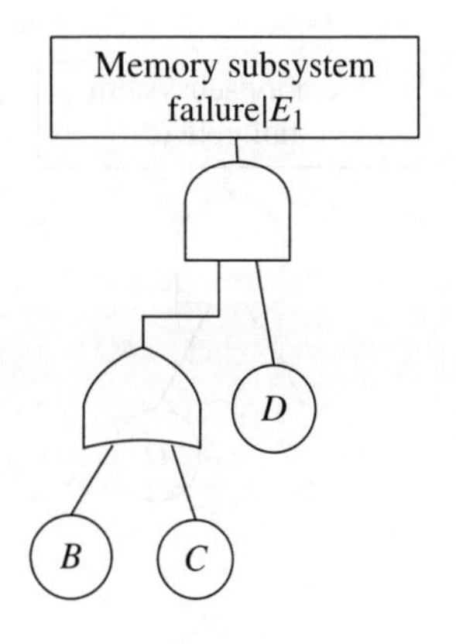

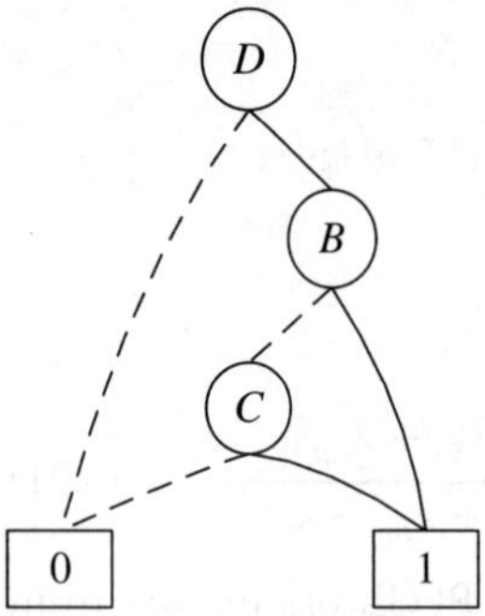

Figure 8.4 BDD model for evaluating $Q(t)$.

Evaluating the BDD of Figure 8.4 using the component conditional failure probabilities computed using (8.25), $Q(t)$ is obtained as

$$\begin{aligned} Q(t) &= q_D(t)q_B(t) + q_D(t)[1 - q_B(t)]q_C(t) \\ &= (1 - e^{-\lambda_{DI}t})(1 - e^{-\lambda_{BI}t}) + (1 - e^{-\lambda_{DI}t})e^{-\lambda_{BI}t}(1 - e^{-\lambda_{CI}t}) \\ &= 1 - e^{-\lambda_{DI}t} - e^{-(\lambda_{BI}+\lambda_{CI})t} + e^{-(\lambda_{BI}+\lambda_{CI}+\lambda_{DI})t}. \end{aligned} \tag{8.26}$$

Based on (8.23), (8.24), and (8.26) are integrated to obtain (8.27).

$$\begin{aligned} &P(\text{system fails}|E_1) \\ &= 1 - P_u(t) + P_u(t)Q(t) \\ &= 1 - e^{-(\lambda_{Bp}+\lambda_{Cp})t}[e^{-\lambda_{DI}t} + e^{-(\lambda_{BI}+\lambda_{CI})t} - e^{-(\lambda_{BI}+\lambda_{CI}+\lambda_{DI})t}]. \end{aligned} \tag{8.27}$$

Under E_2, since the global failure propagation effect takes place. Therefore,

$$P(\text{system fails}|E_2) = 1. \tag{8.28}$$

Under E_3, the failure isolation effect takes place. Figure 8.5 shows the reduced FT generated for evaluating $P(\text{system fails}|E_3)$. Thus,

$$P(\text{system fails}|E_3) = P(Y_D) = 1 - e^{-\lambda_{DI}t}. \tag{8.29}$$

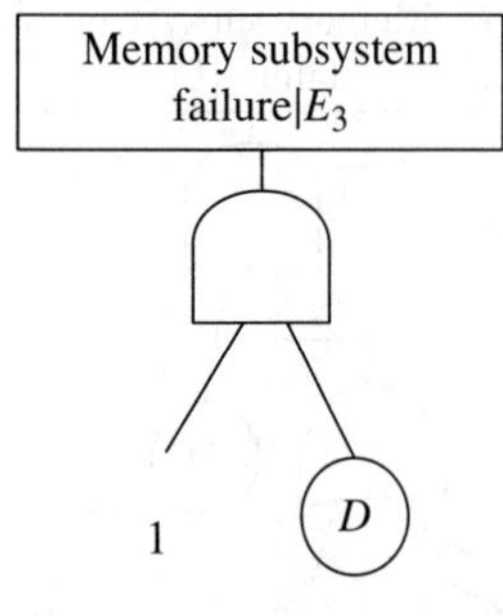

Figure 8.5 Reduced FT for $P(\text{system fails}|E_3)$.

Table 8.1 Unreliability of the example memory sub-system.

Mission time t (*hrs*)	1000	5000	10 000
s-dependent	0.0943	0.6417	0.8949
s-independent	0.0889	0.6128	0.8757
disjoint	0.0894	0.6207	0.8799

Step 3: Integrate for final system unreliability. Based on (8.14), the unreliability of the example memory subsystem is obtained as

$$\begin{aligned} UR_{system}(t) &= \sum_{i=1}^{3}[P(\text{system fails}|E_i)\cdot P(E_i)] \\ &= 1 - e^{-(\lambda_{Al}+\lambda_{Bl}+\lambda_{Bp}+\lambda_{Cl}+\lambda_{Cp})t} - \frac{\lambda_{Al}}{\lambda_{Al}+\lambda_{Bp}+\lambda_{Cp}}e^{-\lambda_{Dl}t} \\ &\quad - \frac{\lambda_{Bp}+\lambda_{Cp}}{\lambda_{Al}+\lambda_{Bp}+\lambda_{Cp}}e^{-(\lambda_{Al}+\lambda_{Bp}+\lambda_{Cp}+\lambda_{Dl})t} + e^{-(\lambda_{Al}+\lambda_{Bl}+\lambda_{Bp}+\lambda_{Cl}+\lambda_{Cp}+\lambda_{Dl})t}. \end{aligned} \tag{8.30}$$

In the case of LF and PFGE of component B or C being s-dependent or disjoint, the combinatorial method presented in Section 8.3.1 can be similarly applied to derive the system unreliability.

Using the given parameter values, the unreliability of the example memory subsystem under the three cases is summarized in Table 8.1. The system unreliability in the s-independent case is lower than that in the disjoint case. This is because that for the same component parameter values, the component reliability in the s-independent case (calculated as $[1-q_{il}(t)]\cdot[1-q_{ip}(t)]$) is higher than that in the disjoint case (calculated as $1-q_{il}(t)-q_{ip}(t)$). The system unreliability in the s-dependent case is higher than that in the s-independent or disjoint case due to the positive dependence assumed in the example input parameters.

8.4 Single-Phase System with Multiple FDEP Groups

This section considers the reliability analysis of a single-phase system subject to competing failures involved in multiple dependent FDEP groups. The method is applicable to any arbitrary *ttf* distributions for the system components.

8.4.1 Combinatorial Method

The combinatorial method contains the following three steps [12]:

Step 1: Construct an event space based on statuses of trigger components. Given m trigger components (denoted by T_i, $i = 1, \dots, m$) involved in FDEPs, an event space consists of 2^m events, each called a combined trigger event (CTE), and is constructed as follows: $CTE_0 = \overline{T_1} \cap \overline{T_2} \cap \ldots \cap \overline{T_m}$, $CTE_1 = \overline{T_1} \cap \overline{T_2} \cap \ldots \cap T_m$, ..., $CTE_{2^m-1} = T_1 \cap T_2 \cap \ldots \cap T_m$. Based on the total probability law [9], the system unreliability is evaluated as

$$UR_{system}(t) = \sum_{i=0}^{2^m-1} [P(\text{system failure}|CTE_i) \cdot P(CTE_i)]. \tag{8.31}$$

Step 2: Evaluate P(system failure|CTE_i)P(CTE_i). Each CTE_i is decomposed into two complementary events defined as follows:

$E_{1,i}$: all *PFGEs* either do not occur or are isolated by failures of corresponding trigger components.

$E_{2,i}$: at least one PFGE is not isolated. It takes place when any PFGE occurs to a dependent component in the FDEP group where the corresponding trigger component does not fail or fails after the PFGE.

As evaluating $P(E_{2,i})$ is more straightforward than evaluating $P(E_{1,i})$, $P(E_{2,i})$ is computed first, $P(E_{1,i})$ is then evaluated as

$$P(E_{1,i}) = P(CTE_i) - P(E_{2,i}). \tag{8.32}$$

$P(\text{system failure}|CTE_i)P(CTE_i)$ can be evaluated as

$$\begin{aligned} &P(\text{system failure}|CTE_i) \cdot P(CTE_i) \\ &\quad = P(\text{system failure}|E_{1,i})P(E_{1,i}) + P(\text{system failure}|E_{2,i})P(E_{2,i}) \\ &\quad = P(\text{system failure}|E_{1,i})P(E_{1,i}) + P(E_{2,i}), \end{aligned} \tag{8.33}$$

where $P(\text{system failure}|E_{2,i}) = 1$ due to the global failure propagation effect.

Under $E_{1,i}$, all PFGEs from dependent components either do not happen or are isolated. A reduced FT is generated for evaluating $P(\text{system failure}|E_{1,i})$ in (8.33). Under each considered CTE_i, the trigger event and corresponding FEDP gate are first removed from the original system FT. If a trigger event occurs, then events of the corresponding dependent components are replaced with constant 1 (TRUE); otherwise, events of the dependent components remain in the FT. Boolean algebra rules are then applied to simplify the FT. The reduced FT generated can be evaluated using the BDD method [13] to find $P(\text{system failure}|E_{1,i})$.

Step 3: Integrate for final system unreliability. Based on (8.31) and (8.33), results of step 2 are integrated to obtain the system unreliability, considering competing failures involved in multiple FDEP groups.

$$\begin{aligned} UR_{system}(t) &= \sum_{i=0}^{2^m-1} [P(\text{system failure}|CTE_i) \cdot P(CTE_i)] \\ &= \sum_{i=0}^{2^m-1} [P(\text{system failure}|E_{1,i}) \cdot P(E_{1,i}) + P(E_{2,i})]. \end{aligned} \tag{8.34}$$

Note that the above three-step procedure does not address PFGEs from nondependent components. In the case of nondependent components undergoing PFGEs, a pre-processing step 0 described below is applied based on the PFGE method (Section 8.2):

Step 0: Separate PFGEs of all nondependent components from the solution combinatorics. Based on the PFGE method, the system unreliability is evaluated as (8.1), where $P_u(t)$ represents the probability that no PFGEs from nondependent components (including trigger components) occur. $Q(t)$ in (8.1) is defined as the conditional system failure probability given that no PFGEs from nondependent components occur, which is evaluated using the above described three-step procedure.

8.4.2 Case Study

Example 8.2 Figure 8.6 shows the FT model of a computer memory system containing three identical memory chips (MC_1, MC_2, MC_3), which are accessible through two identical memory interface units (MIU_1, MIU_2). Specifically, MC_1 is accessible through MIU_1, MC_3 is accessible through MIU_2, MC_2 is accessible through either MIU_1 or MIU_2. Thus, $FDG_{MIU_1} = \{MC_1\}$, $FDG_{MIU_2} = \{MC_3\}$, $FDG_{MIU_1 \cap MIU_2} = \{MC_2\}$. The system functions when at least two of the three memory chips function correctly.

Input Parameters. The exponential distribution is assumed for this illustrative example. The *pdf* and *cdf* of the exponential distribution with failure rate λ are given in (8.35).

$$f(t) = \lambda e^{-\lambda t}, \quad F(t) = 1 - e^{-\lambda t}. \tag{8.35}$$

The three MCs undergo both LFs and PFGEs with constant rates given in Table 8.2. The LF and PFGE of the same MC are *s*-independent. The two MIUs only experience LFs with rates also given in Table 8.2.

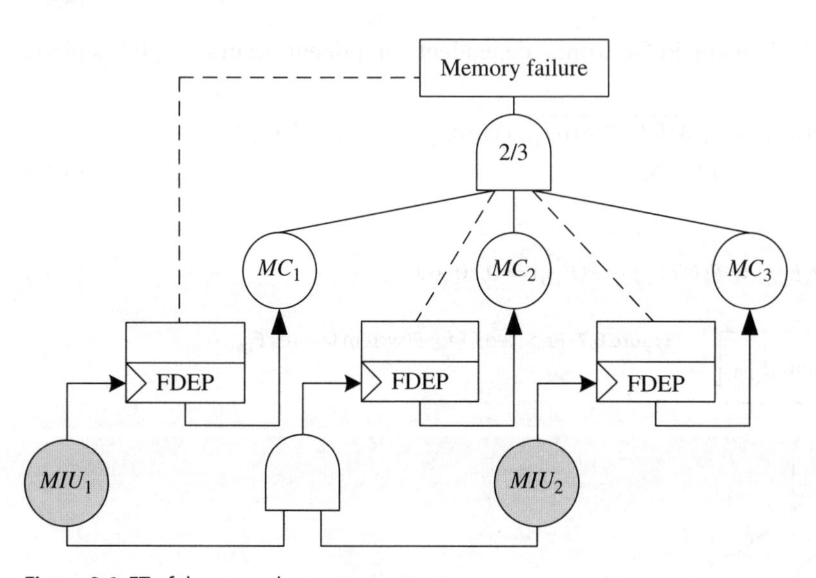

Figure 8.6 FT of the example memory system.

Table 8.2 Failure rates of the example memory system components (/hr).

Component	PFGE rate	LF rate
MC_i	0.00005	0.0002
MIU_i	0	0.0001

Example Analysis. The unreliability of the example memory system at time $t = 1000$ hours is analyzed using the method of Section 8.4.1 as follows.

Step 1: Construct an event space based on statuses of trigger components. The two trigger components lead to an event space with 4 CTEs defined in (8.36).

$$\begin{aligned} CTE_0 &= \overline{MIU_{1l}} \cap \overline{MIU_{2l}}, \\ CTE_1 &= \overline{MIU_{1l}} \cap MIU_{2l}, \\ CTE_2 &= MIU_{1l} \cap \overline{MIU_{2l}}, \\ CTE_3 &= MIU_{1l} \cap MIU_{2l}. \end{aligned} \tag{8.36}$$

Step 2: Evaluate P(system failure|CTE_i)P(CTE_i)

1. Evaluate $P(\text{system } failure|CTE_0)P(CTE_0)$.
   Under CTE_0, no trigger components fail. Thus,

$$P(CTE_0) = P(\overline{MIU_{1l}} \cap \overline{MIU_{2l}}) = (1 - F_{MIU_1}(t))(1 - F_{MIU_2}(t)) = 0.81873. \tag{8.37}$$

   Under CTE_0, if any PFGE from a dependent component occurs, $E_{2,0}$ takes place. So,

$$\begin{aligned} P(E_{2,0}) &= P[(\overline{MIU_{1l}} \cap \overline{MIU_{2l}}) \cap (MC_{1p} \cup MC_{2p} \cup MC_{3p})] \\ &= 0.11404. \end{aligned} \tag{8.38}$$

   Thus,

$$P(E_{1,0}) = P(CTE_0) - P(E_{2,0}) = 0.70469. \tag{8.39}$$

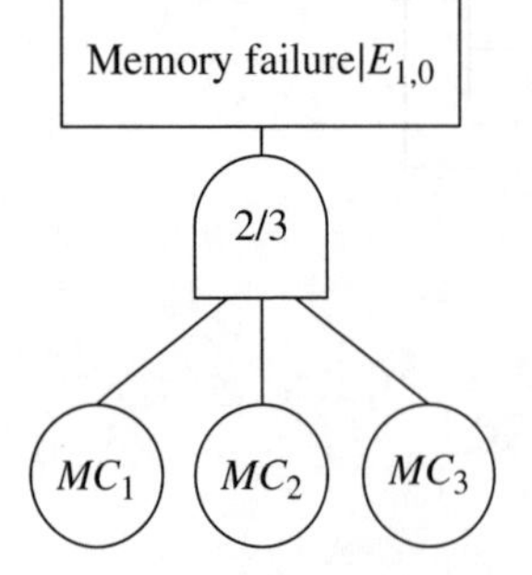

Figure 8.7 Reduced FT for $P(\text{system failure} \mid E_{1,0})$.

Figure 8.8 BDD for P(system failure | $E_{1,0}$).

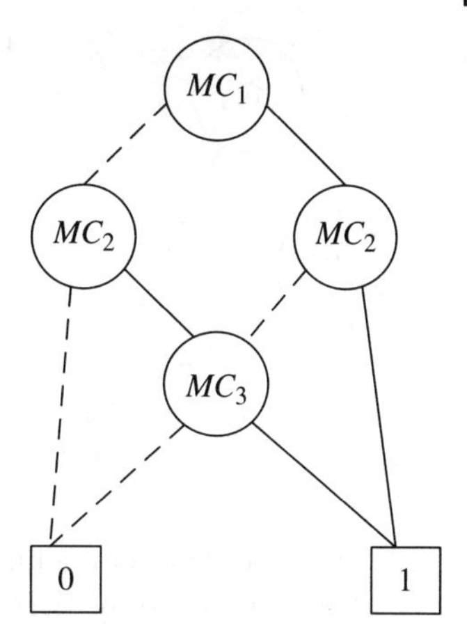

Figure 8.7 shows the reduced FT for evaluating P(system failure | $E_{1,0}$). Figure 8.8 shows the BDD model generated from the FT. The evaluation of the BDD model in Figure 8.8 gives

$$\begin{aligned} & P(\text{system fails}|E_{1,0}) \\ & \quad = P(MC_{1l} \cap MC_{2l}) + P(MC_{1l} \cap \overline{MC_{2l}} \cap MC_{3l}) \\ & \qquad + P(\overline{MC_{1l}} \cap MC_{2l} \cap MC_{3l}) \\ & \quad = 0.08666. \end{aligned} \tag{8.40}$$

2. Evaluate P(system failure$|CTE_1)P(CTE_1)$.
   Under CTE_1, only MIU_2 fails locally. Thus,

$$P(CTE_1) = P(\overline{MIU_{1l}} \cap MIU_{2l}) = 0.08611.$$

Under CTE_1, if any PFGE from MC_1 or MC_2 happens, or PFGE from MC_3 happens before MIU_2 fails, then event $E_{2,1}$ takes place. So

$$\begin{aligned} P(E_{2,1}) &= P\{[(MC_{1p} \cup MC_{2p}) \cap MIU_{2l}] \cup (MC_{3p} \rightarrow MIU_{2l})\} \cdot P(\overline{MIU_{1l}}) \\ &= 0.01008. \end{aligned}$$

The evaluation of $P(E_{2,1})$ involves an sequential event, which can be evaluated using (8.10). With $P(E_{2,1})$, one obtain $P(E_{1,1}) = P(CTE_1) - P(E_{2,1}) = 0.07603$. Figure 8.9 shows the reduced FT for evaluating P(system failure | $E_{1,1}$). Figure 8.10 shows the BDD model generated from the FT. The evaluation of the BDD model in Figure 8.10 gives P(systemfailure | $E_{1,1}$) $= 0.32968$.

Figure 8.9 Reduced FT for P(system failure | $E_{1,1}$).

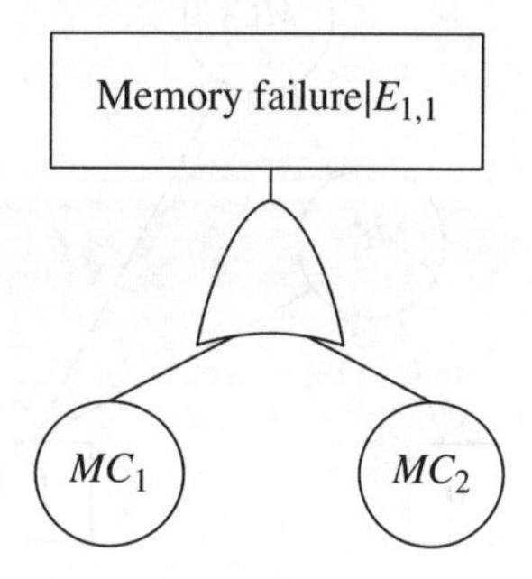

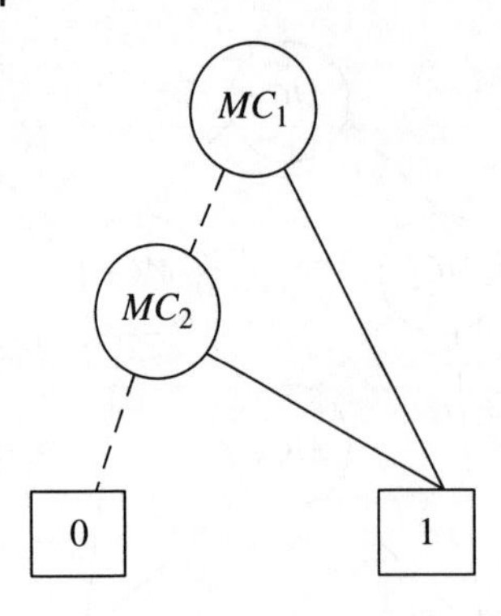

Figure 8.10 BDD for P(system failure | $E_{1,1}$).

3. Evaluate $P(\text{system failure}|CTE_2)P(CTE_2)$.
   Under CTE_2, only MIU_1 fails locally. Thus,

$$P(CTE_2) = P(MIU_{1l} \cap \overline{MIU_{2l}}) = 0.08611.$$

   Under CTE_2, if any PFGE from MC_2 or MC_3 happens, or PFGE from MC_1 happens before MIU_1 fails, then event $E_{2,2}$ takes place. So,

$$\begin{aligned} P(E_{2,2}) &= P\{[(MC_{2p} \cup MC_{3p}) \cap MIU_{1l}] \cup (MC_{1p} \rightarrow MIU_{1l})\} \cdot P(\overline{MIU_{2l}}) \\ &= 0.01008. \end{aligned}$$

   Thus, $P(E_{1,2}) = P(CTE_2) - P(E_{2,2}) = 0.07603$.
   Figure 8.11 shows the reduced FT for evaluating P(system failure | $E_{1,2}$). Figure 8.12 shows the BDD model generated from the FT. The evaluation of the BDD model in Figure 8.12 gives P(systemfailure | $E_{1,2}$) $= 0.32968$.

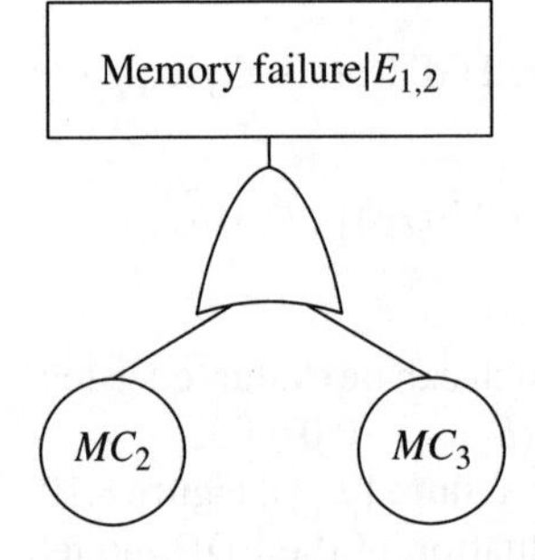

Figure 8.11 Reduced FT for P(system failure | $E_{1,2}$).

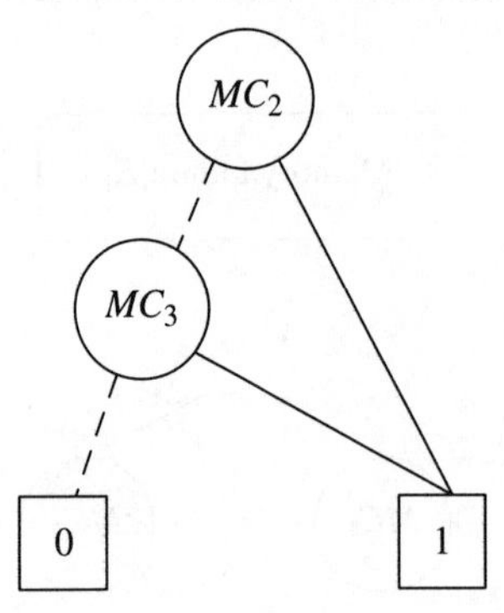

Figure 8.12 BDD for evaluating P(system failure | $E_{1,2}$).

4. Evaluate $P(\text{system failure}|CTE_3)P(CTE_3)$.
   Under CTE_3, both trigger components fail. Thus,

$$P(CTE_3) = P(MIU_{1l} \cap MIU_{2l}) = 0.00906.$$

   Under CTE_3, if at least one PFGE from the three MCs happens before the corresponding trigger component fails, then event $E_{2,3}$ takes place. So,

$$P(E_{2,3}) = P\left\{ \begin{array}{l} [(MC_{1p} \rightarrow MIU_{1l}) \cap MIU_{2l}] \cup [(MC_{3p} \rightarrow MIU_{2l}) \cap MIU_{1l}] \\ \cup[(MC_{2p} \rightarrow MIU_{1l}) \cap MIU_{2l}] \cup [(MC_{2p} \rightarrow MIU_{2l}) \cap MIU_{1l}] \end{array} \right\}$$

$$= 0.00071.$$

   Thus,

$$P(E_{1,3}) = P(CTE_3) - P(E_{2,3}) = 0.00835.$$

   When both of the trigger components fail, the entire system fails. Therefore $P(\text{system failure} \mid E_{1,3}) = 1$.

Step 3: Integrate for final system unreliability. According to (8.34), results obtained at step 2 are integrated to obtain the final system unreliability as

$$UR_{system}(t) = \sum_{i=0}^{3} [P(\text{system failure}|E_{1,i}) \cdot P(E_{1,i}) + P(E_{2,i})] = 0.25446.$$

8.5 Single-Phase System with PFs Having Global and Selective Effects

A PF that originates from a system component causes extensive damages to the rest of the system. A PFGE occurs when the PF causes the entire system to fail. There also exist a propagated failure with selective effect (PFSE), which takes place when the PF causes failure of only a subset of system components. This section presents a combinatorial reliability analysis method for single-phase systems subject to competing failures considering both global and selective propagation effects [14].

8.5.1 Combinatorial Method

The combinatorial reliability analysis method can be described as a seven-step procedure:

Step 1: Define *events representing states of the trigger component.* Two disjoint events are defined:
E_1: the trigger component is functioning correctly.
E_2: the trigger component is failed.
Based on the total probability law, the system unreliability is evaluated as

$$UR_{system} = P(\text{system fails}|E_1)P(E_1) + P(\text{system fails}|E_2)P(E_2). \quad (8.41)$$

Step 2: Evaluate occurrence probability of E_1. $P(E_1)$ in (8.41) is simply the reliability of the trigger component.

Step 3: Evaluate P(system fails$|E_1$*).* Given that E_1 happens, no failure isolation effect takes place. The PFGE method (Section 8.2) is applied to evaluate P(system fails | E_1) as

$$P(\text{system fails}|E_1) = 1 - P_u(t) + Q(t) \cdot P_u(t). \tag{8.42}$$

where $P_u(t) = P$(no PFGEs) and $Q(t) = P$(system fails|no PFGEs). While the PFGEs are separated from the solution combinatorics via the PFGE method, the PFSEs have to be addressed in the evaluation of $Q(t)$ as follows.

Given that up to m independent PFSEs may occur when the trigger component functions, an event space with 2^m events (denoted by SE_i) is constructed, each being a combination of occurrence or nonoccurrence of these m PFSEs. Based on the total probability law, $Q(t)$ in (8.42) is computed as

$$Q(t) = \sum_{i=0}^{2^m-1} [P(SE_i) \cdot P(\text{system fails}|SE_i)], \tag{8.43}$$

where P(system fails $|SE_i$) can be obtained through the BDD-based evaluation of a reduced FT. The reduced FT is generated by the following procedure:

1. Remove the FDEP gate and its trigger component from the original system FT.
2. Replace events of components affected by SE_i with constant 1.
3. Apply Boolean algebra rules to simplify the FT.

Step 4: Define two disjoint event cases given that E_2 takes place. Given that E_2 happens (i.e. the trigger component fails), two disjoint cases are considered for evaluating P(system fails$|E_2$) $P(E_2)$ in (8.41):

Case a: At least one PFGE from any dependent components occurs before E_2 happens. Under this case, the global failure propagation effect destroys the entire system, i.e. P(system fails|Case a) =1.

Case b: No PFGEs occur before E_2 happens.

Based on the total probability law, P(system fails$|E_2$)$P(E_2)$ is calculated as

$$\begin{aligned}P(\text{system fails}|E_2)P(E_2) &= P(\text{system fails} \cap E_2)\\ &= P(\text{system fails}|\text{Case a})P(\text{Case a}) + P(\text{system fails}|\text{Case b})P(\text{Case b})\\ &= P(\text{Case a}) + P(\text{system fails}|\text{Case b})P(\text{Case b})\\ &= P(\text{Case a}) + P(\text{system fails} \cap \text{Case b}).\end{aligned} \tag{8.44}$$

Step 5: Evaluate P(Case a). Assume that the trigger component A has n dependent components $D_1, D_2, \ldots D_n$, whose PFGE events are denoted as $Y_{D1pg}, Y_{D2pg}, \ldots, Y_{Dnpg}$, respectively. Thus, $P(\text{Case a}) = P[(Y_{D1pg} \cup Y_{D2pg} \cup \ldots \cup Y_{Dnpg}) \rightarrow Y_A]$. Similar to the evaluation of (8.8), according to (8.11) and (8.12), P(Case a) is evaluated as

$$P(\text{Case a}) = \int_0^t \int_{\tau_1}^t f_{(Y_{D1pg} \cup Y_{D2pg} \cup \ldots \cup Y_{Dnpg})}(\tau_1) f_{Y_A}(\tau_2) d\tau_2 d\tau_1, \tag{8.45}$$

where,

$$f_{(Y_{D1pg} \cup Y_{D2pg} \cup \ldots \cup Y_{Dnpg})}(t) = \frac{d[P(Y_{D1pg} \cup Y_{D2pg} \cup \ldots \cup Y_{Dnpg})]}{dt}.$$

Step 6: Evaluate P(system fails $\cap$ Case b) in (8.44). Under Case b, while no PFGEs occur before the failure of the trigger component, PFSEs can take place before or after the trigger component failure. Assume there are n PFSEs under Case b. If all of the PFSEs occur after the trigger failure, they are isolated (i.e. the isolation effect takes place).

If at least one PFSE occurs before the trigger failure, the selective failure propagation effect takes place. To handle those effects, an event space with 2^n events (denoted by SE_i') is constructed, each being a combination of occurrence or nonoccurrence of the n PFSE events before the trigger failure. Based on the total probability law, $P(\text{systemfails} \cap \text{Case b})$ is evaluated as

$$P(\text{system fails} \cap \text{Case b}) = \sum_{i=0}^{2^n-1} [P(SE_i') \times P(\text{system fails}|SE_i')], \tag{8.46}$$

where $P(\text{system fails} \mid SE_i')$ can be obtained through the procedure similar to that for evaluating $P(\text{system fails} \mid SE_i)$ in step 3.

Step 7: Integrate *for final system unreliability.* Based on (8.41) and (8.44), the system unreliability is calculated as

$$UR_{system} = P(E_1) \times P(\text{system fails}|E_1) + P(\text{Case a}) + P(\text{system fails} \cap \text{Case b}), \tag{8.47}$$

where $P(E_1)$ is computed at step 2, $P(\text{system fails}|E_1)$ is evaluated at step 3, $P(\text{Case a})$ is computed at step 5, and $P(\text{systemfails} \cap \text{Case b})$ is evaluated at step 6.

The above seven-step procedure assumes that any nondependent components (including the trigger component) only undergo LFs. In the following, the procedure is extended to consider (1) PFGEs or (2) PFSEs or (3) both types of PFs for nondependent components.

(1) If nondependent components undergo PFGEs, the PFGE method (Section 8.2) is applied to separate effects of PFGEs originating from the nondependent components from the overall solution combinatorics. Particularly, an additional step denoted by (8.48) is added before step 1:

$$UR_{system} = 1 - P_u'(t) + Q'(t) \cdot P_u'(t), \tag{8.48}$$

where $P'_u(t) = P(\text{no PFGEs from nondependent components})$, $Q'(t) = P(\text{system fails}|\text{no PFGEs from nondependent components})$. $Q'(t)$ is then evaluated using the seven-step procedure.

(2) If nondependent components undergo PFSEs, these PFSEs can be handled using a method similar to (8.43) before step 1. Specifically, given w independent PFSEs originating from nondependent components, an event space with 2^w events is constructed, each being a combination of occurrence or nonoccurrence of the w PFSEs. The conditional system failure probability conditioned on the occurrence of each event is then evaluated using the seven-step procedure.

(3) If nondependent components undergo both PFSEs and PFGEs, the process below is applied:
   a) As in (1), apply the PFGE method to separate effects of PFGEs originating from all the nondependent components.
   b) As in (2), construct an event space with 2^w event combinations based on PFSEs of the nondependent components.
   c) Evaluate the conditional system failure probability given that each event combination occurs using the seven-step procedure.
   d) Integrate the conditional system failure probabilities computed in c) to obtain $Q'(t)$.
   e) Compute the final system unreliability using (8.48).

8.5.2 Case Study

Example 8.3 Consider a computer memory system illustrated in Figure 8.13, which consists of an embedded memory block and an external memory block (EMB). The former further contains two memory chips (MC_1, MC_2) and an independent MM. The two memory chips are accessible by the CPU through an MIU. Figure 8.14 gives the FT model of the memory system, where components MIU, MC_1, MC_2, MM, EMB are respectively denoted by A, B, C, D, E to simplify the representation. The FDEP gate models the FDEP relationship between the MIU and the dependent MCs.

All of the memory components undergo LFs; only components B and C undergo PFGEs and PFSEs. Define Tl, Tpg, and Tps as events respectively representing LF, PFGE, and PFSE originating from component $T \in \{A, B, C, D, E\}$. It is assumed that the PFGE, PFSE, and LF of the same component are *s*-independent. Also, a PFSE causes only LFs of other system components; the PFGEs originating from the affected components can still happen.

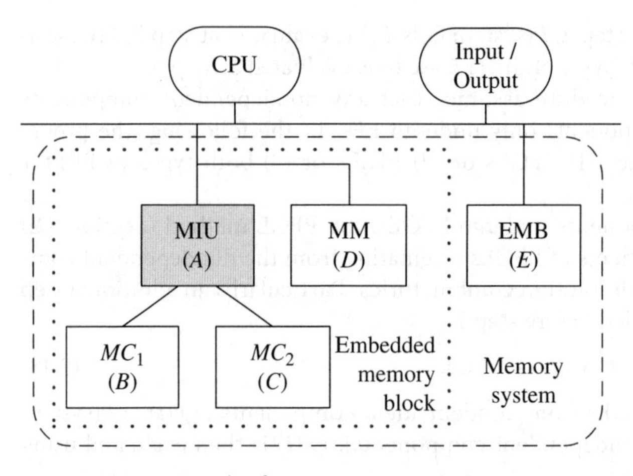

Figure 8.13 An example of a memory system [14].

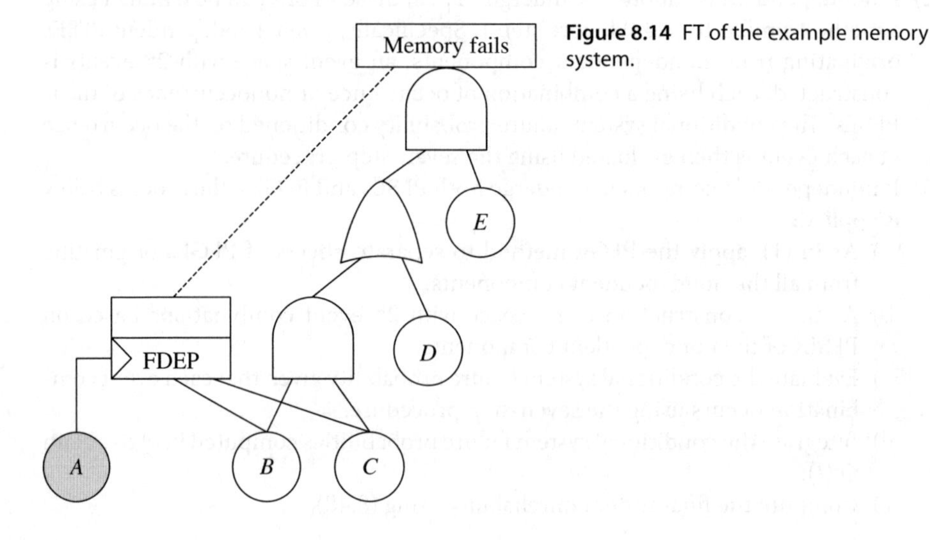

Figure 8.14 FT of the example memory system.

Table 8.3 Component failure events and failure rates.

Component	Failure event	Failure rate	Components affected
A	Al	λ_{Al}	$\{A\}$
B	Bl	λ_{Bl}	$\{B\}$
	Bpg	λ_{Bpg}	All
	$Bps1$	λ_{Bps1}	$\{B, C, D\}$
	$Bps2$	λ_{Bps2}	$\{B, C, E\}$
C	Cl	λ_{Cl}	$\{C\}$
	Cpg	λ_{Cpg}	All
	$Cps1$	λ_{Cps1}	$\{B, C, D\}$
	$Cps2$	λ_{Cps2}	$\{B, C, E\}$
D	Dl	λ_{Dl}	$\{D\}$
E	El	λ_{El}	$\{E\}$

Input Parameters. For illustration purpose, the exponential distribution is assumed for all the system components. Table 8.3 gives possible failure events associated with each component of the example memory system, their constant rates, and set of components affected by each failure event (including the component from which the failure originates).

Example Analysis. Applying the seven-step procedure, the unreliability of the example memory system is analyzed.

Step 1: Define events *representing states of the trigger component.* Two disjoint events are defined regarding states of the trigger component A.
E_1: A functions correctly, i.e. Al does not occur.
E_2: A fails, i.e. Al occurs.

Step 2: Evaluate $P(E_1)$. In this example, $P(E_1)$ is the probability that Al does not occur: $P(E_1) = P(\overline{Al}) = e^{-\lambda_{Al}t}$.

Step 3: Evaluate $P(system\ fails|E_1)$. According to (8.42),

$$P(\text{system fails}|E_1) = 1 - P_u(t) + Q(t)P_u(t).$$

There are two PFGEs: Bpg and Cpg. Hence, $P_u(t)$ is computed as

$$P_u(t) = P(\overline{Bpg} \cap \overline{Cpg}) = P(\overline{Bpg})P(\overline{Cpg}) = [1 - q_{Bpg}(t)][1 - q_{Cpg}(t)] = e^{-(\lambda_{Bpg}+\lambda_{Cpg})t}.$$

To evaluate $Q(t)$, a conditional component failure probability given that no PFGE occurs is computed for components B and C using (8.3). To handle PFSEs an event space with 16 events is constructed in Table 8.4. The last column of Table 8.4 shows the set of active components (components not affected) given the occurrence of each event.

Next, the evaluation of occurrence probability of each event $P(SE_i)$ and the conditional system failure probability given the occurrence of each event $P(\text{system fails}|SE_i)$ is illustrated using SE_0, SE_4, and SE_8.

The occurrence probability of SE_0 is

$$P(\text{SE}_0) = P(\overline{Bps1})P(\overline{Bps2})P(\overline{Cps1})P(\overline{Cps2}) = e^{-(\lambda_{Bps1}+\lambda_{Bps2}+\lambda_{Cps1}+\lambda_{Cps2})t}.$$

Table 8.4 Event space for addressing PFSEs.

i	Definition of SE$_i$	Set of active components
0	$\overline{Bps1} \cap \overline{Bps2} \cap \overline{Cps1} \cap \overline{Cps2}$	$\{B, C, D, E\}$
1	$\overline{Bps1} \cap \overline{Bps2} \cap \overline{Cps1} \cap Cps2$	$\varnothing$
2	$\overline{Bps1} \cap \overline{Bps2} \cap Cps1 \cap \overline{Cps2}$	$\{E\}$
3	$\overline{Bps1} \cap \overline{Bps2} \cap Cps1 \cap Cps2$	$\varnothing$
4	$\overline{Bps1} \cap Bps2 \cap \overline{Cps1} \cap \overline{Cps2}$	$\varnothing$
5	$\overline{Bps1} \cap Bps2 \cap \overline{Cps1} \cap Cps2$	$\varnothing$
6	$\overline{Bps1} \cap Bps2 \cap Cps1 \cap \overline{Cps2}$	$\varnothing$
7	$\overline{Bps1} \cap Bps2 \cap Cps1 \cap Cps2$	$\varnothing$
8	$Bps1 \cap \overline{Bps2} \cap \overline{Cps1} \cap \overline{Cps2}$	$\{E\}$
9	$Bps1 \cap \overline{Bps2} \cap \overline{Cps1} \cap Cps2$	$\varnothing$
10	$Bps1 \cap \overline{Bps2} \cap Cps1 \cap \overline{Cps2}$	$\{E\}$
11	$Bps1 \cap \overline{Bps2} \cap Cps1 \cap Cps2$	$\varnothing$
12	$Bps1 \cap Bps2 \cap \overline{Cps1} \cap \overline{Cps2}$	$\varnothing$
13	$Bps1 \cap Bps2 \cap \overline{Cps1} \cap Cps2$	$\varnothing$
14	$Bps1 \cap Bps2 \cap Cps1 \cap \overline{Cps2}$	$\varnothing$
15	$Bps1 \cap Bps2 \cap Cps1 \cap Cps2$	$\varnothing$

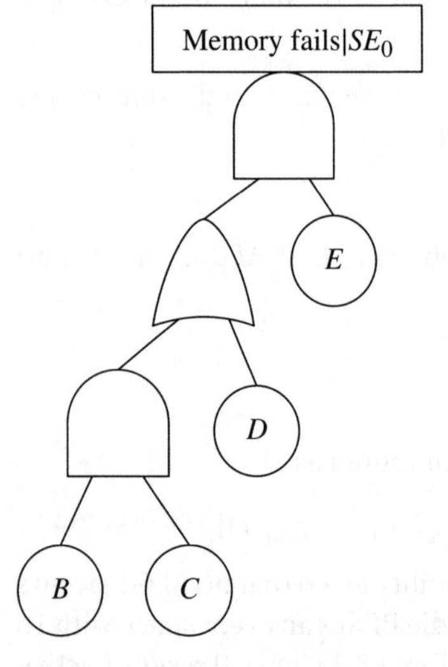

Figure 8.15 Reduced FT for P(system fails|SE_0).

To evaluate P(system fails|SE_0), a reduced FT containing failure events of active components $\{B, C, D, E\}$ is generated in Figure 8.15. Figure 8.16 shows the BDD generated from the reduced FT.

The evaluation of the BDD model in Figure 8.16 gives

$$\begin{aligned} P(\text{system fails}|SE_0) &= q_{EI}(t)\{q_{DI}(t) + [1 - q_{DI}(t)]q_{BI}(t)q_{CI}(t)\} \\ &= (1 - e^{-\lambda_{EI}t})[(1 - e^{-\lambda_{DI}t}) + e^{-\lambda_{DI}t}(1 - e^{-\lambda_{BI}t})(1 - e^{-\lambda_{CI}t})] \\ &= (1 - e^{-\lambda_{EI}t})[1 - e^{-(\lambda_{BI}+\lambda_{DI})t} - e^{-(\lambda_{CI}+\lambda_{DI})t} + e^{-(\lambda_{BI}+\lambda_{CI}+\lambda_{DI})t}]. \end{aligned}$$

Figure 8.16 BDD for evaluating $P(\text{system fails}|SE_0)$.

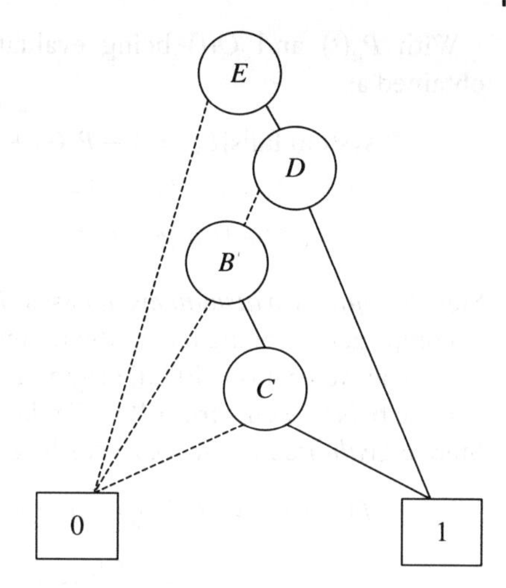

Consider SE_4. Its occurrence probability is

$$P(SE_4) = P(\overline{Bps1})P(Bps2)P(\overline{Cps1})P(\overline{Cps2}) = (1 - e^{-\lambda_{Bps2}t})e^{-(\lambda_{Bps1}+\lambda_{Cps1}+\lambda_{Cps2})t}.$$

To evaluate $P(\text{system fails}|SE_4)$, a reduced FT is generated by replacing events of components affected by SE_4 (including B, C, E) with constant 1 in the FT of Figure 8.15, which leads to $P(\text{system fails}|SE_4) = 1$.

For another example, consider SE_8. Its occurrence probability is

$$P(SE_8) = P(Bps1)P(\overline{Bps2})P(\overline{Cps1})P(\overline{Cps2}) = (1 - e^{-\lambda_{Bps1}t})e^{-(\lambda_{Bps2}+\lambda_{Cps1}+\lambda_{Cps2})t}.$$

To evaluate $P(\text{system fails}|SE_8)$, a reduced FT is generated by replacing events of components affected by SE_8 (including B, C, D) with constant 1 in the FT of Figure 8.15, which leads to an FT containing failure events of active components $\{E\}$ as shown in Figure 8.17. The evaluation of reduced FT gives

$$P(\text{system fails}|SE_8) = q_{El}(t) = (1 - e^{-\lambda_{El}t}).$$

Using the similar procedure, all of $P(SE_i)$ and $P(\text{system fails}|SE_i)$ ($i = 0, \ldots, 15$) can be evaluated. According to (8.43),

$$\begin{aligned} Q(t) &= \sum_{i=0}^{15}[P(SE_i) \cdot P(\text{system fails}|SE_i)] \\ &= 1 - e^{-(\lambda_{Bps2}+\lambda_{Cps2}+\lambda_{El})t} - (1 - e^{-\lambda_{El}t})e^{-(\lambda_{Bps1}+\lambda_{Bps2}+\lambda_{Cps1}+\lambda_{Cps2}+\lambda_{Dl})t} \\ &\quad \cdot [e^{-\lambda_{Bl}t} + e^{-\lambda_{Cl}t} - e^{-(\lambda_{Bl}+\lambda_{Cl})t}]. \end{aligned}$$

Figure 8.17 Reduced FT for $P(\text{system fails}|SE_8)$.

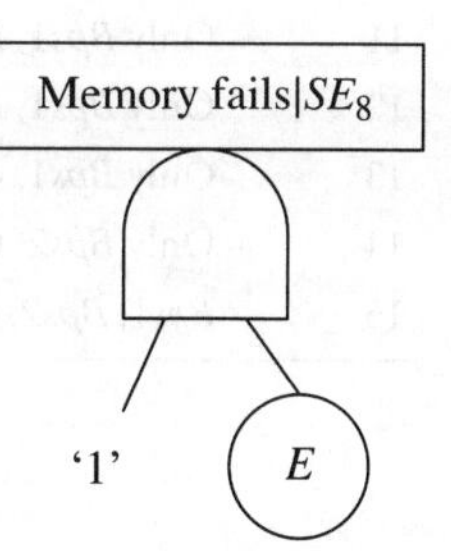

With $P_u(t)$ and $Q(t)$ being evaluated, according to (8.42), $P(\text{system fails}|E_1)$ is obtained as

$$P(\text{system fails}|E_1) = 1 - P_u(t) + Q(t) \cdot P_u(t)$$
$$= 1 - (1 - e^{-\lambda_{El}t})[e^{-\lambda_{Bl}t} + e^{-\lambda_{Cl}t} - e^{-(\lambda_{Bl}+\lambda_{Cl})t}]e^{-(\lambda_{Bps1}+\lambda_{Bps2}+\lambda_{Cps1}+\lambda_{Cps2}+\lambda_{Bpg}+\lambda_{Cpg}+\lambda_{Dl})t}$$
$$- e^{-(\lambda_{Bps2}+\lambda_{Cps2}+\lambda_{Bpg}+\lambda_{Cpg}+\lambda_{El})t}.$$

Step 4: *Define two disjoint event cases given that E_2 takes place.* In the event of the trigger component A being failed, the following two cases are considered:
Case a: At least one PFGE originating from B or C occurs before A fails.
Case b: No PFGEs from B or C take place before A fails.
Step 5: *Evaluate P(Case a).* According to (8.45), one obtains

$$P(\text{Case a}) = P[(Bpg \cup Cpg) \rightarrow Al]$$
$$= \frac{\lambda_{Al}}{\lambda_{Al} + \lambda_{Bpg} + \lambda_{Cpg}} e^{-(\lambda_{Al}+\lambda_{Bpg}+\lambda_{Cpg})t} - e^{-\lambda_{Al}t} + \frac{\lambda_{Bpg} + \lambda_{Cpg}}{\lambda_{Al} + \lambda_{Bpg} + \lambda_{Cpg}}.$$

Step 6: *Evaluate P(system fails $\cap$ Case b).* An event space with 16 events is constructed as shown in Table 8.5.

Next the evaluation of $P(SE_i')$ and $P(\text{system fails} \mid SE_i')$ is illustrated using SE_0' and SE_1'.

Table 8.5 Event space for addressing PFSEs.

i	Definition of SE_i'	Set of active components
0	No PFs happen before Al occurs	$\{E\}$
1	Only $Bps1$ occurs before Al occurs	$\{E\}$
2	Only $Bps2$ occurs before Al occurs	$\emptyset$
3	Only $Cps1$ occurs before Al occurs	$\{E\}$
4	Only $Cps2$ occurs before Al occurs	$\emptyset$
5	Only $Bps1$ and $Bps2$ occur before Al occurs	$\emptyset$
6	Only $Bps1$ and $Cps1$ occur before Al occurs	$\{E\}$
7	Only $Bps1$ and $Cps2$ occur before Al occurs	$\emptyset$
8	Only $Bps2$ and $Cps1$ occur before Al occurs	$\emptyset$
9	Only $Bps2$ and $Cps2$ occur before Al occurs	$\emptyset$
10	Only $Cps1$ and $Cps2$ occur before Al occurs	$\emptyset$
11	Only $Bps1$, $Bps2$ and $Cps1$ occur before Al occurs	$\emptyset$
12	Only $Bps1$, $Bps2$ and $Cps2$ occur before Al occurs	$\emptyset$
13	Only $Bps1$, $Cps1$ and $Cps2$ occur before Al occurs	$\emptyset$
14	Only $Bps2$, $Cps1$ and $Cps2$ occur before Al occurs	$\emptyset$
15	$Bps1$, $Bps2$, $Cps1$ and $Cps2$ all occur before Al occurs	$\emptyset$

Under SE_0', no PF occurs before Al occurs. Thus,

$$P(SE_0') = P(Al) - P(\text{at least one PF occurs before } Al \text{ happens})$$
$$= P(Al) - P[(Bps1 \cup Bps2 \cup Cps1 \cup Cps2 \cup Bpg \cup Cpg) \to Al].$$

where, according to (8.45)

$$P[(Bps1 \cup Bps2 \cup Cps1 \cup Cps2 \cup Bpg \cup Cpg) \to Al]$$

$$= \int_0^t \int_{\tau_1}^t \begin{bmatrix} (\lambda_{Bps1} + \lambda_{Bps2} + \lambda_{Cps1} + \lambda_{Cps2} + \lambda_{Bpg} + \lambda_{Cpg}) \\ e^{-(\lambda_{Bps1}+\lambda_{Bps2}+\lambda_{Cps1}+\lambda_{Cps2}+\lambda_{Bpg}+\lambda_{Cpg})\tau_1} \lambda_{Al} e^{-\lambda_{Al}\tau_2} \end{bmatrix} d\tau_2 d\tau_1$$

$$= \frac{\lambda_{Al}}{\lambda_{Al} + \lambda_{Bps1} + \lambda_{Bps2} + \lambda_{Cps1} + \lambda_{Cps2} + \lambda_{Bpg} + \lambda_{Cpg}} \cdot e^{-(\lambda_{Al}+\lambda_{Bps1}+\lambda_{Bps2}+\lambda_{Cps1}+\lambda_{Cps2}+\lambda_{Bpg}+\lambda_{Cpg})t}$$
$$- e^{-\lambda_{Al}t} + \frac{\lambda_{Bps1} + \lambda_{Bps2} + \lambda_{Cps1} + \lambda_{Cps2} + \lambda_{Bpg} + \lambda_{Cpg}}{\lambda_{Al} + \lambda_{Bps1} + \lambda_{Bps2} + \lambda_{Cps1} + \lambda_{Cps2} + \lambda_{Bpg} + \lambda_{Cpg}}.$$

Thus,

$$P(SE_0') = \frac{\lambda_{Al}}{\lambda_{Al} + \lambda_{Bps1} + \lambda_{Bps2} + \lambda_{Cps1} + \lambda_{Cps2} + \lambda_{Bpg} + \lambda_{Cpg}}$$
$$- \frac{\lambda_{Al}}{\lambda_{Al} + \lambda_{Bps1} + \lambda_{Bps2} + \lambda_{Cps1} + \lambda_{Cps2} + \lambda_{Bpg} + \lambda_{Cpg}} \cdot e^{-(\lambda_{Al}+\lambda_{Bps1}+\lambda_{Bps2}+\lambda_{Cps1}+\lambda_{Cps2}+\lambda_{Bpg}+\lambda_{Cpg})t}.$$

The reduced FT for evaluating $P(\text{system fails} \mid SE_0')$ is same as that in Figure 8.17. Thus, $P(\text{system fails}|SE_0') = q_{El}(t) = 1 - e^{-\lambda_{El}t}$.

Under SE_1', only $Bps1$ occurs before Al occurs. Thus,

$$P(SE_1') = P(Bps1 \to Al) - P[Bps1 \cap (Bps2 \cup Cps1 \cup Cps2 \cup Bpg \cup Cpg) \to Al]$$
$$= P\{\{Bps1 - [Bps1 \cap (Bps2 \cup Cps1 \cup Cps2 \cup Bpg \cup Cpg)]\} \to Al\}$$
$$= P\{[Bps1 \cap \overline{Bps2} \cap \overline{Cps1} \cap \overline{Cps2} \cap \overline{Bpg} \cap \overline{Cpg}] \to Al\}.$$

The reduced FT for evaluating $P(\text{system fails} \mid SE_1')$ is same as that in Figure 8.17. Thus, $P(\text{system fails}|SE_1') = q_{El}(t) = 1 - e^{-\lambda_{El}t}$.

Using the similar procedure, all of $P(SE_i')$ and $P(\text{system fails} \mid SE_i')$ $(i = 0, \ldots, 15)$ can be evaluated. $P(\text{systemfails} \cap \text{Case b})$ can thus be evaluated using (8.46). Then, according to (8.44), one obtains

$$P(\text{system fails}|E_2)P(E_2) = P(\text{Case a}) + P(\text{system fails}|\text{Case b})P(\text{Case b})$$

$$= 1 - e^{-\lambda_{Al}t} - \frac{\lambda_{Al}}{\lambda_{Al} + \lambda_{Bps2} + \lambda_{Cps2} + \lambda_{Bpg} + \lambda_{Cpg}} e^{-\lambda_{El}t}[1 - e^{-(\lambda_{Al}+\lambda_{Bps2}+\lambda_{Cps2}+\lambda_{Bpg}+\lambda_{Cpg})t}].$$

Step 7: Integrate for final system unreliability. According to (8.41), with $P(E_1)$ evaluated at step 2, $P(\text{system fails}|E_1)$ evaluated at step 3, and $P(E_2)P(\text{system fails}|E_2)$ evaluated at step 6, the system unreliability considering competing failures as well as effects of both PFGEs and PFSEs is computed as:

$$
\begin{aligned}
UR_{system} &= P(\text{system fails}|E_1)P(E_1) + P(\text{system fails}|E_2)P(E_2)\\
&= 1 - \frac{\lambda_{AI}}{\lambda_{AI} + \lambda_{Bps2} + \lambda_{Cps2} + \lambda_{Bpg} + \lambda_{Cpg}} e^{-\lambda_{EI}t}\\
&\quad - \frac{\lambda_{Bps2} + \lambda_{Cps2} + \lambda_{Bpg} + \lambda_{Cpg}}{\lambda_{AI} + \lambda_{Bps2} + \lambda_{Cps2} + \lambda_{Bpg} + \lambda_{Cpg}} e^{-(\lambda_{AI}+\lambda_{Bps2}+\lambda_{Cps2}+\lambda_{Bpg}+\lambda_{Cpg}+\lambda_{EI})t}\\
&\quad - e^{-(\lambda_{AI}+\lambda_{BI}+\lambda_{Bps1}+\lambda_{Bps2}+\lambda_{Cps1}+\lambda_{Cps2}+\lambda_{Bpg}+\lambda_{Cpg}+\lambda_{DI})t}\\
&\quad - e^{-(\lambda_{AI}+\lambda_{Bps1}+\lambda_{Bps2}+\lambda_{CI}+\lambda_{Cps1}+\lambda_{Cps2}+\lambda_{Bpg}+\lambda_{Cpg}+\lambda_{DI})t}\\
&\quad + e^{-(\lambda_{AI}+\lambda_{BI}+\lambda_{CI}+\lambda_{Bps1}+\lambda_{Bps2}+\lambda_{Cps1}+\lambda_{Cps2}+\lambda_{Bpg}+\lambda_{Cpg}+\lambda_{DI})t}\\
&\quad + e^{-(\lambda_{AI}+\lambda_{BI}+\lambda_{Bps1}+\lambda_{Bps2}+\lambda_{Cps1}+\lambda_{Cps2}+\lambda_{Bpg}+\lambda_{Cpg}+\lambda_{DI}+\lambda_{EI})t}\\
&\quad + e^{-(\lambda_{AI}+\lambda_{Bps1}+\lambda_{Bps2}+\lambda_{CI}+\lambda_{Cps1}+\lambda_{Cps2}+\lambda_{Bpg}+\lambda_{Cpg}+\lambda_{DI}+\lambda_{EI})t}\\
&\quad - e^{-(\lambda_{AI}+\lambda_{BI}+\lambda_{CI}+\lambda_{Bps1}+\lambda_{Bps2}+\lambda_{Cps1}+\lambda_{Cps2}+\lambda_{Bpg}+\lambda_{Cpg}+\lambda_{DI}+\lambda_{EI})t}.
\end{aligned}
$$

8.6 Multi-Phase System with Single FDEP Group

Previous sections focus on single-phase systems. However, many real-world systems are PMSs, involving multiple, consecutive, and nonoverlapping phases of operations or tasks. Consideration of competing failures in PMSs is a challenging task because PMSs exhibit dynamics in system configuration and component behavior, as well as statistical dependencies across phases for a given component.

This section presents a combinatorial method to address the competing failure effects in reliability analysis of nonrepairable binary-state PMSs, where only one mission phase is subject to the FDEP behavior. As an example of such a PMS, a set of computers work together to accomplish an M-phase mission task. In $M - 1$ of these phases, only local computing is needed (no FDEPs are involved), while in one of the phases, some computers need to access the Internet to access external data. Thus, in this particular phase these computers have FDEP on the router.

The phase with FDEP is referred to as an FDEP phase; other phases are referred to as non-FDEP phases. All PFs have the global effect, i.e. only PFGEs are considered. The LF and PFGE of the same component are s-independent. Also, failure events of different components are s-independent. There is only one FDEP group existing in the system. Thus, the PMS considered undergoes no cascading failure propagation process.

8.6.1 Combinatorial Method

The reliability analysis method for PMSs subject to competing failure isolation and propagation effects involves the following five-step procedure [15]:

Step *1: Separate PFGEs of all nondependent components.* Any component that cannot be isolated by the failure of another component is referred to as a nondependent component (NDC), e.g. the trigger component of an FDEP group or a component not belonging to any FDEP group. A PFGE originating from an NDC causes the failure of the entire system. Thus, the PFGE method described in Section 8.2 can be applied to separate PFGEs of all the NDCs from the solution combinatorics.

$$\begin{aligned} UR_{PMS} &= P(\text{PMS fails}|\text{at least one PFGE from NDCs}) \\ &\quad \cdot P(\text{at least one PFGE from NDCs}) \\ &\quad + P(\text{PMS fails}|\text{no PFGEs from NDCs}) \cdot P(\text{no PFGEs from NDCs}) \\ &= 1 - P_u(t) + Q(t) \cdot P_u(t). \end{aligned} \tag{8.49}$$

Let N denote the set of NDCs undergoing PFGEs. $P_u(t)$ in (8.49) is evaluated as

$$P_u(t) = \prod_{i \in N} P(\text{NDC } i \text{ has no PFGEs during the mission}). \tag{8.50}$$

As explained in Section 8.2, the evaluation of $Q(t)$ requires the use of a conditional LF probability for all the NDCs given that no PFGEs occur to the component during the mission. Since the LF and PFGE of the same component are *s*-independent, (8.3) is applied to compute the conditional LF probability. The evaluation of $Q(t)$ is conducted in steps 2–4.

Step 2: Define three disjoint events E_1, E_2, and E_3 and evaluate their probabilities. Given that no PFGEs from NDCs happen, three disjoint events are defined:

E_1: No PFGEs occur to dependent components (DCs) during the mission. Let D denote the set of DCs involved in the FDEP that undergo PFGEs. $P(E_1)$ is evaluated as

$$P(E_1) = \prod_{i \in D} P(\text{DC } i \text{ has no PFGEs during the mission}). \tag{8.51}$$

E_2: At least one PFGE from DCs takes place in a non-FDEP phase. $P(E_2)$ can be evaluated as

$$\begin{aligned} P(E_2) &= P(\text{At least one PFGE from DCs occurs in a non-FDEP phase}) \\ &= 1 - P(\text{No PFGEs from DCs occurs in any non-FDEP phase}) \\ &= 1 - \prod_{i \in D} P(\text{DC } i \text{ undergoes no PFGEs in non-FDEP phases}). \end{aligned} \tag{8.52}$$

E_3: At least one PFGE from DCs occurs in the FDEP phase and no PFGEs from DCs occur in non-FDEP phase. Because $P(E_1) + P(E_2) + P(E_3) = 1$, $P(E_3)$ is evaluated as

$$P(E_3) = 1 - P(E_1) - P(E_2). \tag{8.53}$$

Based on the total probability law, $Q(t)$ is evaluated as

$$Q(t) = P(E_1) \cdot Q_1^C + P(E_2) \cdot Q_2^C + P(E_3) \cdot Q_3^C, \tag{8.54}$$

where Q_i^C ($i = 1, 2, 3$) is the conditional system failure probability given that E_i occurs and no PFGEs from the NDCs occur.

Step 3: Evaluate Q_1^C and Q_2^C. Under E_1, the PMS can be analyzed as a system without any component undergoing PFGEs. Thus, to evaluate Q_1^C, a reduced FT is first generated by replacing the FDEP gate with an OR gate and by considering only LFs for all the system components. The reduced FT is then evaluated using the PMS BDD method described in Section 3.5.3 [16] to obtain Q_1^C.

Under E_2, the global failure propagation effect takes place, causing the entire system failure. Thus, $Q_2{}^C = 1$.

Step 4: Evaluate $P(E_3) \cdot Q_3^C$. Under E_3, competing failures can take place in the FDEP phase and the occurrence sequence of the trigger LF and PFGEs of DCs needs to be considered. Two disjoint subcases are defined under E_3:

Case 1: Failure isolation effect takes place.

The trigger LF occurs before any PFGEs from the DCs. Since under E_3, all PFGEs from the DCs occur in the FDEP phase, the trigger LF occurs either in phases before the FDEP phase or in FDEP phase but before any PFGEs occurs from the DCs.

Case 2: Failure propagation effect takes place.

The trigger LF either occurs after any PFGE from the DCs occurs or does not occur at all. Under E_3, if the trigger LF occurs, the failure can only occur either in phases after the FDEP phase or in the FDEP phase but after any PFGE from the DCs occurs.

Let $Q_{3,i}{}^C$ $(i = 1, 2)$ denote the conditional system failure probability, given that Case i occurs. Based on the total probability law, one obtains

$$P(E_3) \cdot Q_3^C = P(\text{Case 1}) \cdot Q_{3,1}^C + P(\text{Case 2}) \cdot Q_{3,2}^C, \tag{8.55}$$

where $Q_{3,2}{}^C = 1$ because the global failure propagation effect takes place under Case 2.

Assume the FDEP phase is phase m. To evaluate $P(\text{Case 1})\ Q_{3,1}{}^C$, an event space with m events is constructed, with event i $(i = 1, \ldots, m-1)$ representing the trigger component fails locally in phase i and E_3 occurs, and event m representing the trigger component fails locally in phase m (FDEP phase) and before any PFGE from the DCs occurs in phase m. Based on these events, (8.55) can be evaluated as

$$\begin{aligned} P(E_3) \cdot Q_3^C &= P(\text{Case 1}) \cdot Q_{3,1}^C + P(\text{Case 2}) \cdot Q_{3,2}^C \\ &= \sum_{i=1}^{m} [P(\text{event } i) \cdot Q_{3,1-i}^C] + P(\text{Case 2}). \end{aligned} \tag{8.56}$$

The occurrence probabilities of the m events are

$$\begin{aligned} &P(\text{event } i) = P(\text{Trigger component fails locally in phase } i) \cdot P(E_3), \\ &\quad \text{for } i = 1, \ldots, m-1; \end{aligned} \tag{8.57}$$

$$\begin{aligned} &P(\text{event } m) \\ &= P\left[\begin{pmatrix} \text{Trigger component fails locally in phase } m \\ \text{and before any DC fails globally} \end{pmatrix} \cap (\text{No DCs fail globally in non-FDEP phase})\right]. \end{aligned} \tag{8.58}$$

The sequential failure probability involved in (8.58) can be evaluated using integral formula (8.10), which can be solved using the MathCAD software [17].

$Q_{3,1-i}^C$ in (8.56) can be computed by evaluating a reduced PMS FT, which is obtained through the following procedure:

1. Events in the original PMS FT in phase j $(j < i)$ representing failure of the trigger component are replaced with constant 0 (FALSE).

2. Events in phase k ($k \geq i$) representing failure of the trigger component are replaced with constant 1 (TURE).
3. Since at least one PFGE occurs in the FDEP phase, all DCs in the same FDEP group are affected and fail although other components outside the FDEP group are not affected by the PFGEs due to the isolation effect. Hence, events representing failures of DCs in the FDEP phase and phases after the FDEP phase are replaced with constant 1 (TURE).
4. Boolean rules ($1+x=1$, $1 \cdot x = x$, $0+x=x$, $0 \cdot x = 0$, where x denotes a Boolean variable) are applied to generate the reduced FT.

   The reduced FT generated is then evaluated using the PMS BDD method described in Section 3.5.3 to obtain $Q^C_{3,1-i}$.

   P(Case 2) in (8.56) can be simply computed as

$$P(\text{Case 2}) = P(E_3) - P(\text{Case 1}) = P(E_3) - \sum_{i=1}^{m} P(\text{event } i). \tag{8.59}$$

Step 5: Integrate for final PMS unreliability. According to (8.54), $P(E_1)$ and $P(E_2)$ evaluated at step 2, $Q_1{}^C$ and $Q_2{}^C$ evaluated at step 3, and $P(E_3) \cdot Q^C_3$ evaluated at step 4 are integrated to obtain $Q(t)$. Further, according to (8.49), $Q(t)$ and $P_u(t)$ evaluated at step 1 are integrated to obtain the final PMS unreliability considering effects of competing failures.

8.6.2 Case Study

Example 8.4 Consider an example of PMS FT illustrated in Figure 8.18, which consists of a router A and four computers B, C, D, E. The four computers work together to complete a three-phase mission task. In phase 1 and phase 3, only local computing is involved. In phase 2, computers B and C need to access data files through router A to complete the task. Thus, B and C have FDEP on A. Computer E holds the required data files in its local memory and can complete the task independently. Phase 2 fails when all the three computers B, C, and E malfunction.

Input Parameters. Assume the phase durations for the three phases are independent of the system state and equal to 10, 30, and 20 hours, respectively. Therefore, the entire mission time is 60 hours.

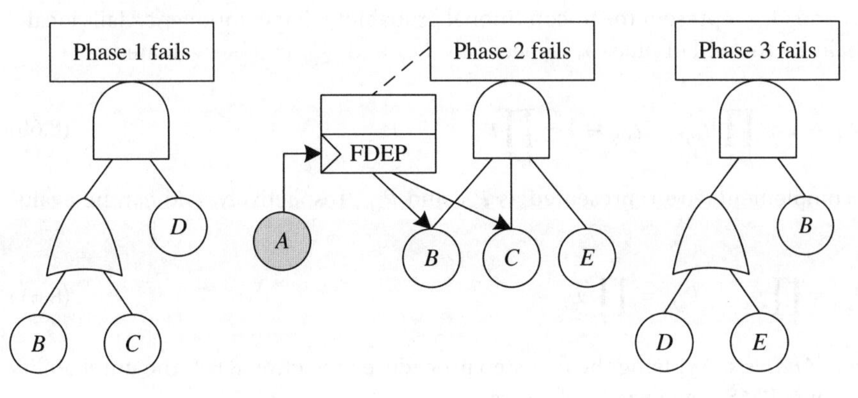

Figure 8.18 An example PMS FT.

Table 8.6 Failure parameters for components *A*, *B*, and *C*.

	Phases 1, 2, 3	
	LF	PFGE
A	$\lambda = 1.5e - 4$	$\lambda = 1e - 4$
B	$\lambda = 2e - 4$	$\lambda = 1e - 4$
C	$\lambda = 2e - 4$	$\lambda = 1e - 4$

Table 8.7 Failure parameters for component *E*.

	Phase 1	Phase 2	Phase 3
λ_W	$2e - 4$	$1e - 4$	$1.5e - 4$
α_W	2	2	2

All five components experience LFs in each of the three phases. Only computers *A*, *B*, and *C* can undergo PFGEs during the mission (e.g. due to computer viruses). Let q_{xl_i} and q_{xp_i} represent the conditional probability that component *x* fails locally and globally at phase *i*, respectively, given that the component has survived the previous phase. Their complements are represented as p_{xl_i} and p_{xp_i}, respectively.

For illustration, three types of *ttf* distributions are considered for evaluating q_{xl_i} and q_{xp_i}:

1. Components *A*, *B*, and *C* have the exponential *ttf* distribution with constant failure rate λ given in Table 8.6. The failure probability is computed as $q_{xl_i/xp_i}(t) = 1 - e^{-\lambda t}$, and *pdf* is $f_{xl_i/xp_i}(t) = \lambda e^{-\lambda t}$.
2. Component *D* has a fixed LF probability $q_{Dl_i} = 0.001$, $i = 1, 2, 3$.
3. Component *E*'s time to LF follows the Weibull distribution with scale parameter λ_W and shape parameter α_W given in Table 8.7. The failure probability is computed as $q_{xl_i/xp_i}(t) = 1 - e^{-(\lambda_W t)^{\alpha_W}}$, and *pdf* is $f_{xl_i/xp_i}(t) = \alpha_W \lambda_W (\lambda_W t)^{\alpha_W - 1} \cdot e^{-(\lambda_W t)^{\alpha_W}}$.

Let Q_{xl_i} and Q_{xp_i} represent the unconditional probability that component *x* fails locally and globally by the end of phase *i*, respectively. Q_{xl_i} and Q_{xp_i} can be evaluated as

$$Q_{xl_i} = 1 - \prod_{j=1}^{i} p_{xl_i}, \quad Q_{xp_i} = 1 - \prod_{j=1}^{i} p_{xp_i}. \tag{8.60}$$

Their complements are represented as P_{xl_i} and P_{xp_i}, respectively, and can be evaluated as

$$P_{xl_i} = \prod_{j=1}^{i} p_{xl_i}, \quad P_{xp_i} = \prod_{j=1}^{i} p_{xp_i}. \tag{8.61}$$

Example Analysis. Applying the five-step procedure in Section 8.6.1, the unreliability of the example PMS is analyzed as follows:

Step 1: Separate PFGEs of all nondependent components. There is only one NDC A, so $Q(t)$ in (8.49) is evaluated as

$$P_u(t) = P(A \text{ has no PFGEs during the mission}) = P_{Ap_3} = \prod_{j=1}^{3} p_{Ap_j} = 0.994018.$$

The evaluation of $Q(t)$ in (8.49) requires the use of a conditional LF probability for A given that no PFGEs occur to A during the mission, which is evaluated based on (8.3).

Step 2: Define three disjoint events E_1, E_2, and E_3 and evaluate their probabilities. There are two DCs: B and C. Both of them can experience PFGEs. Thus, the three events are defined as:

E_1: no PFGEs occur to B and C during the entire mission. Thus,

$$P(E_1) = P_{Bp_3} \cdot P_{Cp_3} = 0.988072.$$

E_2: at least one PFGE from B or C takes place in non-FDEP phases (phase 1 and phase 3). Note that component C does not appear in the phase 3 FT, meaning that its LF makes no contribution to phase 3 failure; however, its PFGE can occur and cause the entire mission failure. $P(E_2)$ is evaluated as

$$\begin{aligned}
P(E_2) &= 1 - P(B \text{ and } C \text{ do not undergo PFGEs in phases 1 and 3})\\
&= 1 - \begin{bmatrix} P(B \text{ does not undergo PFGEs in phases 1 and 3}) \\ \cdot P(C \text{ does not undergo PFGEs in phases 1 and 3}) \end{bmatrix}\\
&= 1 - \begin{bmatrix} P\begin{pmatrix} B \text{ undergoes PFGEs in phase 2} \\ \cup\, B \text{ does not undergo PFGEs during the mission} \end{pmatrix} \\ \cdot P\begin{pmatrix} C \text{ undergoes PFGEs in phase 2} \\ \cup\, C \text{ does not undergo PFGEs during the mission} \end{pmatrix} \end{bmatrix}\\
&= 1 - (p_{Bp_1} q_{Bp_2} + P_{Bp_3})(p_{Cp_1} q_{Cp_2} + P_{Cp_3}) = 0.00597.
\end{aligned}$$

E_3: B or C undergoes PFGEs in the FDEP phase (phase 2), and both B and C do not undergo PFGEs in phase 1 and phase 3. $P(E_3)$ is evaluated as $P(E_3) = 1 - P(E_1) - P(E_2)$.

Step 3: Evaluate $Q_1{}^C$ and $Q_2{}^C$. Figure 8.19 shows the reduced FT for evaluating $Q_1{}^C$. The reduced FT is then evaluated using the PMS BDD method described in Section 3.5.3. Figure 8.20 shows the PMS BDD generated using the variable order of $Al_2 < Bl_3 < Bl_2 < Bl_1 < Cl_2 < Cl_1 < Dl_3 < Dl_1 < El_3 < El_2$. The evaluation of the PMS BDD model gives

$$\begin{aligned}
Q_1^C &= Q_{Al_2}\{Q_{Bl_3}(Q_{Dl_3} + P_{Dl_3}Q_{El_3}) + P_{Bl_3}[Q_{Cl_1}(Q_{Dl_1} + P_{Dl_1}Q_{El_2}) + P_{Cl_1}Q_{El_2}]\}\\
&\quad + P_{Al_2}\{Q_{Bl_3}(Q_{Dl_3} + P_{Dl_3}Q_{El_3}) + P_{Bl_3}Q_{Cl_1}Q_{Dl_1}\} = 0.000038.
\end{aligned}$$

Under E_2, the global failure propagation effect from B or C occurs causing the PMS failure. Thus, $Q_2{}^C = 1$.

Step 4: Evaluate $P(E_3) \cdot Q_3^C$. The two sub-cases under E_3 are

Case 1: LF of trigger A occurs in phase 1, or occurs in phase 2 and before B or C fails globally in phase 2.

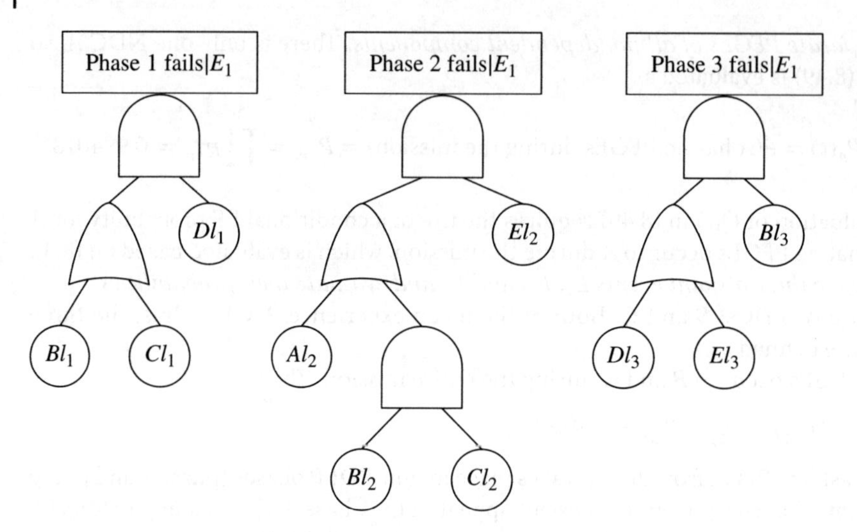

Figure 8.19 Reduced FT for evaluating $Q_1{}^C$.

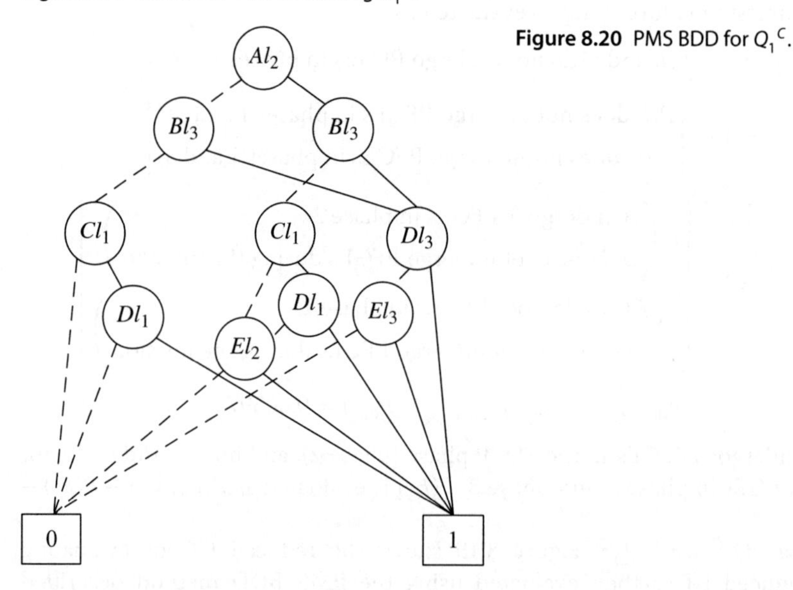

Figure 8.20 PMS BDD for $Q_1{}^C$.

Case 2: LF of trigger A occurs in phase 3, or occurs in phase 2 and after B or C fails globally in phase 2, or does not occur at all during the entire mission.

Under Case 1, the following two events are defined: event 1 – trigger A fails locally in phase 1 and E_3 happens, event 2 – A fails locally in phase 2 and before B or C fails globally in phase 2. P(event 1) is evaluated as

$$P(\text{event 1}) = Q_{Al_1} P(E_3) = 0.000009.$$

To evaluate $Q^C_{3,1-i}$ in (8.56), a reduced FT is generated in Figure 8.21.

Figure 8.22 shows the PMS BDD generated in the PMS BDD method using the order of $Bl_1 < Cl_1 < Dl_3 < Dl_1 < El_3 < El_2$. The evaluation of the PMS BDD model gives

$$Q^C_{3,1-1} = Q_{Dl_3} + P_{Dl_3} Q_{El_3} = 0.003019.$$

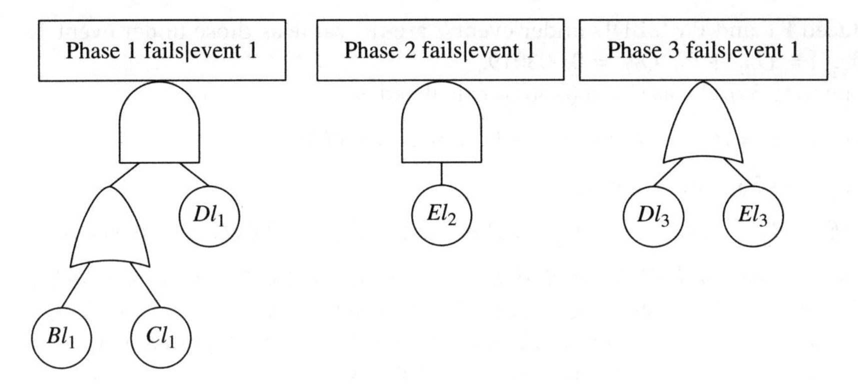

Figure 8.21 Reduced FT under event 1.

Figure 8.22 PMS BDD under event 1.

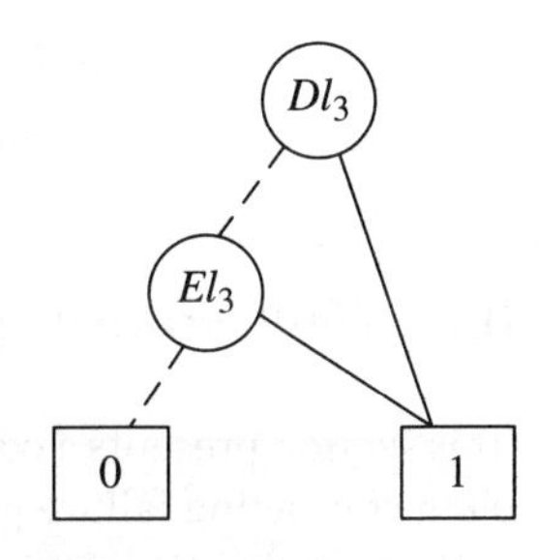

Let T_2 represent the duration of phase 2. P(event 2) is evaluated as

$$
\begin{aligned}
P(\text{event 2}) &= P\left[\begin{array}{l}(\text{LF of } A \text{ occurs in phase 2 and before PFGEs of } B \text{ and } C)\\ \cap\ (\text{PFGEs of } B \text{ and } C \text{ do not occur in phases1 and 3})\end{array}\right]\\
&= P\left[\begin{array}{l}(\text{LF of } A \text{ occurs in phase 2 and before PFGE of } B)\\ \cap\ (C \text{ does not fail globally during the mission})\end{array}\right]\\
&\quad + P\left[\begin{array}{l}(\text{LF of } A \text{ occurs in phase 2 and before PFGE of } C)\\ \cap\ (B \text{ does not fail globally during the mission})\end{array}\right]\\
&\quad + P(\text{LF of } A \text{ occurs in phase 2 and before PFGEs of } B \text{ and } C)\\
&= p_{Al_1} p_{Bp_1} P_{Cp_3} \int_0^{T_2}\int_{\tau_1}^{T_2} f_{Al_2}(\tau_1) f_{Bp_2}(\tau_2)\, d\tau_2\, d\tau_1\\
&\quad + p_{Al_1} p_{Cp_1} P_{Bp_3} \int_0^{T_2}\int_{\tau_1}^{T_2} f_{Al_2}(\tau_1) f_{Cp_2}(\tau_2)\, d\tau_2\, d\tau_1\\
&\quad + p_{Al_1} p_{Bp_1} p_{Cp_1} \int_0^{T_2}\int_{\tau_1}^{T_2}\int_{\tau_2}^{T_2} f_{Al_2}(\tau_1) f_{Bp_2}(\tau_2) f_{Cp_2}(\tau_3)\, d\tau_3\, d\tau_2\, d\tau_1\\
&\quad + p_{Al_1} p_{Bp_1} p_{Cp_1} \int_0^{T_2}\int_{\tau_1}^{T_2}\int_{\tau_2}^{T_2} f_{Al_2}(\tau_1) f_{Cp_2}(\tau_2) f_{Bp_2}(\tau_3)\, d\tau_3\, d\tau_2\, d\tau_1\\
&= 0.000013.
\end{aligned}
$$

The reduced FT and PMS BDD under event 2 are the same as those under event 1. Thus, $Q^C_{3,1-2} = Q_{DI_3} + P_{DI_3} Q_{EI_3} = 0.003019$.
According to (8.59), $P(\text{Case } 2)$ in (8.56) is calculated as

$$P(\text{Case } 2) = P(E_3) - P(\text{event } 1) - P(\text{event } 2) = 0.0059359.$$

According to (8.56), one obtains

$$P(E_3) \cdot Q^C_3 = P(\text{event } 1) \cdot Q^C_{3,1-1} + P(\text{event } 2) \cdot Q^C_{3,1-2} + P(\text{Case } 2) = 0.005936.$$

Step 5: Integrate for final PMS unreliability. According to (8.54) and (8.49), $P(E_1)$ and $P(E_2)$ evaluated at step 2, $Q_1{}^C$ and $Q_2{}^C$ evaluated at step 3, and $P(E_3) \cdot Q^C_3$ evaluated at step 4 are integrated to obtain $Q(t)$, which is then integrated with $P_u(t)$ evaluated at step 1 to obtain the final PMS unreliability.

$$\begin{aligned} UR_{PMS} &= 1 - P_u(t) + Q(t) \cdot P_u(t) \\ &= 1 - P_u(t) + [P(E_1) \cdot Q^C_1 + P(E_2) \cdot Q^C_2 + P(E_3) \cdot Q^C_3] \cdot P_u(t) \\ &= 0.017854. \end{aligned}$$

8.7 Multi-Phase System with Multiple FDEP Groups

This section presents a continuous time Markov chain (CTMC)-based method for modeling competing failure propagation and isolation effects in reliability analysis of PMSs with multiple FDEP groups [18]. The exponential *ttf* distribution is assumed for system components. The LF, PFGE, PFSE of the same component are *s*-independent.

A trigger component failure in one phase, if occurring first, only makes dependent components belonging to the same FDEP group inaccessible in that phase; these dependent components are still available to use in other phases if they are accessible directly by the system function without involving the trigger component in those phases. Both PFGEs and PFSEs from the dependent component can be isolated by the trigger failure. An isolated PFGE or PFSE only affects the component itself. An isolated PFGE or PFSE in a previous phase may still propagate to other components in a later phase that does not involve operation of the related FDEP group.

8.7.1 CTMC-Based Method

In [19] a CTMC-based method was developed for the reliability analysis of PMSs without FDEP and related competing failures. This section presents an extension of the CTMC-based method for considering the competing failure effects in reliability analysis of PMSs with multiple FDEP groups.

The extended CTMC-based method involves the following three-step procedure:

Step 1: Develop a separate CTMC for each phase. In the traditional CTMC model [5], each component corresponds to one failure event. To consider different types of failures, each component in the extended CTMC-based method corresponds to up to three failure events representing the occurrence of LF, PFGE, or PFSE, respectively. The initial state is a state where none of the component failure events takes place or a state where all the system components are good. Each of the subsequent states represents a combination of component LF, PFGE, and PFSE that may occur. As the failure events occur one by one, the system transits from one state to another state until the absorbing state (mission failure in a particular phase) is reached. The transition is characterized by the occurrence rate of the related component failure event.

Note that a component not appearing in a phase FT means that the LF of this component does not contribute to the mission failure in the phase. However, the PFGE or PFSE of the component can still affect the mission failure and thus should be considered for constructing the CTMC of the phase.

Step2: Solve the CTMC of phase 1. According to (2.26), state equations of the CTMC in phase 1 are constructed. Using the initial state probabilities (1 for the initial state and 0 for all other states), the state equations can be solved using the Laplace transform method [11].

Step 3: Solve the CTMC for later phases. Starting from phase 2, the state probabilities evaluated for the previous phase are mapped as the initial state probabilities of the current phase CTMC for evaluation. Note that since compact CTMCs are used in each phase, some states may not exist in all of the mission phases. Consider such an example where an FDEP group exists in phase 1. The CTMC of phase 1 contains a system operation state (state x) where a PFGE has taken place in a dependent component but is isolated by the corresponding trigger failure. However, if this FDEP group does not exist in the later phase 2, then state x does not appear in the CTMC of phase 2. In this case, the system operation state x in phase 1 is mapped to the failure state in phase 2. Therefore, due to the competing failure behavior, special treatment should be taken during the phase-to-phase mapping process.

Repeat this step until the CTMC of all the mission phases are analyzed. The analysis of the final phase gives the final PMS unreliability. In particular, the failure state probability of the final phase is the failure probability of the entire PMS.

8.7.2 Case Study

Two examples are presented to illustrate the CTMC-based method. The PMS in Example 8.5 contains dependent FDEP groups in different phases with dependent components undergoing LFs and PFGEs. The PMS in Example 8.6 contains dependent FDEP groups with dependent components undergoing LFs, PFGEs, and PFSEs.

Example 8.5 Figure 8.23 illustrates the FT of an example PMS, where two FDEP groups from two different phases share a common trigger component. In this example system, two computers B and C work together to complete a three-phase computation

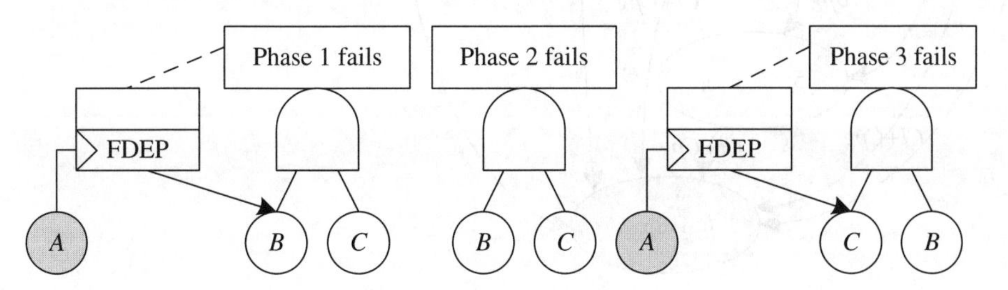

Figure 8.23 FT of an example three-phase PMS.

Table 8.8 Phase durations and component failure parameters.

	Phase 1 (4000 hrs)		Phase 2 (2000 hrs)		Phase 3 (4000 hrs)	
	LF	PFGE	LF	PFGE	LF	PFGE
A	$5e-6$	0	$2e-6$	0	$2e-6$	0
B	$3e-5$	$1e-6$	$3e-5$	$1e-6$	$1e-5$	$1e-6$
C	$1e-5$	$1e-6$	$2e-5$	$1e-6$	$3e-5$	$1e-6$

task. In phase 1, computer B needs to access the network through router A; in phase 3, computer C needs to accesses network through router A; in phase 2, only files in the local memory are needed. Table 8.8 gives the phase duration and component constant failure rates in each phase. Component A only undergoes LFs throughout the mission, while components B and C can experience LFs and PFGEs during the mission.

Figure 8.24 shows the CTMC model for phase 1, where state 0 is the initial state and state 6 is the absorbing state. Under the initial state, five different component failure events (*Al*, *Bl*, *Cl*, *Bpg*, and *Cpg*) can take place. If *Bpg* occurs and is isolated by the LF of component A (states 1, 4, and 5), it only affects component B itself instead of the entire system. Under states 2 and 3, the occurrence of *Bpg* fails the system because the trigger component A does not fail, thus the global failure propagation effect takes place. Under any state, the occurrence of *Cpg* causes the entire system failure because component

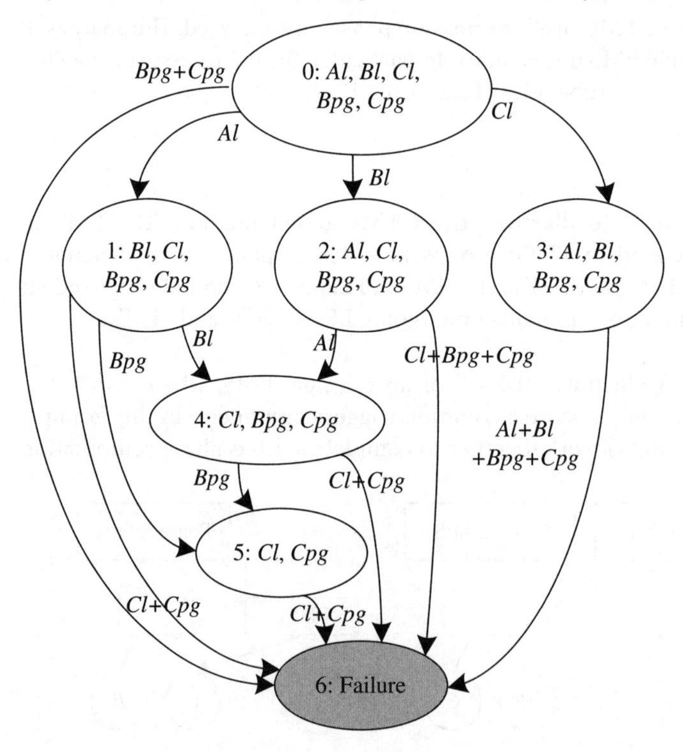

Figure 8.24 CTMC for phase 1.

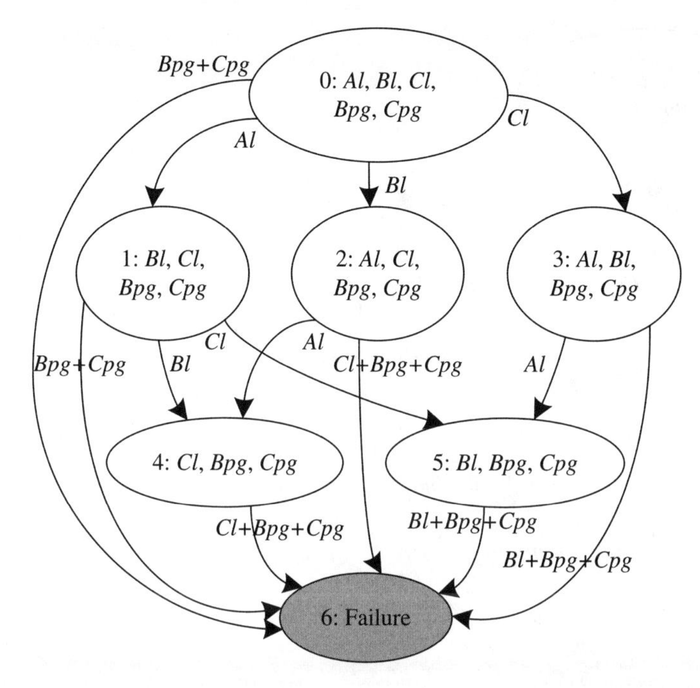

Figure 8.25 CTMC for phase 2.

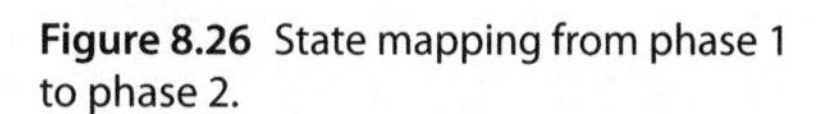
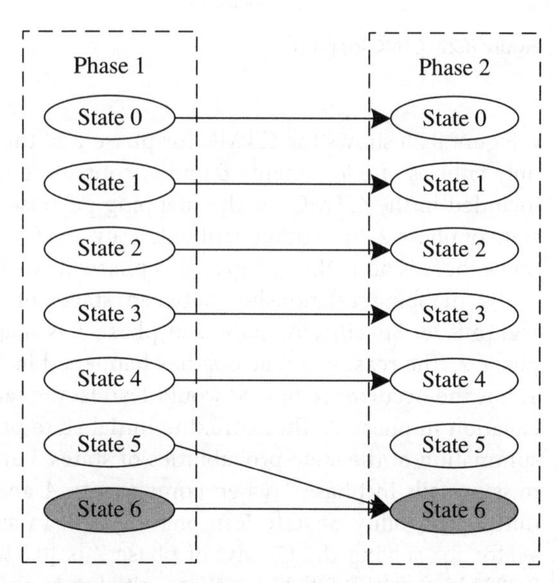

Figure 8.26 State mapping from phase 1 to phase 2.

C does not belong to any FDEP group in this phase and any PFGE from component C cannot be isolated.

Using the initial state probabilities [1, 0, 0, 0, 0, 0, 0], the state equations of phase 1 in the form of (2.26) are solved. The solution contains state probabilities at the end of phase 1 (i.e. mission time = 4000 hrs) as [0.828615, 0.016739, 0.105646, 0.033816, 0.002134, 0.000038, 0.013012].

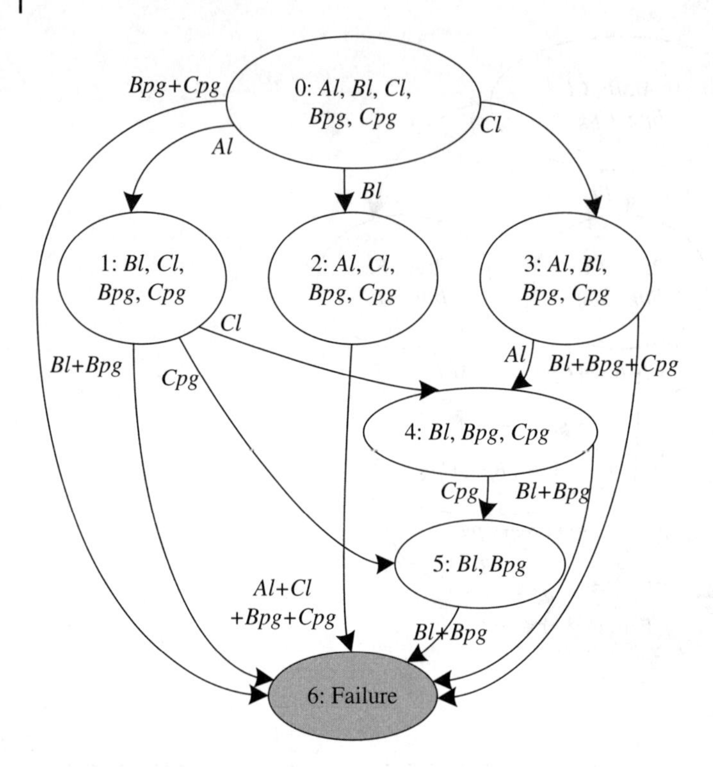

Figure 8.27 CTMC for phase 3.

Figure 8.25 shows the CTMC for phase 2 of the example PMS. In phase 2, although only failures of components *B* and *C* contribute to the system failure, event *Al* is also included in the CTMC for the mapping purpose. Therefore, under state 0, the initial state of phase 2, five component failure events (*Al*, *Bl*, *Cl*, *Bpg*, and *Cpg*) can take place. Since there is no FDEP group in this phase, if any PFGE happens, the system fails.

The mapping relationship between states of phase 1 and phase 2 is shown in Figure 8.26. Specifically, state 5 of phase 1 is mapped to the failure state (state 6) of phase 2. The reason is that *Bpg* has happened in this state and been isolated in phase 1, but the occurrence of *Bpg* would lead to the failure state because of the absence of isolation in phase 2. Therefore, the initial state probability of state 6 of phase 2 is the summation of the state probabilities of states 5 and 6 at the end of phase 1. Since the mission fails in phase 1 when components *A* and *C* both fail locally in phase 1, the initial probability of state 5 in phase 2 is 0. In summary, the initial state probability vector for solving the CTMC of phase 2 is [0.828615, 0.016739, 0.105646, 0.033816, 0.002134, 0, 0.013050]. Using these initial state probabilities, the solution of the CTMC of phase 2 in Figure 8.25 gives the state probabilities at the end of phase 2 (i.e. mission time = 2000 hrs) as [0.743787, 0.018067, 0.146688, 0.061948, 0.003563, 0.000864, 0.025083].

Figure 8.27 shows the CTMC for phase 3 of the example system. The mapping relationship between states of phase 2 in Figure 8.25 and states of phase 3 in Figure 8.27 is shown in Figure 8.28.

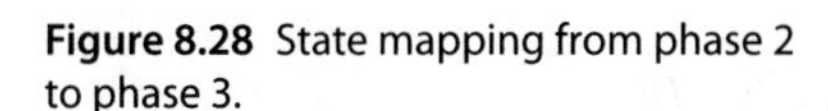
Figure 8.28 State mapping from phase 2 to phase 3.

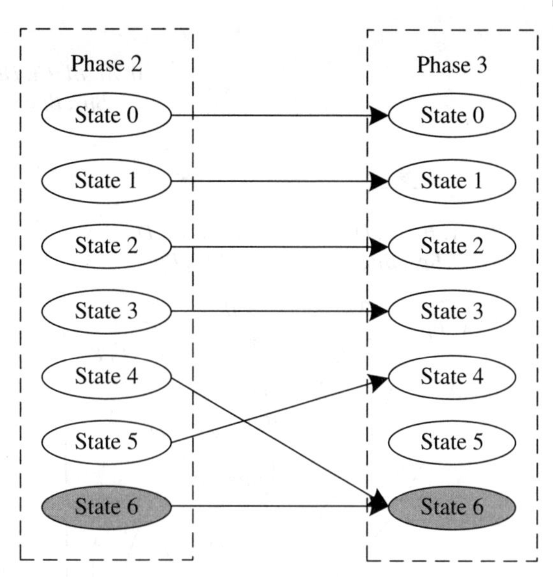

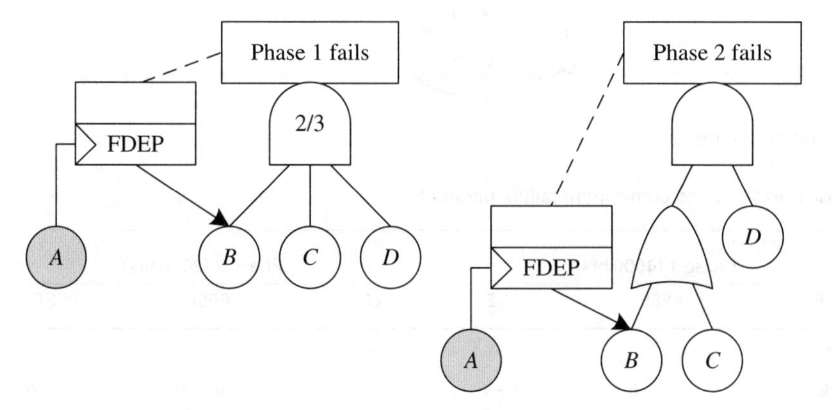

Figure 8.29 FT of the example PMS.

Based on the mapping, the initial state probability vector for solving the CTMC of phase 3 is obtained as [0.743787, 0.018067, 0.146688, 0.061948, 0.000864, 0, 0.028646]. By solving the CTMC in Figure 8.27, the state probability vector at the end of phase 3, i.e. the entire mission is obtained as [0.623754, 0.020283, 0.153491, 0.138101, 0.003880, 0.000085, 0.060407]. Thus, the unreliability of the entire PMS is 0.060407, which is the probability of state 6 at the end of the mission.

Example 8.6 Figure 8.29 shows the FT of an example two-phase computer system where computer B is suffering both PFGEs and PFSEs (e.g. caused by different types of viruses: type I and type II, respectively) during a specific mission task. The router A, computers C and D only undergo LFs during this task execution. Computer C has no protection mechanism against these two types of viruses. Computer D has a firewall that can protect it against type II viruses. Therefore, the type I virus (PFGE) originating from computer B during the mission may infect both computer C and computer D; the type

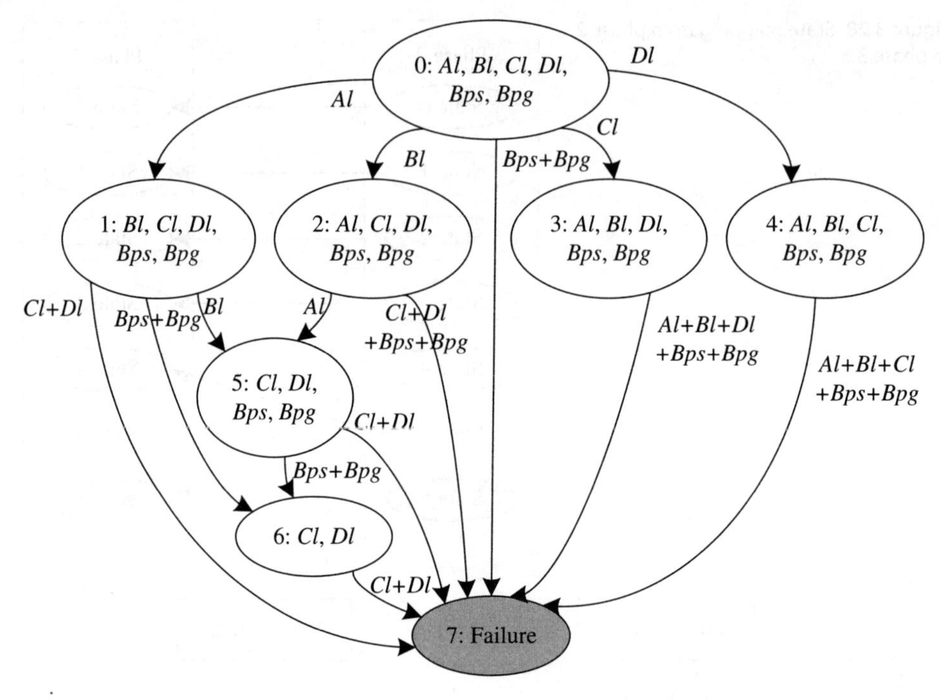

Figure 8.30 CTMC for phase 1.

Table 8.9 Phase durations and component failure parameters.

	Phase 1 (4000 hrs)			Phase 2 (2000 hrs)		
	LF	PFSE	PFGE	LF	PFSE	PFGE
A	5*e* − 6	—	—	2*e* − 6	—	—
B	3*e* − 5	3*e* − 6	1*e* − 6	3*e* − 5	3*e* − 6	1*e* − 6
C	2*e* − 5	—	—	2*e* − 5	—	—
D	1*e* − 5	—	—	2*e* − 5	—	—

II virus (PFSE) originating from computer *B* can only infect computer *C*. Table 8.9 gives failure rates of each component in the two phases.

Figure 8.30 shows the CTMC of phase 1. If *Bpg* or *Bps* happens but is isolated, it can only affect component *B* itself. Like *Bpg*, the occurrences of *Bps* (if not being isolated) causes the failure of the entire system. The reason is that *Bps* affects components *B* and *C*, which crashes the system in phase 1 (where the system fails if two of the three components fail). Specifically, under states 1 and 5, the isolation effect takes place because the trigger component *A* fails first. Under states 2, 3, and 4, the failure propagation effect takes place and the entire system fails when either *Bps* or *Bpg* occurs. Using the initial state probability vector [1, 0, 0, 0, 0, 0, 0, 0] and parameter values given in Table 8.9,

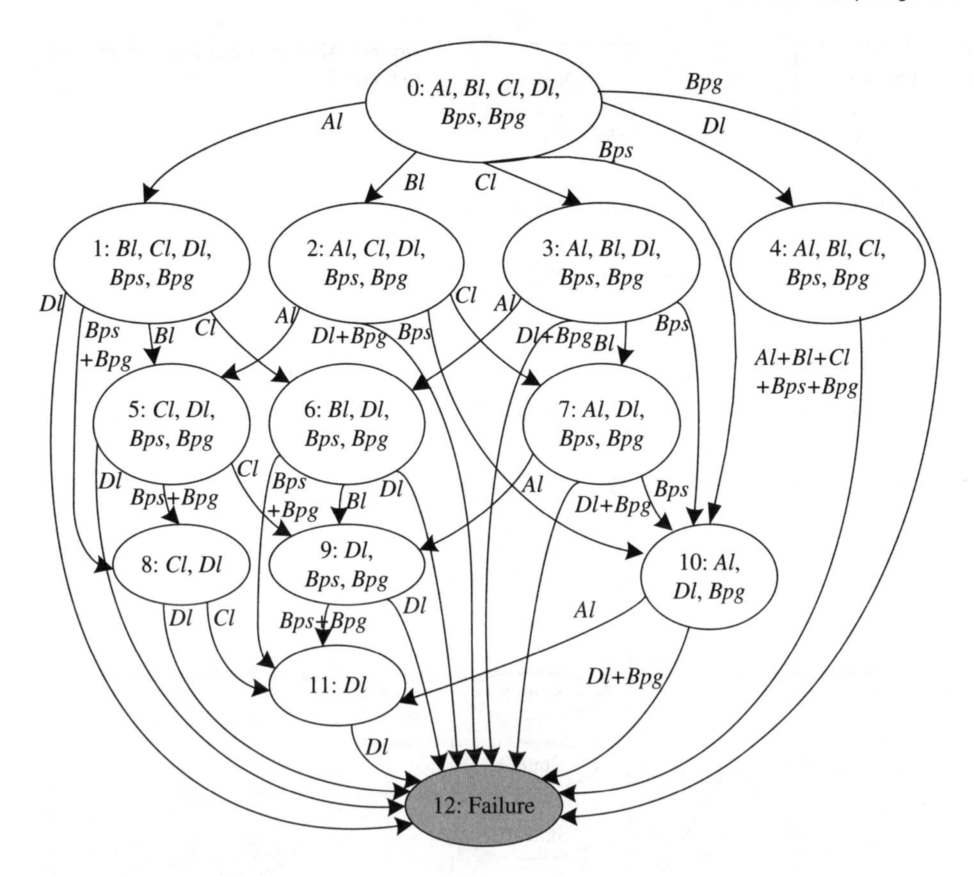

Figure 8.31 CTMC for phase 2.

the CTMC of phase 1 in Figure 8.30 is solved. The state probability vector at the end of phase 1 is obtained as [0.758813, 0.015329, 0.096746, 0.063199, 0.030968, 0.001954, 0.000139, 0.032851].

Figure 8.31 shows the CTMC of phase 2. Under states 0, 2, 3, 4, 7, and 10, the occurrence of *Bpg* causes the system failure because the trigger component *A* is still functioning and the failure isolation effect does not take place. Under states 1, 5, 6, and 9, because the LF of trigger component *A* occurs first, *Bps* and *Bpg* cannot affect other system components except component *B* itself because the isolation effect takes place.

The mapping relationship between states of phase 1 and phase 2 is illustrated in Figure 8.32. The initial state probability vector for solving the CTMC of phase 2 is thus [0.758813, 0.015329, 0.096746, 0.063199, 0.030968, 0.001954, 0, 0, 0.000139, 0, 0, 0, 0.032851]. The solution gives the state probability vector at the end of the mission as [0.651811, 0.015833, 0.128549, 0.083104, 0.054287, 0.003122, 0.000873, 0.008740, 0.000268, 0.000141, 0.005249, 0.000022, 0.048000]. Therefore, the unreliability of the PMS is 0.048.

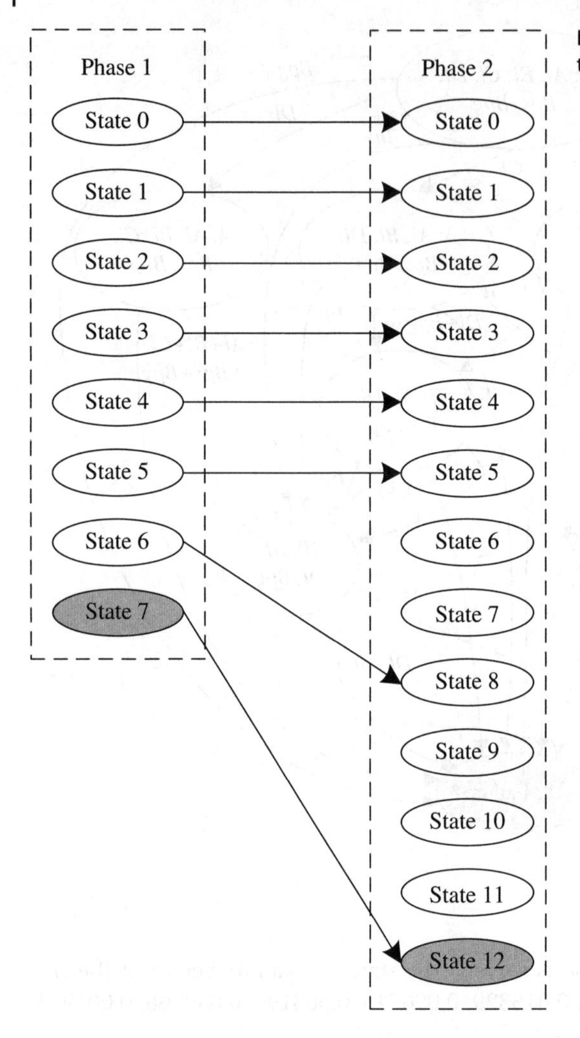

Figure 8.32 State mapping from phase 1 to phase 2.

8.8 Summary

In systems subject to the FDEP behavior, there exist competitions in the time domain between the failure isolation effect (caused by the trigger failure) and the failure propagation effect (caused by the PFGE/PFSE of the dependent components). As different occurrence sequences of the trigger failure and the dependent component PFGE/PFSE can lead to different system statuses, it is significant to address the competing effects in the system reliability analysis. Combinatorial methods are presented for reliability analysis of single-phase systems with a single FDEP group or multiple dependent FDEP groups, and for multi-phase systems with a single FDEP group. A CTMC-based method is discussed for addressing the competing effects in the reliability analysis of multi-phase systems with multiple FDEP groups.

References

1 Levitin, G. and Xing, L. (2010). Reliability and performance of multi-state systems with propagated failures having selective effect. *Reliability Engineering & System Safety* 95 (6): 655–661.
2 Amari, S.V., Dugan, J.B., and Misra, R.B. (1999). A separable method for incorporating imperfect coverage in combinatorial model. *IEEE Transactions on Reliability* 48 (3): 267–274.
3 Levitin, G. and Amari, S.V. (2007). Reliability analysis of fault tolerant systems with multi-fault coverage. *International Journal of Performability Engineering* 3 (4): 441–451.
4 Xing, L. and Dugan, J.B. (2002). Analysis of generalized phased-mission systems reliability, performance and sensitivity. *IEEE Transaction on Reliability* 51 (2): 199–211.
5 Misra, K.B. (2008). *Handbook of Performability Engineering*. London: Springer-Verlag.
6 Myers, A.F. (2007). *k*-out-of-*n*: G system reliability with imperfect fault coverage. *IEEE Transactions on Reliability* 56 (3): 464–473.
7 Shrestha, A. and Xing, L. (2008). Quantifying application communication reliability of wireless sensor networks. *International Journal of Performability Engineering, Special Issue on Reliability and Quality in Design* 4 (1): 43–56.
8 Xing, L., Wang, C., and Levitin, G. (2012). Competing failure analysis in non-repairable binary systems subject to functional dependence. *Proc IMechE, Part O: Journal of Risk and Reliability* 226 (4): 406–416.
9 Papoulis, A. (1984). *Probability, Random Variables, and Stochastic Processes*, 2e. New York, NY: McGraw-Hill.
10 Xing, L. and Levitin, G. (2010). Combinatorial analysis of systems with competing failures subject to failure isolation and propagation effects. *Reliability Engineering & System Safety* 95 (11): 1210–1215.
11 Rausand, M. and Hoyland, A. (2003). *System Reliability Theory: Models and Statistical Methods*, Wiley Series in Probability and Mathematical Statistics. Wiley.
12 Wang, C., Xing, L., and Levitin, G. (2013). Reliability analysis of multi-trigger binary systems subject to competing failures. *Reliability Engineering & System Safety* 111: 9–17.
13 Dugan, J.B. and Doyle, S.A. (1996). New results in fault-tree analysis. In: *Tutorial Notes of Annual Reliability and Maintainability Symposium*, Las Vegas, Nevada, USA.
14 Wang, C., Xing, L., and Levitin, G. (2012). Propagated failure analysis for non-repairable systems considering both global and selective effects. *Reliability Engineering & System Safety* 99: 96–104.
15 Wang, C., Xing, L., and Levitin, G. (2012). Competing failure analysis in phased-mission systems with functional dependence in one of phases. *Reliability Engineering & System Safety* 108: 90–99.
16 Zang, X., Sun, H., and Trivedi, K.S. (1999). A BDD-based algorithm for reliability analysis of phased-mission systems. *IEEE Transactions on Reliability* 48 (1): 50–60.
17 Maxfield, B. (2009). *Essential Mathcad for Engineering, Science, and Math*, 2e. Academic Press.

18 Wang, C., Xing, L., Peng, R., and Pan, Z. (2017). Competing failure analysis in phased-mission systems with multiple functional dependence groups. *Reliability Engineering & System Safety* 164: 24–33.

19 Somani, A.K., Ritcey, J.A., and Au, S.H.L. Computationally efficient phased-mission reliability analysis for systems with variable configurations. *IEEE Transactions on Reliability* 41 (4): 504–511.

9

Probabilistic Competing Failure

9.1 Overview

Chapter 8 focuses on reliability analysis of systems subject to deterministic competing failures, where the local failure (LF) of a trigger component, if it happens first, can cause a *deterministic* or *certain* isolation effect to propagated failure with global effect (PFGEs) originating from its corresponding dependent components within the same functional dependence (FDEP) group. However, in some real-world systems, such an isolation effect can be probabilistic or uncertain.

For example, consider a relay-assisted wireless sensor network (WSN) system where wireless signal attenuations can degrade the system performance significantly. Some sensors preferably deliver their sensed information to the sink device through a relay node [1]. Each component can undergo LFs (e.g. due to disabled transmissions) and PFGEs (e.g. due to jamming attacks). When the relay fails locally, each sensor may increase its transmission power to be wirelessly connected to the sink device with a certain probability related to the percentage of remaining energy. A sensor is isolated from the rest of the WSN only when its remaining energy is insufficient to enable the direct transmission to the sink device. In this case, the global propagation effect of the PFGE originating from the isolated sensor is prevented. Therefore, sensors have the probabilistic functional dependence (PFD) on the relay, and the relay LF (if happening first) causes probabilistic failure isolation effects to the corresponding sensors. The relay is referred to as the trigger component; the sensors are referred to as probabilistic-dependent (PDEP) components.

Figure 9.1 illustrates the fault tree (FT) gate introduced for modeling the PFD behavior [2]. A single trigger event T serves as the input, which corresponds to LF of one or multiple trigger components. The gate also has one or multiple PDEP components (D_i, $i = 1, \ldots, n$). The occurrence of the trigger event causes the PDEP components to become isolated (unusable or inaccessible) with certain and maybe different probabilities (q_{TDi}), modeled by the *switch symbol* in the gate.

Similar to systems with the deterministic FDEP behavior discussed in Chapter 8, time-domain competitions existing between the trigger event (probabilistic failure isolation effect) and PFGEs of the corresponding PDEP components (failure propagation effect) can lead to different system statuses. If the trigger event takes place first, the PDEP components can be isolated with different probabilities. Such an isolation effect may prevent the system function from being compromised by the PFGEs of those PDEP components. However, if any PFGE originating from the PDEP components takes place

Dynamic System Reliability: Modeling and Analysis of Dynamic and Dependent Behaviors,
First Edition. Liudong Xing, Gregory Levitin and Chaonan Wang.

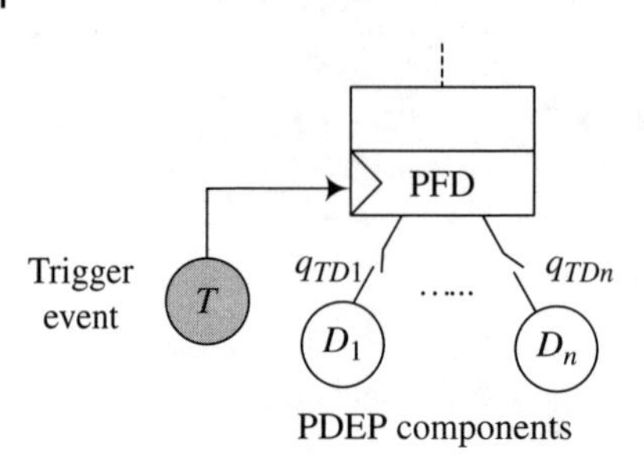

Figure 9.1 Structure of a PFD gate [2].

before the occurrence of the trigger event, the whole system fails because of the global failure propagation effect.

This chapter presents methods for modeling the probabilistic competing failure effects in the reliability analysis of different types of nonrepairable single-phase systems.

9.2 System with Single Type of Component Local Failures

A combinatorial methodology is presented for reliability analysis of nonrepairable single-phase systems subject to the PFD behavior. Each system component undergoes PFGEs and a single or identical type of LFs.

9.2.1 Combinatorial Method

The method involves a five-step procedure described as follows.

Step 1: Separate effects of PFGEs from the trigger component. According to the PFGE approach (Section 8.2), particularly (8.1), the system unreliability is evaluated as

$$UR_{system}(t) = 1 - P_u(t) + Q(t) \cdot P_u(t), \tag{9.1}$$

where $P_u(t) = P$ (no PFGEs from the trigger component T take place during the mission) $= 1 - q_{Tp}(t)$. $Q(t)$ in (9.1) is evaluated through the following steps.

Step 2: Define probabilistic functional dependence cases (PFDCs). PFDCs are identified to cover all possible isolation relationships between the trigger component and related PDEP-components belonging to the same PFD group. Their occurrence probabilities are calculated, and the set of components affected by each PFDC is also identified in this step.

Given that n PDEP components exist in the PFD group, there are 2^n disjoint PFDCs, denoted by $PFDC_i$, $i = 0, 1, \ldots, 2^n - 1$. The occurrence probability of each $PFDC_i$, denoted by $P(PFDC_i)$, is computed using parameters modeling the PFD relationships between the trigger and PDEP components, i.e. q_{TDi} in Figure 9.1. The set of components affected by each $PFDC_i$, denoted by $S(PFDC_i)$, is a subset of all PDEP components belonging to the group.

Step 3: Define failure competition events (FCEs) and evaluate their probabilities. Based on the state of the trigger component and competitions in the time domain between the trigger LF and PFGEs of the PDEP components, three FCEs are defined as follows:

FCE_1: the trigger component is functioning during the entire mission.
FCE_2: at least one PFGE originating from PDEP components occurs before the trigger LF, i.e. the global failure propagation effect takes place.
FCE_3: the trigger LF occurs before all the PFGEs originating from the PDEP components. Under this event, the probabilistic isolation effect takes place.
$P(FCE_1)$ is calculated as

$$P(FCE_1) = 1 - q_T(t). \tag{9.2}$$

$q_T(t)$ in (9.2) is the conditional failure probability of trigger T given that T undergoes no PFGEs. It is computed according to the s-relationships between the LF and PFGE of the trigger component, particularly, using (8.3) for s-independent, (8.4) for s-dependent, and (8.6) for disjoint LF and PFGE.

Based on the PFDCs defined in step 2 and the total probability law, $P(FCE_2)$ is calculated as

$$P(FCE_2) = \sum_{i=0}^{2^n-1} [P(FCE_2 \mid PFDC_i) \cdot P(PFDC_i)]. \tag{9.3}$$

Assume $S(PFDC_i)$ contains k components, whose unconditional PFGE events are represented by X_{D1p}, X_{D2p}, …, and X_{Dkp}. Let X_T denote the conditional LF event of the trigger component given that T undergoes no PFGEs. Thus according to (8.11), one obtains

$$\begin{aligned} &P(FCE_2 \mid PFDC_i) \\ &\quad = P[(X_{D1p} \to X_T) \cup (X_{D2p} \to X_T) \cup \ldots \cup (X_{Dkp} \to X_T)] \\ &\quad = P[(X_{D1p} \cup X_{D2p} \cup \ldots \cup X_{Dkp}) \to X_T] \\ &\quad = \int_0^t \int_{\tau_1}^t f_{(X_{D1p} \cup X_{D2p} \cup \ldots \cup X_{Dkp})}(\tau_1) f_T(\tau_2) d\tau_2 d\tau_1, \end{aligned}$$

where

$$\begin{aligned} f_{(X_{D1p} \cup X_{D2p} \cup \ldots \cup X_{Dkp})}(t) &= \frac{d[P(X_{D1p} \cup X_{D2p} \cup \ldots \cup X_{Dkp})]}{dt} \\ &= \frac{d[1 - P(\overline{X}_{D1p})P(\overline{X}_{D2p}) \ldots P(\overline{X}_{Dkp})]}{dt}, \end{aligned}$$

$$f_T(t) = dq_T(t)/dt. \tag{9.4}$$

The multiple integrals in (9.4) can be computed using the Mathcad tool, which performs the evaluation numerically using the Romberg algorithm [3].
When $S(PFDC_i)$ is an empty set, $P(FCE_2|PFDC_i)$ is simply 0.

Since the three FCEs are mutually exclusive and form a complete event space, $P(FCE_3)$ can be calculated as

$$P(FCE_3) = 1 - P(FCE_1) - P(FCE_2). \tag{9.5}$$

Step 4: Evaluate P(system fails|FCE_i), i = 1, 2, 3. P(system fails|FCE_1) is evaluated by applying the PFGE approach in Section 8.2 to a reduced system model, which is obtained by removing the PFD gate and its trigger input from the original system FT model.

$P(\text{system fails}|FCE_2)$ simply equals to one since the global failure propagation effect takes place, causing the entire system failure.

Under FCE_3, the probabilistic isolation effect occurs. Based on the PFDCs defined in step 2 and the total probability law, $P(\text{system fails}|FCE_3)$ is calculated as

$$P(\text{system fails}|FCE_3) = \sum_{i=0}^{2^n-1} [P(\text{system fails}|PFDC_i) \cdot P(PFDC_i)]. \tag{9.6}$$

To evaluate $P(\text{system fails}|PFDC_i)$ in (9.6), a reduced FT model is generated using the following procedure.

1. Remove the PFD gate and its trigger input from the original system FT model.
2. Replace events of PDEP components affected by $PFDC_i$, i.e. components in the set $S(PFDC_i)$ with constant 1 (TRUE).
3. Apply Boolean algebra rules to simplify the FT model.

The PFGE approach in Section 8.2 can then be applied to the reduced system model for computing $P(\text{system fails}|PFDC_i)$.

Step 5: Integrate for the final system unreliability. Based on the three FCEs and the total probability law, $Q(t)$ in (9.1) is evaluated as

$$Q(t) = \sum_{j=1}^{3} [P(\text{system fails}|FCE_j) \cdot P(FCE_j)]. \tag{9.7}$$

The final system unreliability can be obtained by integrating $Q(t)$ and $P_u(t)$ evaluated in step 1 using (9.1).

The above five-step procedure practices the "divide-and-conquer" approach, which decomposes the original reliability problem into a set of independent reduced problems. Given adequate computing resources, these reduced problems can be solved in parallel. While the method is presented for systems with one PFD group, it is applicable to systems with multiple independent PFD groups, which can be analyzed separately.

9.2.2 Case Study

Example 9.1 Consider a WSN system used for condition monitoring of a critical production room, illustrated in Figure 9.2. There are five sensors (A, B, C, D, E) and a relay node (R). Sensors A and B measure temperature, while C, D, and E measure humidity of the room. Both types of physical information (temperature and humidity) are required for a successful monitoring. The five sensors are settled on the wall at different positions. The relay node is positioned in the center of the room on the ceiling and is used to transit data acquired by sensors A and C to the sink device located lower beside the right wall. The sink device gathers all the data and communicates with the users through an external gateway.

Figure 9.3 shows the FT model of the example WSN system. Sensors A and C have PFD on the relay R. If an LF of the relay R takes place first, sensors A and C become isolated to the rest of the WSN system with conditional probabilities q_{RA} and q_{RC}, respectively. These conditional probabilities are referred to as isolation factors.

All the WSN components (sensors and the relay) undergo LFs due to transmission malfunction and PFGEs due to jamming attacks. The transmission LF and PFGE of the

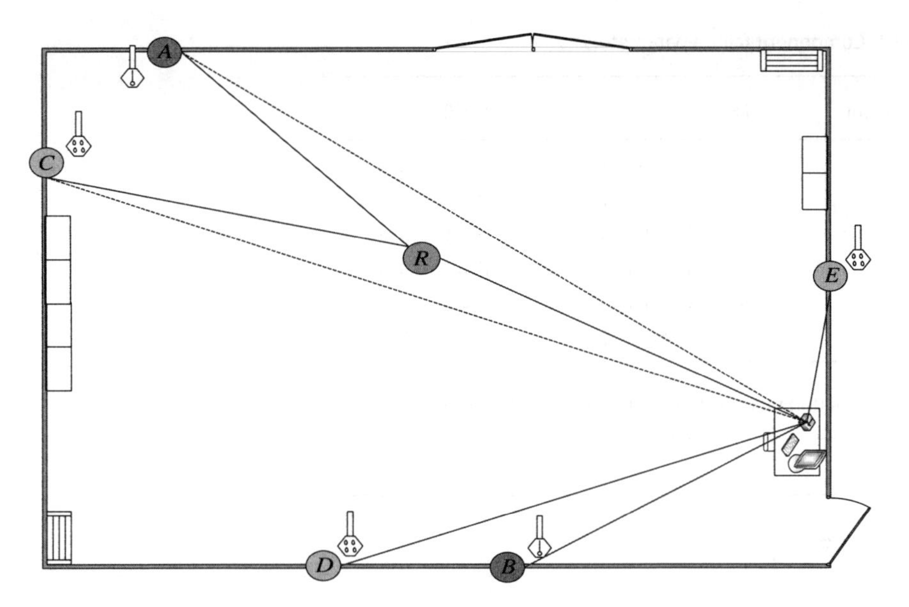

Figure 9.2 Example of a WSN for condition monitoring.

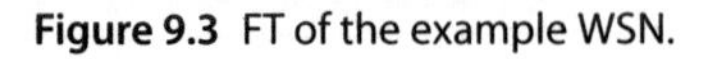

Figure 9.3 FT of the example WSN.

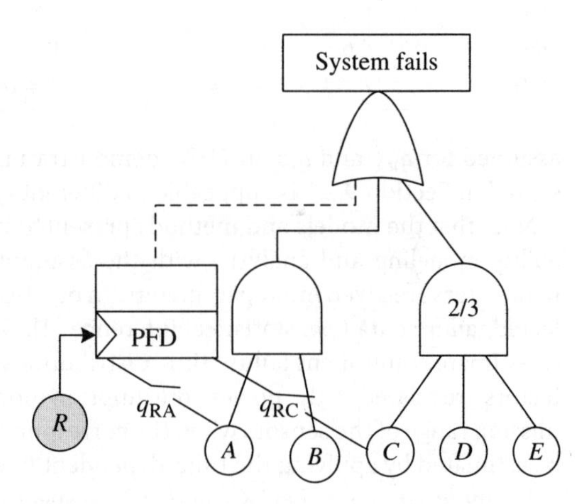

same component are disjoint or mutually exclusive because the jamming attacks are based on the transmission function (they are launched by continually transmitting interference signals from a compromised system component).

Input Parameters: Table 9.1 gives the component time to failure distribution parameters given that for all the system components these distributions are Weibull with shape parameter α and scale parameter β.

Table 9.2 gives values of isolation factors in different groups considered for illustrating effects of the probabilistic failure isolation on the system reliability. IFG_1 corresponds to the system with the deterministic FDEP. IFG_2 actually corresponds to the system without any FDEP behavior. Under IFG_3 and IFG_4, only one component has deterministic FDEP on the relay. IFG_5 and IFG_6 correspond to general cases where the two components have PFD on the relay with different probabilities. The Weibull distribution is

Table 9.1 Component failure parameters.

Component	LF	PFGE
A	$\beta = 2e-4, \alpha = 1$	$\beta = 1e-5, \alpha = 2$
B	$\beta = 2e-4, \alpha = 1$	$\beta = 1e-5, \alpha = 2$
C	$\beta = 1.5e-4, \alpha = 2$	$\beta = 1e-5, \alpha = 2$
D	$\beta = 1.5e-4, \alpha = 2$	$\beta = 1e-5, \alpha = 2$
E	$\beta = 1.5e-4, \alpha = 2$	$\beta = 1e-5, \alpha = 2$
R	$\beta = 1e-4, \alpha = 1$	$\beta = 1e-5, \alpha = 1$

Table 9.2 Isolation factor groups (IFGs) of the example WSN.

	q_{RA}	q_{RC}
IFG_1	1	1
IFG_2	0	0
IFG_3	1	0
IFG_4	0	1
IFG_5	0.1	0.9
IFG_6	0.6	0.6
IFG_7	$\beta = 1.5e-4, \alpha = 2$	$\beta = 1e-4, \alpha = 1$

assumed for q_{RA} and q_{RC} in IFG_7, demonstrating that the combinatorial method presented in Section 9.2.1 is applicable to different types of distributions.

Note that the models and methods presented in this book focus on system-level reliability modeling and analysis, with the assumption that the component-level failure parameters are given input parameters. In practice, estimation approaches based on collected failure data (e.g. statistical inference [4], Bayesian estimation [5]) can be applied to estimate component failure time distributions and related parameters. The isolation factors are related to the power consumption model of sensors and the maximum transmission range of the sensors when the corresponding relay failure takes place, which can be estimated by applying the time-dependent link failure model in [6].

Example Analysis: Let X_{il} and X_{ip}, respectively, represent the unconditional LF and PFGE events of component i; $f_{il}(t)$ and $f_{ip}(t)$, respectively, represent the probability density function (*pdf*) of time to LF and time to PFGE. As the LF and PFGE of the same component are mutually exclusive, the conditional component failure probability $q_i(t)$ given that no PFGE originating from component i occurs is computed using (8.6). The example WSN system is analyzed as follows:

Step 1: Separate effects of PFGEs from the trigger component. According to (9.1), the unreliability of the WSN system is

$$UR_{system}(t) = 1 - P_u(t) + Q(t) \cdot P_u(t), \tag{9.8}$$

where

$$P_u(t) = 1 - q_{Rp}(t). \tag{9.9}$$

Step 2: Define PFDCs, calculate their occurrence probabilities, and identify $S(PFDC_i)$. There are two PDEP components (sensors A and C). Thus, four PFDCs are defined as shown in Table 9.3.

Step 3: Define FCEs and evaluate their probabilities. The three FCEs are defined as follows:

FCE_1: relay R is operational during the entire mission. According to (8.6), $P(FCE_1)$ is calculated as

$$P(FCE_1) = P(\overline{X}_R) = 1 - q_R(t) = 1 - \frac{q_{Rl}(t)}{1 - q_{Rp}(t)} = \frac{1 - \int_0^t f_{Rp}(\tau)d\tau - \int_0^t f_{Rl}(\tau)d\tau}{1 - \int_0^t f_{Rp}(\tau)d\tau}. \tag{9.10}$$

FCE_2: at least one PFGE from A or C occurs before the LF of R. $P(FCE_2)$ is calculated using (9.3).

Specifically, under $PFDC_0$, since $S(PFDC_0) = \emptyset$, one obtains $P(FCE_2|PFDC_0) = 0$. Under $PFDC_1$, $S(PFDC_1) = \{C\}$. Thus, $P(FCE_2|PFDC_1)$ is calculated as

$$P(FCE_2|PFDC_1) = P(X_{Cp} \to X_R) = \int_0^t \int_{\tau_1}^t f_{Cp}(\tau_1) f_R(\tau_2) d\tau_2 d\tau_1. \tag{9.11}$$

Under $PFDC_2$, $S(PFDC_2) = \{A\}$. Thus, $P(FCE_2|PFDC_2)$ is calculated as

$$P(FCE_2|PFDC_2) = P(X_{Ap} \to X_R) = \int_0^t \int_{\tau_1}^t f_{Ap}(\tau_1) f_R(\tau_2) d\tau_2 d\tau_1. \tag{9.12}$$

Under $PFDC_3$, $S(PFDC_3) = \{A, C\}$. Thus, PFGEs of either A or C occur before the LF of R. $P(FCE_2|PFDC_3)$ is computed as

$$\begin{aligned} P(FCE_2|PFDC_3) &= P\{(X_{Ap} \cup X_{Cp}) \to X_R\} \\ &= \int_0^t \int_{\tau_1}^t f_{Ap}(\tau_1) f_R(\tau_2) d\tau_2 d\tau_1 + \int_0^t \int_{\tau_1}^t f_{Cp}(\tau_1) f_R(\tau_2) d\tau_2 d\tau_1 \\ &\quad - \int_0^t \int_{\tau_1}^t \int_{\tau_2}^t f_{Ap}(\tau_1) f_{Cp}(\tau_2) f_R(\tau_3) d\tau_3 d\tau_2 d\tau_1 \\ &\quad - \int_0^t \int_{\tau_1}^t \int_{\tau_2}^t f_{Cp}(\tau_1) f_{Ap}(\tau_2) f_R(\tau_3) d\tau_3 d\tau_2 d\tau_1, \end{aligned} \tag{9.13}$$

Table 9.3 PFDCs for the example WSN.

$PFDC_i$	Definition	$P(PFDC_i)$	$S(PFDC_i)$
0	$\overline{A} \cap \overline{C}$	$(1 - q_{RA})(1 - q_{RC})$	$\emptyset$
1	$\overline{A} \cap C$	$(1 - q_{RA})q_{RC}$	$\{C\}$
2	$A \cap \overline{C}$	$q_{RA}(1 - q_{RC})$	$\{A\}$
3	$A \cap C$	$q_{RA}q_{RC}$	$\{A, C\}$

where

$$f_R(\tau) = dq_R(t)/dt = d\left[\frac{\int_0^t f_{Rl}(\tau)d\tau}{1-\int_0^t f_{Rp}(\tau)d\tau}\right]/dt.$$

According to (9.3), $P(FCE_2)$ can then be computed as

$$\begin{aligned}P(FCE_2) &= \sum_{i=0}^{3}[P(FCE_2|PFDC_i)\cdot P(PFDC_i)]\\&= q_{\mathrm{RA}}\int_0^t\int_{\tau_1}^t f_{Ap}(\tau_1)f_R(\tau_2)d\tau_2 d\tau_1 + q_{RC}\int_0^t\int_{\tau_1}^t f_{Cp}(\tau_1)f_R(\tau_2)d\tau_2 d\tau_1\\&\quad -q_{\mathrm{RA}}q_{\mathrm{RC}}\int_0^t\int_{\tau_1}^t\int_{\tau_2}^t f_{Ap}(\tau_1)f_{Cp}(\tau_2)f_R(\tau_3)d\tau_3 d\tau_2 d\tau_1\\&\quad -q_{\mathrm{RA}}q_{\mathrm{RC}}\int_0^t\int_{\tau_1}^t\int_{\tau_2}^t f_{Cp}(\tau_1)f_{Ap}(\tau_2)f_R(\tau_3)d\tau_3 d\tau_2 d\tau_1,\end{aligned} \tag{9.14}$$

where

$$f_R(t) = dq_R(t)/dt = d\left[\frac{\int_0^t f_{Rl}(\tau)d\tau}{1-\int_0^t f_{Rp}(\tau)d\tau}\right]/dt.$$

FCE_3: the LF of trigger R occurs before PFGEs originating from any PDEP components. $P(FCE_3)$ can be calculated as (9.5).

Step 4: Evaluate P(system fails|FCE_i), i = 1, 2, 3. P(system fails|FCE_1): under FCE_1, no failure isolation effect exists. Figure 9.4 shows the reduced FT, which is obtained by removing the PFD gate and trigger R from the FT in Figure 9.3. Figure 9.5 shows the BDD model of the reduced FT.

Applying the PFGE method (Section 8.2), $P(\text{system fails}|FCE_1)$ is evaluated as

$$P(\text{system fails}|FCE_1) = 1 - P_{FCE_1}(t) + P_{FCE_1}(t)\cdot Q_{FCE_1}(t). \tag{9.15}$$

In (9.15), according to (8.2)

$$P_{FCE_1}(t) = (1-q_{Ap}(t))(1-q_{Bp}(t))(1-q_{Cp}(t))(1-q_{Dp}(t))(1-q_{Ep}(t)), \tag{9.16}$$

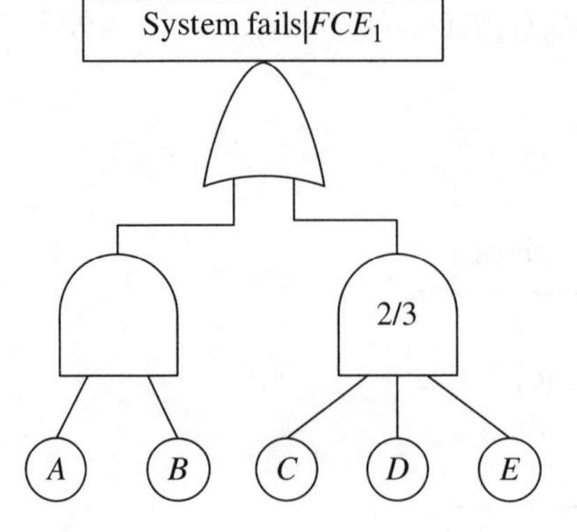

Figure 9.4 Reduced FT under FCE_1.

Figure 9.5 BDD for evaluating $Q_{FCE_1}(t)$.

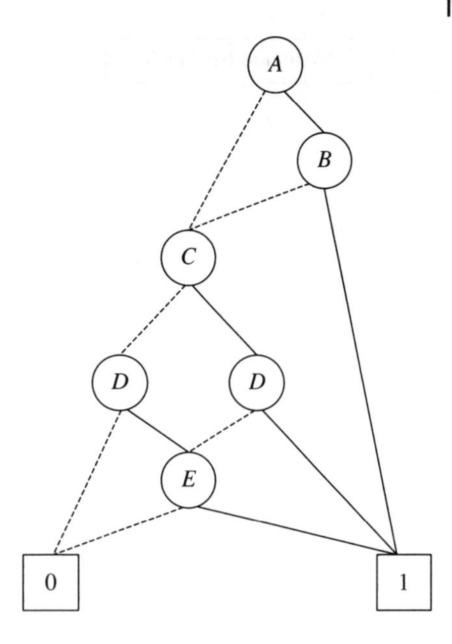

where

$$q_{ip}(t) = \int_0^t f_{ip}(\tau)d\tau, \quad i \in \{A, B, C, D, E\}. \tag{9.17}$$

Evaluating the BDD in Figure 9.5, one obtains

$$Q_{FCE_1}(t) = q_A(t)q_B(t) + [1 - q_A(t)q_B(t)] \cdot \begin{bmatrix} q_C(t)q_D(t) + q_C(t)q_E(t) \\ +q_D(t)q_E(t) - 2q_C(t)q_D(t)q_E(t) \end{bmatrix}, \tag{9.18}$$

where

$$q_i(t) = \frac{q_{il}(t)}{1 - q_{ip}(t)} = \frac{\int_0^t f_{il}(\tau)d\tau}{1 - \int_0^t f_{ip}(\tau)d\tau}. \tag{9.19}$$

P(system fails$|FCE_2$): due to the global failure propagation effect, P(system fails$|FCE_2$).

P(system fails$|FCE_3$): under FCE_3, the probabilistic isolation effect occurs. P(system fails$|FCE_3$) is calculated as (9.6).

Specifically, under $PFDC_0$, neither A nor C is isolated by the LF of R. The reduced FT model generated under this case is the same as that in Figure 9.4. Thus, P(system fails$|PFDC_0$) is the same as P(system fails$|FCE_1$) in (9.15).

Under $PFDC_1$, only C is isolated by the LF of R. The reduced FT model generated under this case is shown in Figure 9.6. The BDD mode of the reduced FT is shown in Figure 9.7. Applying the PFGE method, P(system fails$|PFDC_1$) is evaluated as

$$P(\text{system fails}|PFDC_1) = 1 - P_{PFDC_1}(t) + P_{PFDC_1}(t) \cdot Q_{PFDC_1}(t), \tag{9.20}$$

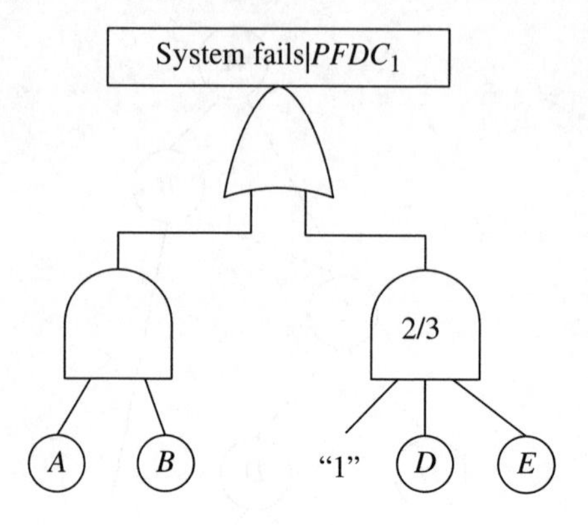

Figure 9.6 Reduced FT under $PFDC_1$.

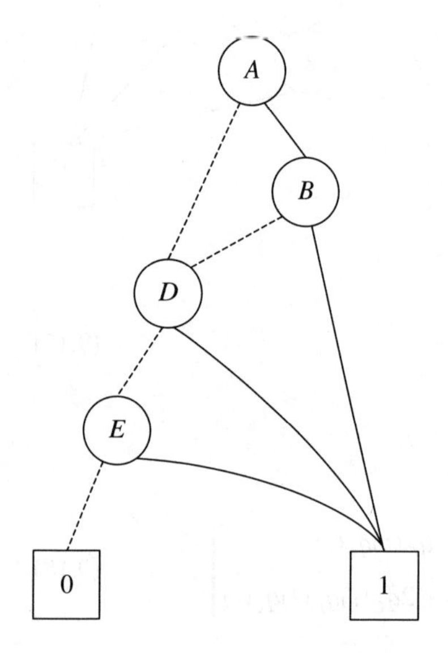

Figure 9.7 BDD for evaluating $Q_{PFDC_1}(t)$.

where

$$P_{PFDC_1}(t) = (1 - q_{Ap}(t))(1 - q_{Bp}(t))(1 - q_{Dp}(t))(1 - q_{Ep}(t)),$$

$$Q_{PFDC_1}(t) = q_A(t)q_B(t) + (1 - q_A(t)q_B(t))(q_D(t) + q_E(t) - q_D(t)q_E(t)).$$

Under $PFDC_2$, only A is isolated by the LF of R. The reduced FT model generated under this case is shown in Figure 9.8. The BDD mode of the reduced FT is shown in Figure 9.9. Applying the PFGE method, $P(\text{system fails}|PFDC_2)$ is evaluated as

$$P(\text{system fails}|PFDC_2) = 1 - P_{PFDC_2}(t) + P_{PFDC_2}(t) \cdot Q_{PFDC_2}(t), \tag{9.21}$$

where

$$P_{PFDC_2}(t) = (1 - q_{Bp}(t))(1 - q_{Cp}(t))(1 - q_{Dp}(t))(1 - q_{Ep}(t)),$$

$$Q_{PFDC_2}(t) = q_B(t) + (1 - q_B(t)) \begin{pmatrix} q_C(t)q_D(t) + q_C(t)q_E(t) + q_D(t)q_E(t) \\ -2q_C(t)q_D(t)q_E(t) \end{pmatrix}.$$

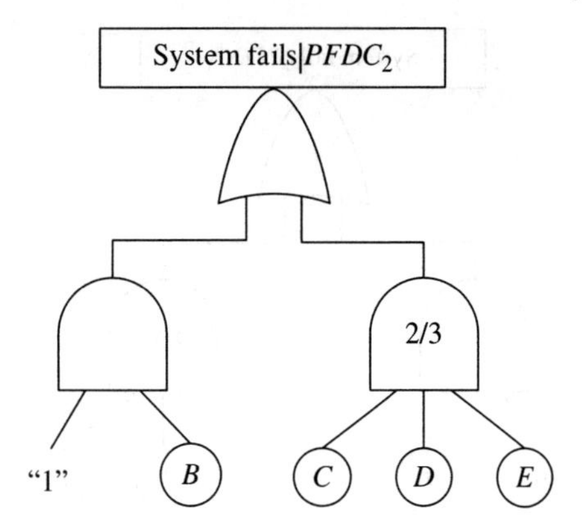

Figure 9.8 Reduced FT under $PFDC_2$.

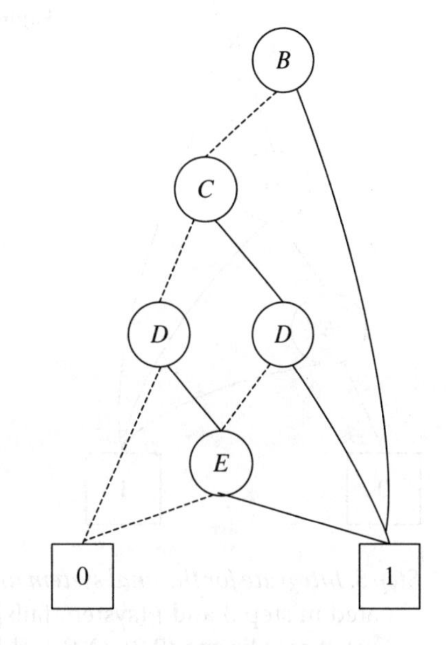

Figure 9.9 BDD for evaluating $Q_{PFDC_2}(t)$.

Under $PFDC_3$, both A and C are isolated by the LF of R. Figure 9.10 shows the reduced FT model generated under this case. Figure 9.11 shows the BDD generated for the reduced FT. Applying the PFGE method (Section 8.2), P(system fails|$PFDC_3$) is evaluated as

$$P(\text{system fails}|PFDC_3) = 1 - P_{PFDC_3}(t) + P_{PFDC_3}(t) \cdot Q_{PFDC_3}(t), \tag{9.22}$$

where

$$P_{PFDC_3}(t) = (1 - q_{Bp}(t))(1 - q_{Dp}(t))(1 - q_{Ep}(t)),$$

$$Q_{PFDC_3}(t) = q_B(t) + [1 - q_B(t)][q_D(t) + q_E(t) - q_D(t)q_E(t)].$$

According to (9.6), P(system fails|FCE_3) can be obtained by integrating $P(PFDC_i)$ in Table 9.3 and P(system fails|$PFDC_i$) calculated using (9.15), (9.20), (9.21), and (9.22).

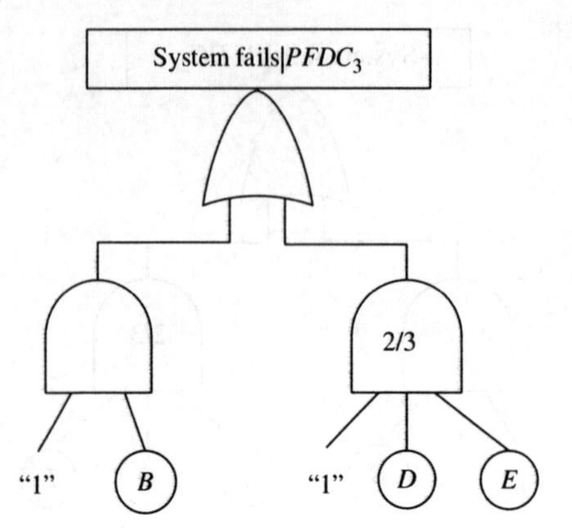

Figure 9.10 Reduced FT under $PFDC_3$.

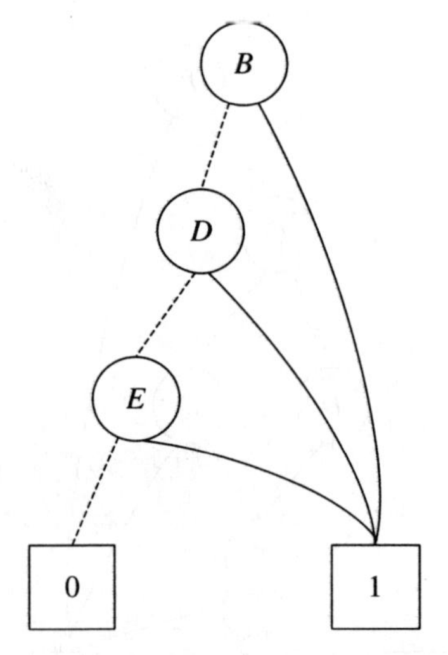

Figure 9.11 BDD for evaluating $Q_{PFDC_3}(t)$.

Step 5: Integrate for the final system unreliability. Finally, according to (9.7), $P(FCE_i)$ evaluated in step 3 and $P(\text{system fails}|FCE_i)$ evaluated in step 4 are integrated to obtain $Q(t)$. According to (9.8), $Q(t)$ and $P_u(t)$ calculated using (9.9) are integrated to obtain the final unreliability of the example WSN system.

Evaluation Results: Table 9.4 gives the unreliability of the example WSN system using parameters listed in Table 9.1 and Table 9.2 for three different values of the mission time.

The LF of a trigger component (if happening first) causes two-sided effects. On one hand, it prevents propagation of PFGEs originating from the dependent components, which could improve the overall system reliability; on the other hand, these dependent components are isolated and become unusable, which could deteriorate the system performance. This deterioration effect is considered by replacing the isolated PDEP components with 1 for generating the reduced system FT.

Comparing unreliability results under IFG_1 (both components are isolated deterministically) and IFG_2 (both components are not isolated at all), the unreliability of the example WSN under IFG_2 is actually lower than that under IFG_1. This is because the deterioration effect of the failure isolation under the given component failure

Table 9.4 Unreliability of the example WSN.

t (h)	1000	5000	10 000
IFG_1	0.061 766	0.755 232	0.997 676
IFG_2	0.044 367	0.661 346	0.994 672
IFG_3	0.058 458	0.715 533	0.996 393
IFG_4	0.048 275	0.726 089	0.997 058
IFG_5	0.049 239	0.722 786	0.996 893
IFG_6	0.054 951	0.723 704	0.996 741
IFG_7	0.045 051	0.705 911	0.997 097

parameter values contributes to the system unreliability more than the improvement effect.

9.3 System with Multiple Types of Component Local Failures

Some system components may experience multiple types of LFs. For example, a sensor in a WSN system can undergo two types of LFs, which respectively disable the transmission and sensing function of the sensor node. Correspondingly, they are referred to as transmission LFs and sensing LFs. A relay is subject to only transmission LFs based on its function performed. Both sensors and relay nodes may undergo PFGEs due to jamming attacks. They are launched by continually transmitting interference signals so as to block other WSN components, crashing the entire system [7]. Since the jamming attacks depend on the transmission function of the compromised component, the PFGE and transmission LF of the same component are disjoint or mutually exclusive. However, the PFGE and sensing LF of the same component are *s*-independent. Figure 9.12 summarizes the *s*-relationship among the different types of component failures.

A combinatorial method, which extends the method of Section 9.2.1, is presented to handle system components undergoing two types of LFs, which may have different statistical relationships with PFGEs originating from the same component.

9.3.1 Combinatorial Method

The combinatorial method involves the following five-step procedure.

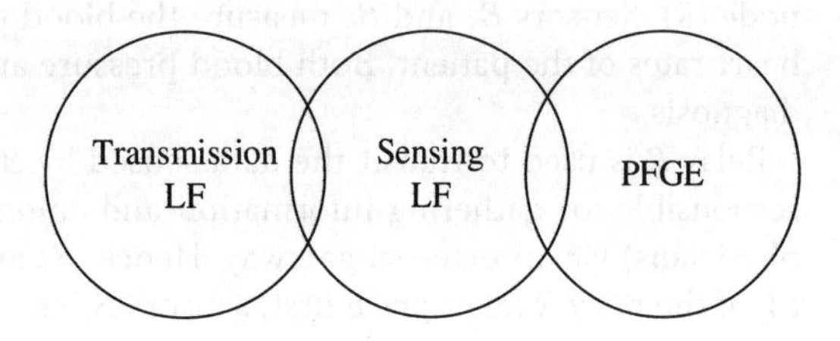

Figure 9.12 *s*-Relationships among PFGE, transmission LF and sensing LF.

Step 1: Separate effects of PFGEs from the trigger component. According to the PFGE approach (Section 8.2), the system unreliability is evaluated as (8.1), i.e.

$$UR_{system}(t) = 1 - P_u(t) + Q(t) \cdot P_u(t), \tag{9.23}$$

where $P_u(t) = P$(no PFGEs from the trigger, e.g. denoted by T take place during the mission) $= 1 - q_{Tp}(t)$. $Q(t)$ is evaluated through the following steps.

Step 2: Define PFDCs. Similar to the PFDC and FCE definitions in Section 9.2.1, 2^n disjoint and complete PFDCs are identified, given that n PDEP components involved in the PFD group.

Step 3: Define FCEs and evaluate their occurrence probabilities. Three FCEs are defined as follows:

FCE_1: the trigger component is functioning during the entire mission.

FCE_2: at least one PFGE originating from PDEP components occurs before the trigger LF.

FCE_3: the trigger LF occurs before all the PFGEs originating from the PDEP components.

$P(FCE_i)$ can be evaluated using the same method presented in Section 9.2.1. Particularly, (9.2) for $P(FCE_1)$, (9.3) for $P(FCE_2)$, and (9.5) for $P(FCE_3)$.

Step 4: Evaluate P(system fails|FCE_i), $i = 1, 2, 3$. For each component undergoing multiple types of LFs, a logic OR gate is added to the system FT, which connects these different types of LFs to represent the total LF of the component. The conditional component LF probabilities are computed based on statistical relationships between each type of LF and the PFGE of the component. Specifically, (8.3) for *s*-independent, (8.4) for *s*-dependent, and (8.6) for disjoint LF and PFGE.

Based on the FT expanded with the logic OR gates for all components undergoing multiple types of LFs, P(system fails|FCE_i) can be evaluated using the same method presented in step 4 in Section 9.2.1.

Step 5: Integrate for the final system unreliability. Based on the three FCEs defined and the total probability law, $Q(t)$ is evaluated as

$$Q(t) = \sum_{j=1}^{3} [P(\text{system fails}|FCE_j) \cdot P(FCE_j)]. \tag{9.24}$$

$Q(t)$ is then combined with $P_u(t)$ computed in step 1 using (9.23) to obtain the final system unreliability.

9.3.2 Case Study

Example 9.2 Figure 9.13 illustrates an example of a body sensor network (BSN) system for patient monitoring. There are four biomedical sensors (B_1, B_2, H_1, H_2) and a relay node (R). Sensors B_1 and B_2 measure the blood pressure, sensors H_1 and H_2 measure heart rates of the patient. Both blood pressure and heart rates are required for a valid diagnosis.

Relay R is used to transit the data sensed by B_1 and H_1 to the sink device, which is responsible for gathering information and communicating with users (e.g. nurses or physicians) via an external gateway. Hence, B_1 and H_1 have PFD on the relay R. If an LF of the relay R takes place first, sensors B_1 and H_1 become isolated to the rest of the

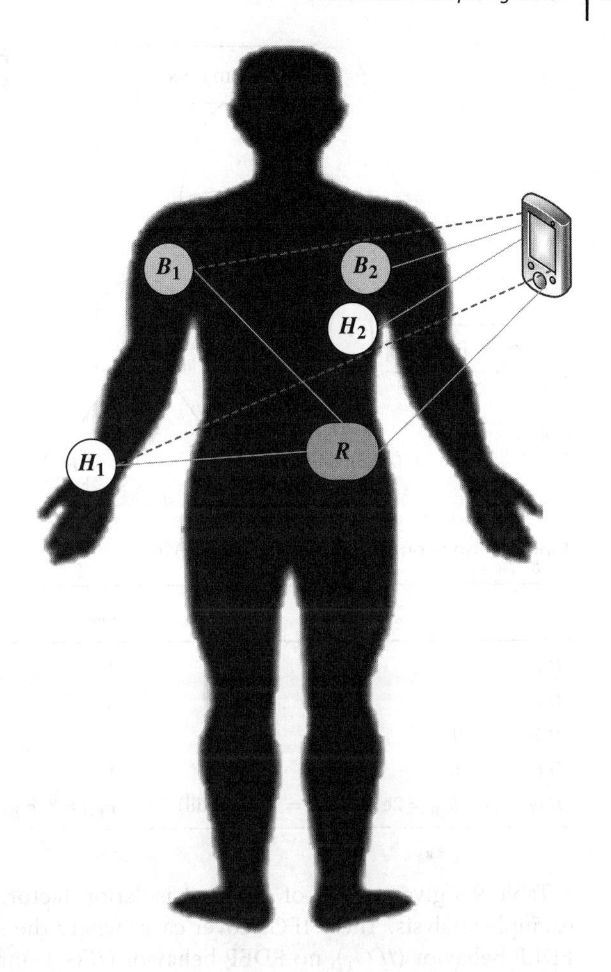

Figure 9.13 An example of a BSN system for patient monitoring.

BSN system with conditional probabilities q_{RB} and q_{RH}, respectively. Figure 9.14 shows the FT model of the example BSN system.

Input Parameters: Table 9.5 lists time to failure distribution parameters given that these distributions for sensing LF (LF_S), transmission LF (LF_T) and PFGE of the BSN components are Weibull with scale parameter β and shape parameter α. The *pdf* of the Weibull distribution is $f(t) = \alpha(\beta)^{\alpha}t^{\alpha-1}e^{-(\beta t)^{\alpha}}$. Note that the Weibull distribution is reduced to an exponential distribution when $\alpha = 1$.

Table 9.5 Component failure parameters for the BSN system.

	LF_S	LF_T	PFGE
H_1	$\beta = 2.5e-4, \alpha = 2$	$\beta = 3.5e-4, \alpha = 2$	$\beta = 3e-5, \alpha = 2$
H_2	$\beta = 2.5e-4, \alpha = 2$	$\beta = 3.5e-4, \alpha = 2$	$\beta = 3e-5, \alpha = 2$
B_1	$\beta = 2.2e-4, \alpha = 1$	$\beta = 3.2e-4, \alpha = 1$	$\beta = 3e-5, \alpha = 2$
B_2	$\beta = 2.2e-4, \alpha = 1$	$\beta = 3.2e-4, \alpha = 1$	$\beta = 3e-5, \alpha = 2$
R	–	$\beta = 2e-4, \alpha = 2$	$\beta = 3e-5, \alpha = 2$

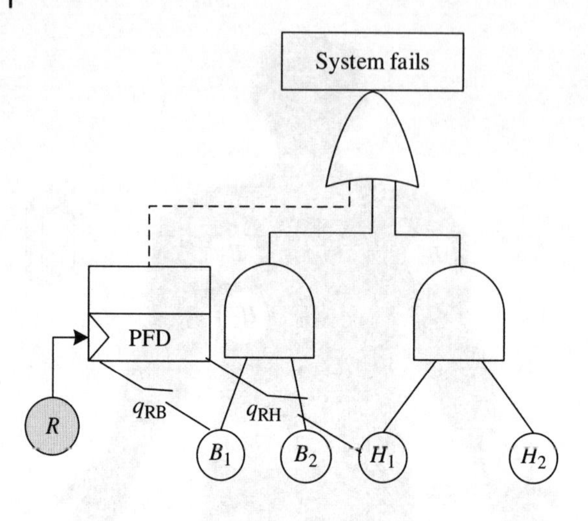

Figure 9.14 FT of the example BSN system.

Table 9.6 Isolation factors of the example BSN.

	q_{RB}	q_{RH}
IFG_1	1	1
IFG_2	0	0
IFG_3	1	0
IFG_4	0	1
IFG_5	$\beta_{RB} = 2e-4$, $\alpha_{RB} = 1$ (Weibull)	$\mu_{RH} = 8$, $\sigma_{RH} = 0.6$ (Lognormal)

Table 9.6 gives values of different isolation factor group (IFGs) considered for the example analysis. These IFGs cover cases where the system is subject to deterministic FDEP behavior (IFG_1), no FDEP behavior (IFG_2), and only one component has deterministic FDEP (IFG_3 and IFG_4). IFG_5 represents a general case with multiple components having PFD on the trigger with different probabilities.

Example Analysis: Let $X_{iLS}(t)$ and $X_{iLT}(t)$, respectively, denote the unconditional sensing and transmission LF events for all of the biomedical sensors, $i \in \{B_1, B_2, H_1, H_2\}$. $X_{RLT}(t)$ denotes the unconditional transmission LF event of relay R. The PFGE event of a system component is represented as $X_{ip}(t), i \in \{B_1, B_2, H_1, H_2, R\}$. The time to occurrence of $X_{iLS}(t)$, $X_{iLT}(t)$, and $X_{ip}(t)$ has *pdf* of $f_{iLS}(t), f_{iLT}(t)$, and $f_{ip}(t)$, respectively. The occurrence probabilities of these different types of failure events are denoted by $q_{iLS}(t)$, $q_{iLT}(t)$, and $q_{ip}(t)$, respectively. $X_i(t), i \in \{B_1, B_2, H_1, H_2, R\}$ denotes the conditional failure event of component i given that no PFGEs occur to the component; its occurrence probability is denoted by $q_i(t)$.

The example BSN system is analyzed as follows:

Step 1: Separate effects of PFGEs from the trigger R. $P_u(t)$ in (9.23) is evaluated as $1 - q_{Rp}(t)$. $Q(t)$ in (9.23) is evaluated through the following steps.

Step 2: Define PFDCs. Since there are two PDEP components (B_1 and H_1), four disjoint PFDCs are defined in Table 9.7. The occurrence probability and set of components isolated under each PFDC are also given in Table 9.7.

Table 9.7 PFDCs for the example BSN.

$PFDC_i$	Definition	$P(PFDC_i)$	$S(PFDC_i)$
0	$\overline{B}_1 \cap \overline{H}_1$	$(1 - q_{RB})(1 - q_{RH})$	$\emptyset$
1	$\overline{B}_1 \cap H_1$	$(1 - q_{RB})q_{RH}$	$\{H_1\}$
2	$B_1 \cap \overline{H}_1$	$q_{RB}(1 - q_{RH})$	$\{B_1\}$
3	$B_1 \cap H_1$	$q_{RB}q_{RH}$	$\{B_1, H_1\}$

Step 3: Define FCEs and evaluate their occurrence probabilities. The three FCEs are defined as:

FCE_1: trigger R is operational during the entire mission.

FCE_2: at least one PFGE originating from the PDEP sensors occurs before the trigger LF.

FCE_3: the trigger LF occurs before PFGEs originating from any PDEP sensors (B_1 and H_1).

Since the LF and PFGE of relay R are mutually exclusive, according to (9.2) and (8.6) the occurrence probability of FCE_1 is

$$P(FCE_1) = 1 - q_R(t) = 1 - \frac{q_{RLT}(t)}{1 - q_{Rp}(t)} = \frac{1 - \int_0^t f_{Rp}(\tau)d\tau - \int_0^t f_{RLT}(\tau)d\tau}{1 - \int_0^t f_{Rp}(\tau)d\tau} \tag{9.25}$$

According to (9.3), $P(FCE_2)$ is computed as

$$P(FCE_2) = \sum_{k=0}^{3}[P(FCE_2|PFDC_k) \cdot P(PFDC_k)], \tag{9.26}$$

where $P(PFDC_k)$ is given in Table 9.7 and $P(FCE_2|PFDC_k)$ is evaluated as follows.

Under $PFDC_0$, there is no dependent component in the PFD group. Thus, $P(FCE_2|PFDC_0)$ equals to 0.

Under $PFDC_1$, FCE_2 happens when the PFGE of the dependent component H_1 occurs before the LF of R. Thus,

$$P(FCE_2|PFDC_1) = P(X_{H_1p} \rightarrow X_R) = \int_0^t \int_{\tau_1}^t f_{H_1p}(\tau_1)f_R(\tau_2)d\tau_2 d\tau_1. \tag{9.27}$$

Under $PFDC_2$, FCE_2 happens when the PFGE of the dependent component B_1 occurs before the LF of R. Thus,

$$P(FCE_2|PFDC_2) = P(X_{B_1p} \rightarrow X_R) = \int_0^t \int_{\tau_1}^t f_{B_1p}(\tau_1)f_R(\tau_2)d\tau_2 d\tau_1 \tag{9.28}$$

Under $PFDC_3$, both B_1 and H_1 are dependent components. $P(FCE_2|PFDC_3)$ is evaluated as

$$\begin{aligned} P(FCE_2|PFDC_3) &= P[(X_{B_1p} \rightarrow X_R) \cup (X_{H_1p} \rightarrow X_R) = P[(X_{B_1p} \cup X_{H_1p}) \rightarrow X_R]] \\ &= \int_0^t \int_{\tau_1}^t f_{(X_{B_1p} \cup X_{H_1p})}(\tau_1)f_{X_R}(\tau_2)d\tau_2 d\tau_1, \end{aligned} \tag{9.29}$$

where

$$f_{(X_{B_1p} \cup X_{H_1p})}(t) = d[1 - (1 - q_{B_1p}(t))(1 - q_{H_1p}(t))]/dt,$$

$$f_R(t) = \frac{dq_R(t)}{dt}.$$

With $P(FCE_1)$ and $P(FCE_2)$, $P(FCE_3)$ is calculated as

$$P(FCE_3) = 1 - P(FCE_1) - P(FCE_2).$$

Step 4: Evaluate P(system fails | FCE_i), $i = 1, 2, 3$. Under FCE_1, the failure isolation does not occur. Figure 9.15 shows the FT for evaluating $P(\text{system fails}|FCE_1)$, which is obtained by expanding the original FT with component logic OR gate connecting two different types of LFs for each biomedical sensor and removing the PFD gate and its trigger from the original FT in Figure 9.14. Particularly, each OR gate at the bottom level models that either a sensing failure or a transmission failure causes the sensor's inability to provide information for diagnosis. Figure 9.16 shows the BDD model generated from the FT of Figure 9.15.

Applying the PFGE approach in Section 8.2, $P(\text{system fails}|FCE_1)$ is computed as

$$P(\text{system fails}|FCE_1) = 1 - P_{FCE_1}(t) + P_{FCE_1}(t) \cdot Q_{FCE_1}(t) \tag{9.30}$$

According to (8.2), $P_{FCE_1}(t)$ in (9.30) is computed as

$$P_{FCE_1}(t) = (1 - q_{B_1p}(t))(1 - q_{B_2p}(t))(1 - q_{H_1p}(t))(1 - q_{H_2p}(t)). \tag{9.31}$$

Because the sensing LF and PFGE of each sensor are *s*-independent, according to (8.3), given that no PFGEs occur, the conditional probability of the sensing LF for sensor i is

$$q_{iS}(t) = q_{iLS}(t) = \int_0^t f_{iLS}(\tau)d\tau, i \in \{B_1, B_2, H_1, H_2\}. \tag{9.32}$$

Because the transmission LF and PFGE of each sensor are mutually exclusive, according to (8.6), given that no PFGEs occur, the conditional probability of the transmission

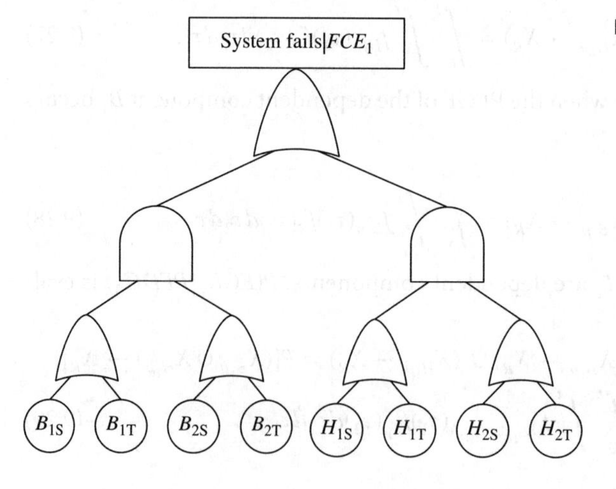

Figure 9.15 FT under FCE_1.

Figure 9.16 BDD under FCE_1.

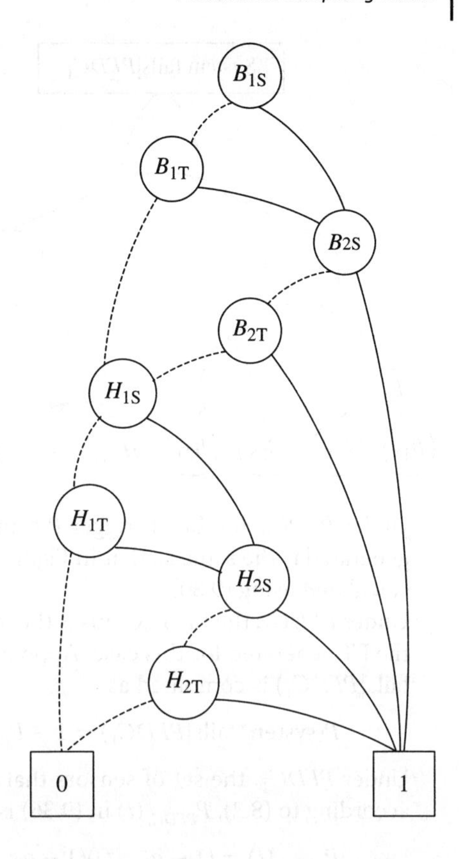

LF for sensor i is

$$q_{iT}(t) = \frac{q_{iLT}(t)}{1 - q_{ip}(t)} = \frac{\int_0^t f_{iLT}(\tau)d\tau}{1 - \int_0^t f_{ip}(\tau)d\tau}, i \in \{B_1, B_2, H_1, H_2\}. \tag{9.33}$$

Using these conditional LF probabilities, the BDD model in Figure 9.16 is evaluated to obtain $Q_{FCE_1}(t)$ in (9.30) as

$$\begin{aligned} Q_{FCE_1}(t) = {} & [q_{B_1S}(t) + (1 - q_{B_1S}(t))q_{B_1T}(t)][q_{B_2S}(t) + (1 - q_{B_2S}(t))q_{B_2T}(t)] \\ & + \{1 - [q_{B_1S}(t) + (1 - q_{B_1S}(t))q_{B_1T}(t)][q_{B_2S}(t) + (1 - q_{B_2S}(t))q_{B_2T}(t)]\} \\ & \cdot [q_{H_1S}(t) + (1 - q_{H_1S}(t))q_{H_1T}(t)][q_{H_2S}(t) + (1 - q_{H_2S}(t))q_{H_2T}(t)]. \end{aligned} \tag{9.34}$$

Under FCE_2, the global failure propagation effect takes place. Hence, $P(\text{system fails}|FCE_2) = 1$.

Under FCE_3, the probabilistic failure isolation effect takes place. According to (9.6), $P(\text{system fails}|FCE_3)$ is evaluated as

$$P(\text{system fails}|FCE_3) = \sum_{k=0}^{3}[P(\text{system fails}|PFDC_k) \cdot P(PFDC_k)]. \tag{9.35}$$

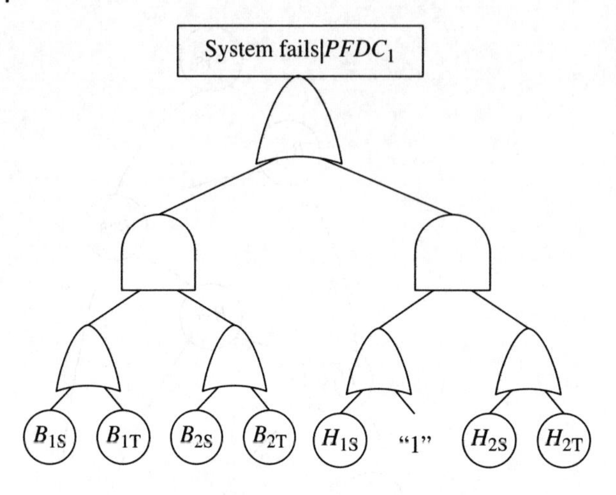

Figure 9.17 Reduced FT under $PFDC_1$.

Under $PFDC_0$, the LF of trigger R cannot cause any isolation effect. The reduced FT generated is the same as that in Figure 9.15. Therefore, $P(\text{system fails}|PFDC_0)$ can be calculated using (9.30).

Under $PFDC_1$, the LF of R causes the failure isolation to sensor H_1. Figure 9.17 shows the FT generated for this case. Applying the PFGE approach in Section 8.2, $P(\text{system fails}|PFDC_1)$ is computed as

$$P(\text{system fails}|PFDC_1) = 1 - P_{PFDC_1}(t) + P_{PFDC_1}(t) \cdot Q_{PFDC_1}(t). \tag{9.36}$$

Under $PFDC_1$, the set of sensors that can cause the PFGE is $I = \{B_1, B_2, H_2\}$. Thus, according to (8.2), $P_{PFDC_1}(t)$ in (9.36) is computed as

$$P_{PFDC_1}(t) = (1 - q_{B_1p}(t))(1 - q_{B_2p}(t))(1 - q_{H_2p}(t)).$$

The BDD evaluation of the FT in Figure 9.17 using the conditional probabilities computed using (9.32) and (9.33) gives $Q_{PFDC_1}(t)$ in (9.36) as

$$\begin{aligned} Q_{PFDC_1}(t) = 1 &- (1 - q_{H_2S}(t))(1 - q_{H_2T}(t)) + (1 - q_{H_2S}(t))(1 - q_{H_2T}(t)) \\ &\cdot [q_{B_1S}(t) + (1 - q_{B_1S}(t))q_{B_1T}(t)][q_{B_2S}(t) + (1 - q_{B_2S}(t))q_{B_2T}(t)]. \end{aligned}$$

Under $PFDC_2$, the LF of R causes the failure isolation to sensor B_1. Figure 9.18 shows the FT generated for this case. Applying the PFGE approach in Section 8.2, $P(\text{system fails}|PFDC_2)$ is computed as

$$P(\text{system fails}|PFDC_2) = 1 - P_{PFDC_2}(t) + P_{PFDC_2}(t) \cdot Q_{PFDC_2}(t). \tag{9.37}$$

Under $PFDC_2$, the set of sensors that can cause the PFGE is $I = \{B_2, H_1, H_2\}$. Thus according to (8.2), $P_{PFDC_2}(t)$ in (9.37) is computed as

$$P_{PFDC_2}(t) = \prod_{\forall i \in I} [1 - q_{ip}(t)] = (1 - q_{B_2p}(t))(1 - q_{H_1p}(t))(1 - q_{H_2p}(t)).$$

The BDD evaluation of the FT in Figure 9.18 using the conditional probabilities computed using (9.32) and (9.33) gives $Q_{PFDC_2}(t)$ in (9.37) as

$$\begin{aligned} Q_{PFDC_2}(t) = 1 &- (1 - q_{B_2S}(t))(1 - q_{B_2T}(t)) + (1 - q_{B_2S}(t))(1 - q_{B_2T}(t)) \\ &\cdot [q_{H_1S}(t) + (1 - q_{H_1S}(t))q_{H_1T}(t)][q_{H_2S}(t) + (1 - q_{H_2S}(t))q_{H_2T}(t)]. \end{aligned}$$

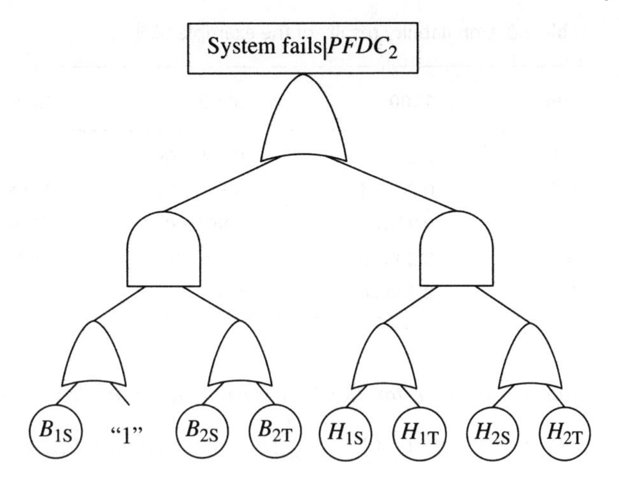

Figure 9.18 Reduced FT model under *PFDC*$_2$.

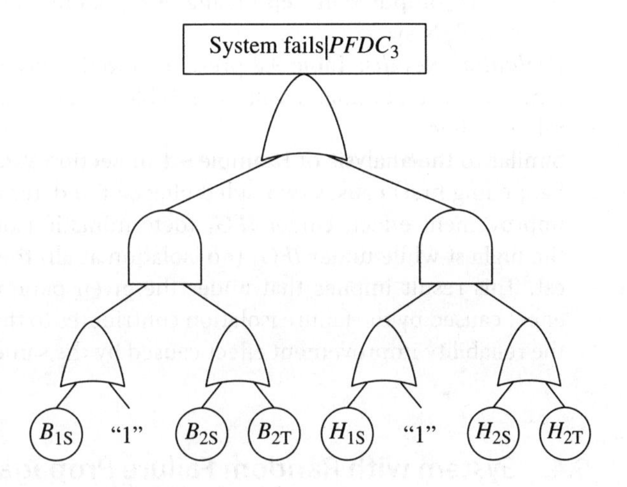

Figure 9.19 Reduced FT model under *PFDC*$_3$.

Under $PFDC_3$, the LF of trigger R can isolate both sensors B_1 and H_1. That is, both sensors B_1 and H_1 cannot get direct wireless connection with the sink device. Figure 9.19 shows the reduced FT generated in this case. Applying the PFGE approach in Section 8.2, $P(\text{system fails}|PFDC_3)$ is computed as

$$P(\text{system fails}|PFDC_3) = 1 - P_{PFDC_3}(t) + P_{PFDC_3}(t) \cdot Q_{PFDC_3}(t). \tag{9.38}$$

Since the PFGEs of B_1 and H_1 are isolated, the set of sensors that can experience PFGEs is $I = \{B_2, H_2\}$ under $PFDC_3$. According to (8.2), $P_{PFDC_3}(t)$ is calculated as

$$P_{PFDC_3}(t) = \prod_{\forall i \in I} [1 - q_{ip}(t)] = (1 - q_{B_2p}(t))(1 - q_{H_2p}(t)).$$

The BDD evaluation of the FT in Figure 9.19 using the conditional probabilities computed via (9.32) and (9.33) gives $Q_{PFDC_3}(t)$ as

$$Q_{PFDC_3}(t) = 1 - (1 - q_{B_2S}(t))(1 - q_{B_2T}(t))(1 - q_{H_2S}(t))(1 - q_{H_2T}(t)).$$

Table 9.8 Unreliability results of the example BSN.

t (h)	1000	3000	5000
IFG_1	0.213 857	0.909 838	0.999 417
IFG_2	0.201 448	0.885 745	0.998 889
IFG_3	0.210 669	0.901 189	0.999 232
IFG_4	0.205 967	0.901 287	0.999 242
IFG_5	0.203 267	0.899 005	0.999 305

Step 5: Integrate for the final system unreliability. According to (9.24), $Q(t)$ in (9.23) is evaluated as $Q(t) = \sum_{j=1}^{3}[P(\text{system fails}|FCE_j) \cdot P(FCE_j)]$. $Q(t)$ is then combined with $P_u(t)$ computed in step 1 using (9.23) to obtain the final failure probability of the example BSN system.

Evaluation Results: Table 9.8 presents unreliability results of the example BSN system using the parameter values of Table 9.5 and Table 9.6 for several values of the mission time.

Similar to the analysis of Example 9.1 in Section 9.2.2, the LF of the trigger R (when happening first) causes two-sided effects: the deterioration effect and the reliability improvement effect. Under IFG_1 (deterministic isolation), the BSN unreliability is the highest while under IFG_2 (no isolation at all) the system unreliability is the lowest. This result implies that under the given parameter settings, the deterioration effect caused by the failure isolation contributes to the system unreliability more than the reliability improvement effect caused by the same failure isolation.

9.4 System with Random Failure Propagation Time

The methods presented in Sections 9.2 and 9.3 assume that any failure propagation originating from a system component instantaneously takes effect, which is often not true in real-world cases [8]. In this section, a combinatorial method is presented for addressing random propagation time (PT) in the reliability analysis of single-phase systems subject to the competing probabilistic failure isolation and failure propagation effects.

9.4.1 Combinatorial Method

The method involves a six-step procedure described as follows:

Step 1: Separate effects of PFGEs from the trigger component. According to the PFGE approach (Section 8.2), the system unreliability is evaluated as

$$UR_{system}(t) = 1 - P_u(t) + Q(t) \cdot P_u(t), \tag{9.39}$$

where $P_u(t) = P(\text{no PFGEs from the trigger, e.g. } R)$, $Q(t) = P(X_{SF}(t)|X_{Rp}^{C}(t))$ is the conditional probability that the system fails at time t (denoted by $X_{SF}(t)$) given that the trigger undergoes no PFGEs (denoted by $X_{Rp}^{C}(t)$), which is evaluated through the following steps.

Step 2: Define PFDCs. Similar to the method in Section 9.2.1, PFDCs are identified to cover all possible relationships between the trigger component and related PDEP components belonging to the same PFD group.

Given that n PDEP components exist in the PFD group, 2^n disjoint PFDCs are defined (denoted by $PFDC_k$, $k = 0, 1, \ldots, 2^n - 1$). $P(PFDC_k)$ and $S(PFDC_k)$ are also determined in this step.

Step 3: Define FCEs and decompose $Q(t)$. Based on the trigger component's status, three FCEs are defined as follows:

FCE_1: the trigger remains working throughout the mission.

FCE_2: the trigger LF occurs after the PFGE of at least one of its PDEP components and the PT is less than the time difference between the trigger LF and the PFGE of the PDEP component. In this case, the global failure propagation effect takes place.

FCE_3: the trigger LF causes the probabilistic isolation effect.

Based on these three events, $Q(t)$ can be evaluated as

$$Q(t) = \sum_{j=1}^{3} P(FCE_j \cap X_{SF}(t) | X_{Rp}^{C}(t))$$
$$= \sum_{j=1}^{3} [P(FCE_j | X_{Rp}^{C}(t)) \cdot P(X_{SF}(t) | X_{Rp}^{C}(t) \cap FCE_j)]. \tag{9.40}$$

Step 4: Evaluate $P(FCE_j | X_{Rp}^{C}(t))$, $j = 1, 2, 3$ *in* (9.40). Based on the PFGE method, $P(FCE_1 | X_{Rp}^{C}(t))$ is evaluated as

$$P(FCE_1 | X_{Rp}^{C}(t)) = 1 - q_R(t), \tag{9.41}$$

where $q_R(t)$ can be evaluated according to the s-relationships between the LF and PFGE of the trigger R, particularly, using (8.3) for s-independent, (8.4) for s-dependent, and (8.6) for disjoint LF and PFGE.

Based on the PFDCs defined in step 2 and the total probability law, $P(FCE_2 | X_{Rp}^{C}(t))$ can be evaluated as

$$P(FCE_2 | X_{Rp}^{C}(t)) = \sum_{k=0}^{3} P(PFDC_k \cap FCE_2 | X_{Rp}^{C}(t))$$
$$= \sum_{k=0}^{3} [P(PFDC_k) \cdot P(FCE_2 | X_{Rp}^{C}(t) \cap PFDC_k)]. \tag{9.42}$$

Based on $S(PFDC_k)$ determined in step 2, $P(FCE_2 | X_{Rp}^{C}(t) \cap PFDC_k)$ in (9.42) can be evaluated using the method similar to (9.4) but with the consideration of the random failure PT. Specifically, assume $S(PFDC_k)$ contains k components, whose unconditional PFGE events are represented by X_{D1p}, X_{D2p}, …, and X_{Dkp}. Thus according to the definition of FCE_2, one obtains

$$P(FCE_2 | X_{Rp}^{C}(t) \cap PFDC_k)$$
$$= P\begin{bmatrix} \{(X_{D1p} \rightarrow X_{Rl}) \cap X_{D1PT}\} \cup \{(X_{D2p} \rightarrow X_{Rl}) \cap X_{D2PT}\} \\ \cup \ldots \cup \{(X_{Dkp} \rightarrow X_{Rl}) \cap X_{DkPT}\} \end{bmatrix}. \tag{9.43}$$

where X_{DjPT} ($j = 1, 2, \ldots, k$) represents event that the PT of the PFGE originating from the jth PDEP component in $S(PFDC_i)$ is less than the time difference between the trigger LF (X_{Rl}) and the PFGE of the PDEP component (X_{Djp}).

$P(FCE_3|X_{Rp}^C(t))$ can be evaluated as

$$P(FCE_3|X_{Rp}^C(t)) = 1 - P(FCE_1|X_{Rp}^C(t)) - P(FCE_2|X_{Rp}^C(t)). \tag{9.44}$$

Step 5: Evaluate $P(X_{SF}(t)|X_{Rp}^C(t) \cap FCE_j)$, *j = 1, 2, 3 in* (9.40), $P(X_{SF}(t)|X_{Rp}^C(t) \cap FCE_j)$ is evaluated using the method similar to that used for evaluating $P(\text{systemfails} \mid FCE_i)$ in step 4 of Section 9.2.1.

Specifically, $P(X_{SF}(t)|X_{Rp}^C(t) \cap FCE_1)$ is evaluated by applying the PFGE method to a reduced FT, which is obtained by removing the PFD gate and its trigger input from the original system FT. In the case of system components undergoing multiple types of LFs, the OR gate expansion used in step 4 described in Section 9.3.1 should also be applied.

$P(X_{SF}(t)|X_{Rp}^C(t) \cap FCE_2)$ simply equals to one due to the global failure propagation effect.

$P(X_{SF}(t)|X_{Rp}^C(t) \cap FCE_3)$ is evaluated based on the PFDCs defined in step 2 as (9.45).

$$P(X_{SF}(t)|X_{Rp}^C(t) \cap FCE_3) = \sum_{k=0}^{2^n-1} P(X_{SF}(t)|X_{Rp}^C(t) \cap FCE_3 \cap PFDC_k) \cdot P(PFDC_k). \tag{9.45}$$

$P(X_{SF}(t)|X_{Rp}^C(t) \cap FCE_3 \cap PFDC_k)$ in (9.45) can be evaluated by applying the PFGE method to a reduced FT, which is obtained by removing the PFD gate and its trigger input from the original system FT model, and then replacing events of components in set $S(PFDC_k)$ with constant 1 (TRUE).

Step 6: Integrate for the final system unreliability. According to (9.40), $Q(t)$ can be obtained by integrating $P(FCE_j|X_{Rp}^C(t))$ evaluated in step 4 and $P(X_{SF}(t)|X_{Rp}^C(t) \cap FCE_j)$ evaluated in step 5. According to (9.39), $Q(t)$ and $P_u(t)$ evaluated at step 1 are integrated to obtain the final system unreliability.

9.4.2 Case Study: WSN System

Example 9.3 Consider a WSN system used in an example smart home power generation system (Figure 9.20). The WSN system contains a relay node R, and a set of sensors ($S1$, $S2$, $E1$, $E2$, $M1$, $M2$, L) for monitoring conditions of the corresponding physical devices and transmitting the sensed data to the home energy management system (EMS), which is referred to as the sink node of the WSN system. The sink node is assumed to be perfectly reliable for the illustrative analysis.

Sensors $S1$ and $S2$ communicate with the sink node through R, while other sensors have direct communication with the sink node. When R malfunctions itself (undergoing an LF), $S1$ and $S2$ may become isolated (with probabilities q_{RS1}, q_{RS2} respectively) or communicate with the sink node directly, depending on the remaining energy of these two sensors. Figure 9.21 illustrates the FT of the WSN system, which models the function of the system requiring the correct function of at least one of ($S1$, $S2$), at least one

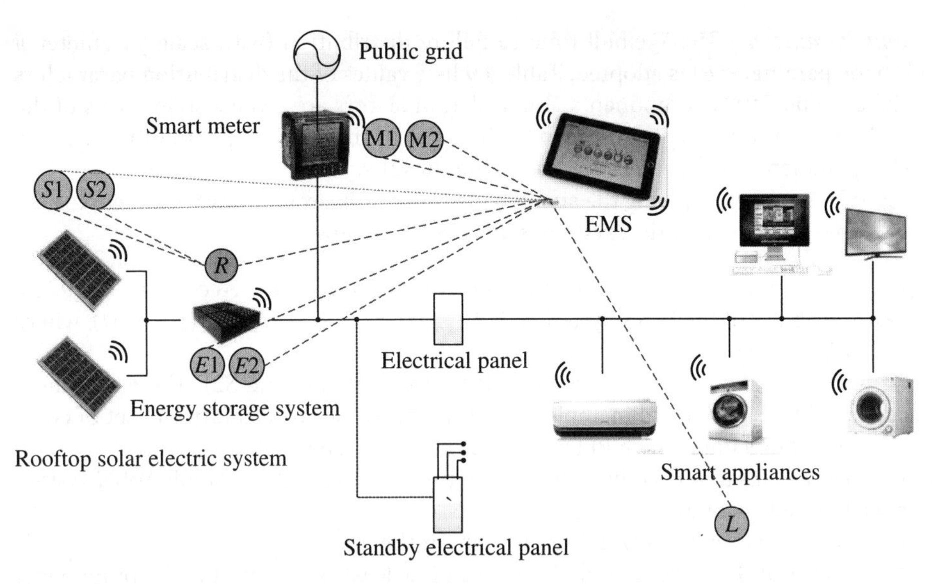

Figure 9.20 An example of a smart home power generation system [9].

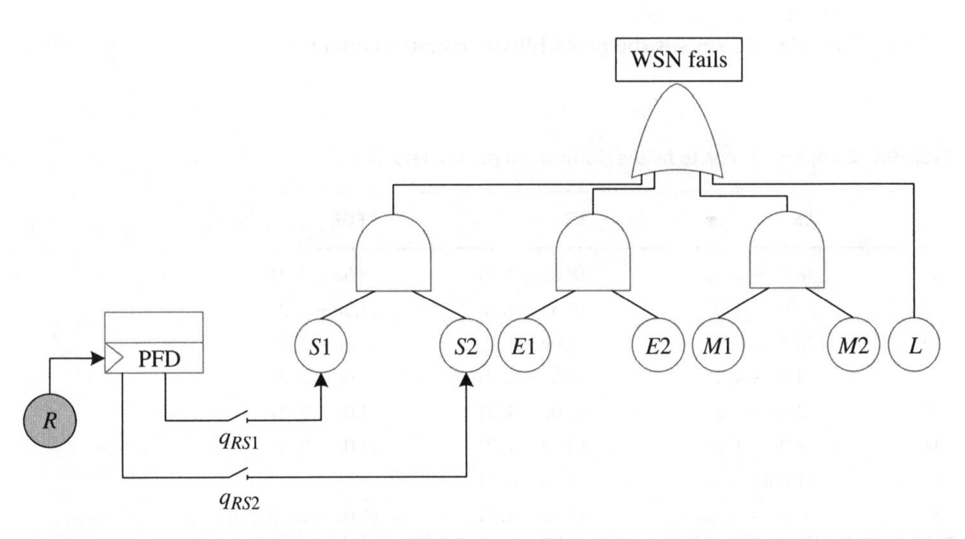

Figure 9.21 FT of the example WSN system.

of ($E1$, $E2$), at least one of ($M1$, $M2$), and L. All the communication links are assumed to be perfectly reliable in this example.

Similar to Example 9.2, each sensor (except L) can undergo PFGEs caused by jamming attacks and two types of LFs: sensing LF (LS) and transmitting LF (LT). Their s-relationship is illustrated in Figure 9.12. The relay R is only subject to the LT and PFGE. The PFGE takes random time to spread its effects. Because jamming power is never infinite, the target location must usually be known by the jammer for launching a jamming attack [10]. Since locations of the indoor smart appliance are not revealed to the jammer, sensor L is assumed to be free of the jamming attack (or PFGE).

Input Parameters: The Weibull time to failure distribution (with scale parameter β and shape parameter α) is adopted. Table 9.9 lists values of the distribution parameters for the example WSN components. The isolation factors are given also in terms of the Weibull distribution: ($\beta_{RS1} = 1.05$ per month, $\alpha_{RS1} = 2$) and ($\beta_{RS2} = 1$ per month, $\alpha_{RS2} = 2$). The PT parameters are given in the Evaluation Results.

Example Analysis: Applying the six-step procedure, the example WSN system for the smart home power generation system is analyzed as follows:

Step 1: Separate effects of PFGEs from the trigger component. According to (9.39), the unreliability of the WSN system is $UR_{system}(t) = 1 - P_u(t) + Q(t) \cdot P_u(t)$, where $P_u(t) = 1 - q_{Rp}(t)$.

Step 2: Define PFDCs. There are two PDEP components ($S1$ and $S2$). Therefore, there are four PFDCs as defined in Table 9.10. The occurrence probability and set of components isolated under each PFDC are also given in Table 9.10.

Step 3: Define FCEs and decompose $Q(t)$. The three FCEs for the example WSN system are elaborated as follows:

FCE_1: the relay R remains working throughout the mission.

FCE_2: the relay LT occurs after the PFGE of at least one of its PDEP components and the PT is less than the time difference between the relay LT and the PFGE of the PDEP component.

FCE_3: the relay LT causes the probabilistic isolation effect.

Table 9.9 Component time to failure distribution parameters (β, α).

	LS	LT	PFGE
$S1$	$(8.0e-4, 2)$	$(8.0e-4, 2)$	$(8.5e-2, 2)$
$S2$	$(4.0e-4, 2)$	$(4.5e-4, 2)$	$(8.5e-2, 2)$
$E1$	$(3.5e-4, 2)$	$(4.0e-4, 2)$	$(5.0e-2, 2)$
$E2$	$(4.0e-4, 2)$	$(4.0e-4, 2)$	$(5.0e-2, 2)$
$M1$	$(2.8e-4, 2)$	$(3.0e-4, 2)$	$(4.0e-2, 2)$
$M2$	$(3.0e-4, 2)$	$(3.5e-4, 2)$	$(4.0e-2, 2)$
L	$(1.666e-4, 2)$	$(3.0e-4, 2)$	–
R	–	$(3.0e-3, 1)$	$(6.0e-2, 2)$

Table 9.10 PFDCs.

k	$PFDC_k$	$P(PFDC_k)$	$S(PFDC_k)$
0	$\overline{S1} \cap \overline{S2}$	$(1 - q_{RS1})(1 - q_{RS2})$	$\emptyset$
1	$\overline{S1} \cap S2$	$(1 - q_{RS1})q_{RS2}$	$\{S2\}$
2	$S1 \cap \overline{S2}$	$q_{RS1}(1 - q_{RS2})$	$\{S1\}$
3	$S1 \cap S2$	$q_{RS1}q_{RS2}$	$\{S1, S2\}$

According to (9.40), $Q(t)$ is evaluated as

$$Q(t) = \sum_{j=1}^{3}[P(FCE_j|X_{Rp}^C(t)) \cdot P(X_{SF}(t)|X_{Rp}^C(t) \cap FCE_j)]$$

Step 4: Evaluate $P(FCE_j|X_{Rp}^C(t))$, *j = 1, 2, 3 in* (9.40). Based on (9.33), $P(FCE_1|X_{Rp}^C(t))$ is evaluated as

$$P(FCE_1|X_{Rp}^C(t)) = 1 - q_{RT}(t) = \frac{q_{RLT}(t)}{1 - q_{Rp}(t)} \tag{9.46}$$

According to (9.42), $P(FCE_2|X_{Rp}^C(t))$ is evaluated as

$$P(FCE_2|X_{Rp}^C(t)) = \sum_{k=0}^{3}[P(PFDC_k) \cdot P(FCE_2|X_{Rp}^C(t) \cap PFDC_k)], \tag{9.47}$$

where according to (9.43)

$$P(FCE_2|X_{Rp}^C(t) \cap PFDC_0) = 0,$$

$$P(FCE_2|X_{Rp}^C(t) \cap PFDC_1) = P((X_{S2p} \rightarrow X_{RT}) \cap X_{S2PT})$$
$$= \int_0^t \int_{\tau_1}^t \int_0^{\tau_2-\tau_1} f_{S2p}(\tau_1) f_{RT}(\tau_2) f_{S2PT}(\tau_3) d\tau_3 d\tau_2 d\tau_1,$$

$$P(FCE_2|X_{Rp}^C(t) \cap PFDC_2) = P((X_{S1p} \rightarrow X_{RT}) \cap X_{S1PT})$$
$$= \int_0^t \int_{\tau_1}^t \int_0^{\tau_2-\tau_1} f_{S1p}(\tau_1) f_{RT}(\tau_2) f_{S1PT}(\tau_3) d\tau_3 d\tau_2 d\tau_1,$$

$$\begin{aligned} &P(FCE_2|X_{Rp}^C(t) \cap PFDC_3) \\ &= P\{(X_{S1p} \rightarrow X_{RT}) \cap X_{S1PT}\} + P\{(X_{S2p} \rightarrow X_{RT}) \cap X_{S2PT}\} \\ &\quad - P\{(X_{S1p} \rightarrow X_{S2p} \rightarrow X_{RT}) \cap X_{S1PT} \cap X_{S2PT}\} \\ &\quad - P\{(X_{S2p} \rightarrow X_{S1p} \rightarrow X_{RT}) \cap X_{S2PT} \cap X_{S1PT}\}. \end{aligned}$$

According to (9.44), $P(FCE_3|X_{Rp}^C(t))$ is evaluated as

$$P(FCE_3|X_{Rp}^C(t)) = 1 - P(FCE_1|X_{Rp}^C(t)) - P(FCE_2|X_{Rp}^C(t))$$

Step 5: Evaluate $P(X_{SF}(t)|X_{Rp}^C(t) \cap FCE_j)$, *j = 1, 2, 3 in* (9.40). Under FCE_1, relay R functions and no sensors are isolated. Figure 9.22 shows the FT generated from the original FT in Figure 9.21 by removing the PFD gate and its trigger input and by adding component OR gates connecting two types of LFs.

Applying the PFGE method, $P(X_{SF}(t)|X_{Rp}^C(t) \cap FCE_1)$ is evaluated as

$$P(X_{SF}(t)|X_{Rp}^C(t) \cap FCE_1) = 1 - P_{FCE_1}(t) + P_{FCE_1}(t) \cdot Q_{FCE_1}(t), \tag{9.48}$$

where $P_{FCE_1}(t) = \prod_{\forall i \in I}(1 - q_{ip}(t))$, $I = \{S1, S2, E1, E2, M1, M2\}$. $Q_{FCE_1}(t)$ can be evaluated as

$$Q_{FCE_1}(t) = 1 - (1 - U_S)(1 - U_E)(1 - U_M)(1 - U_L),$$

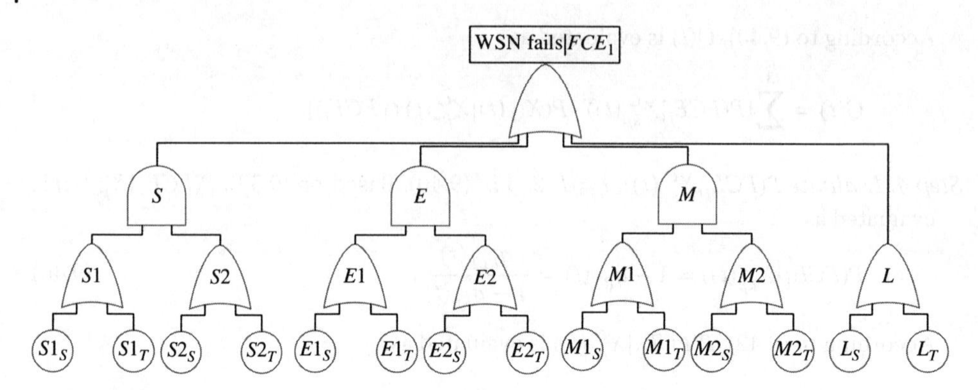

Figure 9.22 FT under FCE_1.

where U_S, U_E, U_M, U_L denote unreliabilities of the sensor subsystem monitoring the solar panels, energy storage system, smart meter, and indoor appliances, respectively. The sub-FTs modeling these subsystems are illustrated in Figure 9.22. Based on the sub-FTs, U_S, U_E, U_M, U_L can be evaluated using the BDD-based method.

Under FCE_2, $P(X_{SF}(t)|X_{Rp}^C(t) \cap FCE_2) = 1$ due to the global failure propagation effect.

Under FCE_3, the probabilistic failure isolation effect takes place. According to (9.45), $P(X_{SF}(t)|X_{Rp}^C(t) \cap FCE_3)$ is evaluated as

$$P(X_{SF}(t)|X_{Rp}^C(t) \cap FCE_3)$$
$$= \sum_{k=0}^{3} [P(X_{SF}(t)|X_{Rp}^C(t) \cap FCE_3 \cap PFDC_k) \cdot P(PFDC_k)]. \tag{9.49}$$

Under $PFDC_0$, the LT of trigger R cannot cause any isolation effect. The FT generated is the same as that in Figure 9.22. Therefore, $P(X_{SF}(t)|X_{Rp}^C(t) \cap FCE_3 \cap PFDC_0)$ can be calculated using (9.48).

Under $PFDC_1$, the LT of R causes the failure isolation to sensor $S2$. Figure 9.23 shows the FT generated for this case. Applying the PFGE approach in Section 8.2, $P(X_{SF}(t)|X_{Rp}^C(t) \cap FCE_3 \cap PFDC_1)$ is computed as

$$P(X_{SF}(t)|X_{Rp}^C(t) \cap FCE_3 \cap PFDC_1) = 1 - P_{PFDC_1}(t) + P_{PFDC_1}(t) \cdot Q_{PFDC_1}(t),$$

where

$$P_{PFDC_1}(t) = \prod_{\forall i \in \{S1,E1,E2,M1,M2\}} (1 - q_{ip}(t)),$$
$$Q_{PFDC_1}(t) = 1 - (1 - U_{S1})(1 - U_E)(1 - U_M)(1 - U_L). \tag{9.50}$$

Under $PFDC_2$, the LT of R causes the failure isolation to sensor $S1$. Figure 9.24 shows the FT generated for this case. Applying the PFGE approach in Section 8.2, $P(X_{SF}(t)|X_{Rp}^C(t) \cap FCE_3 \cap PFDC_2)$ is computed as

$$P(X_{SF}(t)|X_{Rp}^C(t) \cap FCE_3 \cap PFDC_2) = 1 - P_{PFDC_2}(t) + P_{PFDC_2}(t) \cdot Q_{PFDC_2}(t),$$

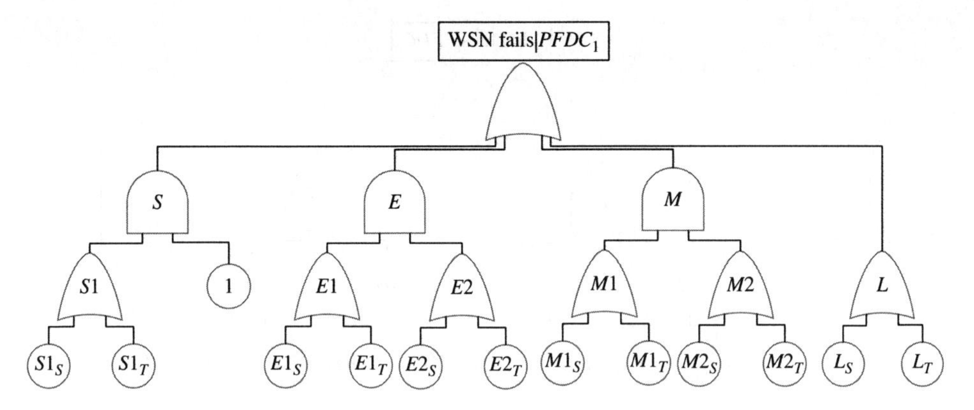

Figure 9.23 FT under $PFDC_1$.

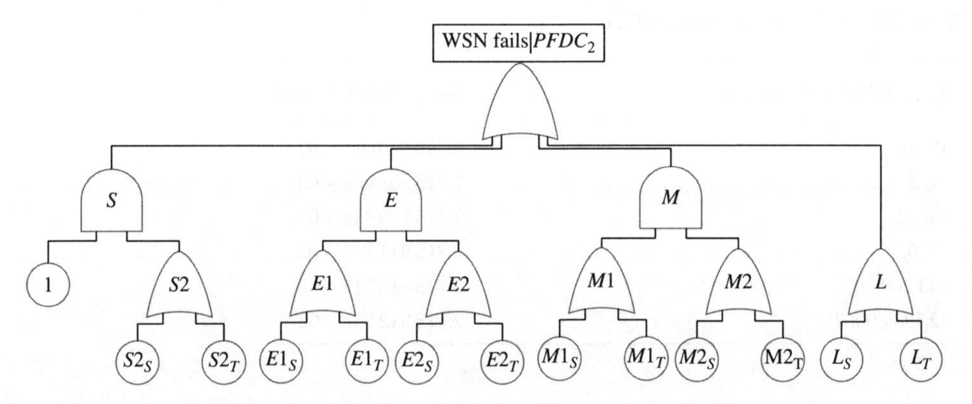

Figure 9.24 FT under $PFDC_2$.

where

$$P_{PFDC_2}(t) = \prod_{\forall i \in \{S2,E1,E2,M1,M2\}} (1 - q_{ip}(t)),$$

$$Q_{PFDC_2}(t) = 1 - (1 - U_{S2})(1 - U_E)(1 - U_M)(1 - U_L). \tag{9.51}$$

Under $PFDC_3$, the LT of R causes the failure isolation to sensors $S1$ and $S2$. Figure 9.25 shows the FT generated for this case, which can be simplified as constant 1. That is $P(X_{SF}(t)|X_{Rp}^C(t) \cap FCE_3 \cap PFDC_3) = 1$.

Step 6: Integrate for the final system unreliability. According to (9.40), $Q(t)$ is obtained by integrating $P(FCE_j|X_{Rp}^C(t))$ evaluated in step 4 and $P(X_{SF}(t)|X_{Rp}^C(t) \cap FCE_j)$ evaluated in step 5. According to (9.39), $Q(t)$ and $P_u(t)$ evaluated at step 1 are integrated to obtain the unreliability of the example WSN system.

Evaluation Results: The PT is assumed to follow the Weibull distribution with parameters (β_{S1PT}, $\alpha_{S1PT} = 2$) and ($\beta_{S2PT} = 0.25$ per month, $\alpha_{S2PT} = 2$). Using the component failure parameters and the isolation factors given in the Input Parameters, Table 9.11 gives the unreliability of the example WSN system at $t = 1$ month using different values of β_{S1PT} to show effects of the failure PT.

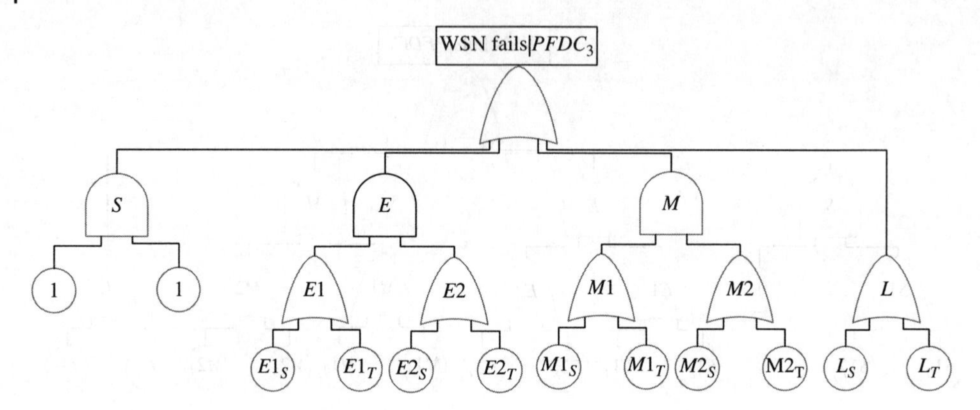

Figure 9.25 FT under $PFDC_3$.

Table 9.11 WSN unreliability for different β_{S1PT}.

(β_{S1PT} [per month], α_{S1PT})	UR_{system} ($t = 1$ month)
(0, 2)	$2.713\,539\,01e-02$
(1, 2)	$2.713\,563\,00e-02$
(30, 2)	$2.713\,789\,42e-02$
(720, 2)	$2.713\,811\,72e-02$
(43 200, 2)	$2.713\,812\,71e-02$
(2 592 000, 2)	$2.713\,812\,73e-02$

The analysis results show that for the given example parameter settings, effects of PT on the WSN unreliability at $t = 1$ day barely take place (the WSN unreliability ranges from 2.916 650 62e-05 when $\beta_{S1PT} = 0$ per month to 2.916 650 66e-05 when $\beta_{S1PT} = 2\,592\,000$ per month). As the mission time proceeds, the effects of PT become more apparent as demonstrated in Table 9.11. As β_{S1PT} increases, the PFGE of $S1$ is more quickly to propagate to the entire WSN, making it less likely to be isolated when relay R fails, thus the WSN unreliability increases.

9.5 Summary

This chapter addresses the probabilistic competing failure behavior, which extends the deterministic competing failure behavior addressed in Chapter 8 by considering uncertain or probabilistic failure isolation effects. As demonstrated by the case studies from the Internet of Things applications (WSNs, body sensor systems, and smart homes), the probabilistic competing failure behavior abounds in systems involving relayed wireless communications. Combinatorial methods are presented for reliability analysis of single-phase systems with a single type of component LFs, multiple different types of component LFs, and random PT. Refer to [11] on a combinatorial procedure for analyzing reliability of single-phase systems subject to multiple dependent PFD groups. Refer to [12] on addressing the probabilistic competing failure behavior in the reliability analysis of phased-mission systems (PMSs).

References

1 Luo, X., Yu, H., and Wang, X. (2013). Energy-aware self-organisation algorithms with heterogeneous connectivity in wireless sensor networks. *International Journal of Systems Science* 44 (10): 1857–1866.
2 Wang, Y., Xing, L., and Wang, H. (2017). Reliability of systems subject to competing failure propagation and probabilistic failure isolation. *International Journal of Systems Science: Operations & Logistics* 4 (3): 241–259.
3 Benker, H. (2012). *Practical Use of Mathcad: Solving Mathematical Problems with a Computer Algebra System*. Springer Science & Business Media.
4 Allen, A. (1990). *Probability, Statistics and Queuing Theory: With Computer Science Applications*, 2e. Academic Press.
5 Dellaportas, P. and Wright, D.E. (1991). Numerical prediction for the 2-parameter Weibull distribution. *The Statistician* 40 (4): 365–372.
6 Zonouz, A.E., Xing, L., Vokkarane, V.M., and Sun, Y. (2014). Reliability-oriented single-path routing protocols in wireless sensor networks. *IEEE Sensors Journal* 14 (11): 4059–4068.
7 Wang, Y., Xing, L., Wang, H., and Levitin, G. (2015). Combinatorial analysis of body sensor networks subject to probabilistic competing failures. *Reliability Engineering & System Safety* 142: 388–398.
8 Levitin, G., Xing, L., Ben-Haim, H., and Dai, Y. (2013). Reliability of series-parallel systems with random failure propagation time. *IEEE Transactions on Reliability* 62 (3): 637–647.
9 Zhao, G., Xing, L., Zhang, Q., and Jia, X. (2018). A hierarchical combinatorial reliability model for smart home systems. *Quality and Reliability Engineering International* 34 (1): 37–52.
10 M. Ståhlberg, "Radio Jamming Attacks Against Two Popular Mobile Networks," *Proceedings of the Helsinki University of Technology Seminar on Network Security and Mobile Security*, http://www.tml.tkk.fi/Opinnot/Tik-110.501/2000/papers/stahlberg.pdf, 2000.
11 Wang, Y., Xing, L., Wang, H., and Coit, D.W. (2018). System reliability modeling considering correlated probabilistic competing failures. *IEEE Transactions on Reliability* 67 (2): 416–431, https://doi.org/10.1109/TR.2017.2716183.
12 Wang, Y., Xing, L., Levitin, G., and Huang, N. (2018). Probabilistic competing failure analysis in phased-mission systems. *Reliability Engineering & System Safety* 176: 37–51.

10

Dynamic Standby Sparing

Standby sparing is a technique in which one or several components are online and operating with some redundant components serving as standby spares [1, 2]. When an online component malfunctions, an available standby component is activated to replace the failed online component and take over the mission task so the entire system can continue to function correctly. The standby sparing technique has been applied in many industries to achieve fault-tolerance and high system reliability or availability. Examples of applications include power systems [3–5], storage systems [6], high-performance computing systems [7], distributed systems [8], and telecommunication systems [9, 10]. The standby sparing technique is especially crucial for communication and computer systems used in mission- or life-critical applications, e.g. flight control [1, 11] and space missions [12, 13] where maintaining or replacing a malfunctioned component through onboard or online manual intervention is difficult or even impossible.

In this chapter, following a discussion on different types of standby sparing systems, the continuous time Markov chain (CTMC)-based method (Section 10.2), decision diagram–based combinatorial method (Section 10.3), approximation method (Section 10.4), and event transition method (Section 10.5) are presented for their reliability modeling and evaluation.

10.1 Types of Standby Systems

Based on the number of components required to be online and functioning for the system operation, standby sparing systems with n components can be classified into 1-out-of-n: G and k-out-of-n: G systems. In a k-out-of-n: G standby system, $k \leq n$ components are functioning with the remaining components waiting in the standby mode [14, 15]. A 1-out-of-n: G standby system is a special case of k-out-of-n: G systems with $k = 1$.

Based on failure characteristics and standby cost associated with standby components, standby sparing systems can be classified into three types: hot, cold, and warm [16–18]. A hot standby component operates concurrently with the online active component [1]; it undergoes the same operational environment and stresses, thus has the same failure rate as the online working component does. The hot standby component is ready to take over the mission task at any time. Hence, a hot standby component

Dynamic System Reliability: Modeling and Analysis of Dynamic and Dependent Behaviors,
First Edition. Liudong Xing, Gregory Levitin and Chaonan Wang.

can provide fast system restoration following a failure, but at the cost of high overhead because it consumes energy and materials as much as the online component does. The hot standby technique is often used for applications where the system recovery time is critical.

A cold standby component is unpowered and fully shielded from the operational stresses. It cannot fail during the standby mode (unless it may be destroyed by external factors such as explosion). It does not consume any energy or materials before being activated to replace a malfunctioned online component. Thus, keeping standby components in the cold mode is nearly costless. However, a long restoration delay or a large startup cost is needed in the case of a cold standby component being required to replace a malfunctioned online component [19–21]. The cold standby technique is typically adopted in applications where the energy consumption is critical, for example, satellites [1], space exploration [13], fielded systems [22], and textile manufacturing systems [4].

As a trade-off between hot standby and cold standby, a warm standby component undergoes milder operational environment and is partially exposed to operational stresses [5, 23–28]. Hence, the failure rate and operation cost of a warm standby component are lower than those of the corresponding hot standby component. On the other hand, as compared to cold standby, less restoration delay and expense are involved to replace a failed online component by a warm standby component. An example of the warm standby systems is a power plant where spinning extra generating units are waiting in a standby mode [17]. The standby units can fail, but their failure rates as well as exploitation costs are less than those for the primary online unit working under the full load and consuming more materials.

The hot, cold, and warm standby behavior can be modeled using the hot spare (HSP), cold spare (CSP), and warm spare (WSP) gates described in the dynamic fault tree (DFT) modeling (Section 2.3.2), respectively. A hot standby system with a perfect switching mechanism is equivalent to an active redundant system, where the system components have static failure behavior during their lifetimes. However, components in systems designed with cold and warm standby sparing exhibit dynamic failure behaviors [18, 25]. Particularly, they have different failure rates before and after they are activated to replace the failed online component. In the following subsections, four different methods are presented to address these dynamic failure rate behaviors in the modeling and evaluation of standby sparing systems with perfect switching mechanisms.

10.2 CTMC-Based Method

As discussed in Section 2.5, system states and state transitions are two essential concepts of the Markov-based method [29]. A system state is defined by a specific combination of component state variables at a given instant of time. A state transition occurs due to the failure or repair of a system component [30]. The CTMC-based methods assume exponential time-to-failure and time-to-repair distributions for the system components.

The solution to a CTMC model with n different states includes the probability of the system being in each state, particularly, $P_j(t)$ that denotes the probability that the system is in state j at time t ($j = 1,\ldots,n$). They can be obtained by solving a set of differential

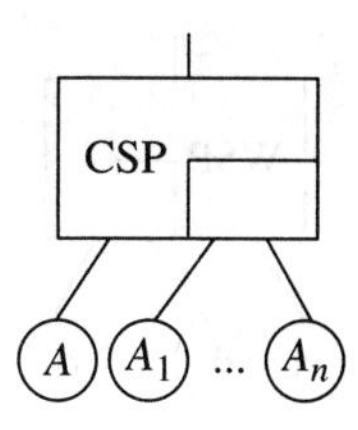

Figure 10.1 DFT model of a cold standby system.

equations in the form of (2.26), which is

$$\begin{bmatrix} -\alpha_{11} & \alpha_{21} & \alpha_{31} & \dots & \alpha_{n1} \\ \alpha_{12} & -\alpha_{22} & \alpha_{32} & \dots & \alpha_{n2} \\ \alpha_{13} & \alpha_{23} & -\alpha_{33} & \dots & \alpha_{n3} \\ \dots & \dots & \dots & \dots & \dots \\ \alpha_{1n} & \alpha_{2n} & \alpha_{3n} & \dots & -\alpha_{nn} \end{bmatrix} \bullet \begin{bmatrix} P_1(t) \\ P_2(t) \\ P_3(t) \\ \dots \\ P_n(t) \end{bmatrix} = \begin{bmatrix} P'_1(t) \\ P'_2(t) \\ P'_3(t) \\ \dots \\ P'_n(t) \end{bmatrix}. \tag{10.1}$$

In (10.1), $\alpha_{jk}, j \neq k$ denotes the transition rate from state j to state k. The diagonal element is obtained as $\alpha_{jj} = \sum_{k=1,k\neq j}^{n} \alpha_{jk}$. To solve (10.1), Laplace transform is commonly applied [31]. The final system unreliability is obtained as the sum of the probability of the system being in each failure state.

In Sections 10.2.1 and 10.2.2, the CTMC solutions to analyzing a cold standby system and a warm standby system are presented, respectively.

10.2.1 Cold Standby System

Figure 10.1 illustrates the DFT model of a cold standby system with a primary component A and n cold standby components A_i $(i = 1, 2, \dots n)$. For simplicity, all system components are assumed to fail exponentially with an identical failure rate of λ.

Example 10.1 Figure 10.2 shows the state transition diagram for the cold standby system with one primary component A and one CSP A_1. The state labeled (A, A_1) is an initial system state in which A is operational and A_1 is in the cold standby state; the state (A_1) is a system operational state where A has failed and A_1 has been activated to replace A and is operational; the state (F) denotes the system failure state where both A and A_1 have failed. The two state transitions are characterized by the failure rates of components A and A_1, respectively.

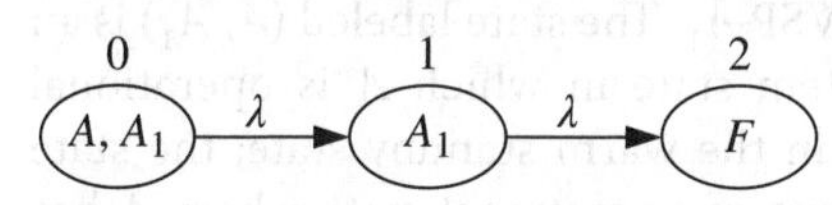

Figure 10.2 Markov model of the cold standby system with one spare.

According to (10.1), the state equations of this example cold standby system are in the form of (10.2).

$$\begin{bmatrix} -\lambda & 0 & 0 \\ \lambda & -\lambda & 0 \\ 0 & \lambda & 0 \end{bmatrix} \bullet \begin{bmatrix} P_0(t) \\ P_1(t) \\ P_2(t) \end{bmatrix} = \begin{bmatrix} P'_0(t) \\ P'_1(t) \\ P'_2(t) \end{bmatrix}. \tag{10.2}$$

Solving (10.2) using the Laplace transform-based method described in [31] and initial probabilities of $P_0(0) = 1, P_1(0) = P_2(0) = 0$, the state probabilities at time t are obtained as:

$$P_0(t) = e^{-\lambda t},$$

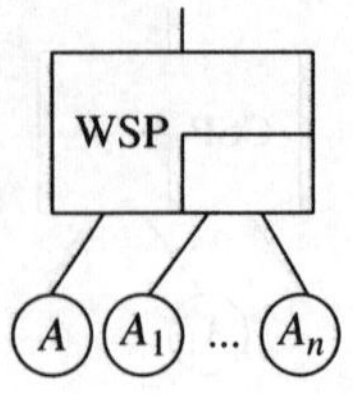

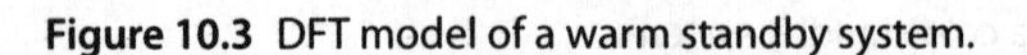
Figure 10.3 DFT model of a warm standby system.

$$P_1(t) = \lambda t e^{-\lambda t},$$

$$P_2(t) = 1 - P_0(t) - P_1(t).$$

Thus, the reliability of the cold standby system with one spare at time t is:

$$R(t) = P_0(t) + P_1(t) = e^{-\lambda t} + \lambda t e^{-\lambda t} = (1 + \lambda t)e^{-\lambda t}. \tag{10.3}$$

In general, the polynomial representing the reliability of a cold standby system with n identical cold standby spares at time t can be derived as [32]:

$$R(t) = \left(\sum_{i=0}^{n} \frac{(\lambda t)^i}{i!}\right) e^{-\lambda t} = \left(1 + \lambda t + \frac{\lambda^2 t^2}{2!} + \ldots + \frac{\lambda^n t^n}{n!}\right) e^{-\lambda t}. \tag{10.4}$$

10.2.2 Warm Standby System

Figure 10.3 illustrates the DFT model of a warm standby system with a primary component A and n warm standby components A_i ($i = 1, 2, \ldots n$). For simplicity, all system components are assumed to fail exponentially with an identical failure rate of λ when they are online and active. For components in the warm standby mode, the failure rate is α ($\alpha \leq \lambda$).

Example 10.2 Figure 10.4 shows the state transition diagram for the warm standby system with one primary component A and one WSP A_1. The state labeled (A, A_1) is an initial system state in which A is operational and A_1 is in the warm standby state; the state (A_1) is a system operational state where A has failed and A_1 has been activated to replace A and is online functioning; the state (A) is a system operational state where A_1 has failed and the primary component A is still online functioning; the state (F) denotes the system failure state. Note that the state transition from (A, A_1) to (A) is characterized by the reduced failure rate α of component A_1.

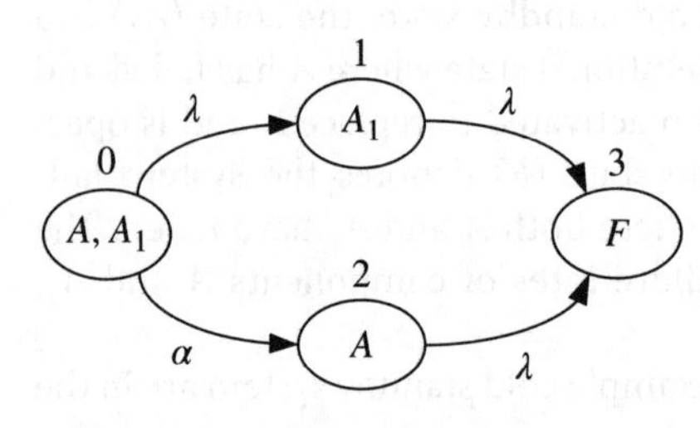

Figure 10.4 Markov model of the warm standby system with one spare.

According to (10.1), the state equations of this example warm standby system are in the form of (10.5).

$$\begin{bmatrix} -(\lambda+\alpha) & 0 & 0 & 0 \\ \lambda & -\lambda & 0 & 0 \\ \alpha & 0 & -\lambda & 0 \\ 0 & \lambda & \lambda & 0 \end{bmatrix} \bullet \begin{bmatrix} P_0(t) \\ P_1(t) \\ P_2(t) \\ P_3(t) \end{bmatrix} = \begin{bmatrix} P_0'(t) \\ P_1'(t) \\ P_2'(t) \\ P_3'(t) \end{bmatrix}. \tag{10.5}$$

Solving (10.5) using the Laplace transform–based method described in [31], and initial probabilities of $P_0(0) = 1$, $P_1(0) = P_2(0) = P_3(0) = 0$, the state probabilities can be obtained. The reliability of the warm standby system with one primary component and one spare at time t is further obtained as [23]:

$$R(t) = \left(1 + \frac{\lambda}{\alpha}\right) e^{-\lambda t} - \frac{\lambda}{\alpha} e^{-(\lambda+\alpha)t}. \tag{10.6}$$

In general, the polynomial representing the reliability of a general k-out-of-n warm standby system at time t can be derived as [23]:

$$R(t) = \frac{1}{(n-k)!\alpha^{n-k}} \sum_{i=0}^{n-k} (-1)^i C_{n-k}^i \cdot \left[\prod_{j=0, j\neq i}^{n-k} (k\lambda + j\alpha)\right] e^{-(k\lambda+i\alpha)t}. \tag{10.7}$$

The reliability of a 1-out-of-n warm standby system is given by setting $k = 1$ in (10.7).

10.3 Decision Diagram–Based Method

The Markov-based method typically requires exponential time to failure (*ttf*) distribution for the system components [34]. In this section, combinatorial methods based on sequential binary decision diagrams (SBDDs) are presented for reliability analysis of cold and warm standby systems, which are applicable to any arbitrary *ttf* distributions. Refer to [34–37] for other decision diagram–based methods for considering effects of imperfect fault coverage or backups in the reliability analysis of standby systems.

10.3.1 Cold Standby System

The SBDD-based combinatorial method for reliability analysis of cold standby systems involves a three-step process [19]: system DFT conversion, system SBDD generation, and system SBDD evaluation, which are presented as follows:

Step 1: System DFT conversion

In this step, CSP gates in the system DFT model are replaced with sequential events. Specifically, each CSP gate, e.g. with primary component P and a spare S is replaced with a *sequential* event $(P \prec S)$, where $\prec$ denotes that the component to the right of the symbol is powered up and starts to work and then fails only after the component to the left of the symbol has failed.

In the case of multiple CSP gates sharing the same cold standby component, an OR gate is used to connect the sequential events corresponding to those CSP gates. The OR gate is used because if the standby component is used to replace the primary component of one CSP gate, it becomes unavailable for replacing primary components of other CSP gates.

Example 10.3 Figure 10.5a shows the DFT of a processor subsystem in a computer system containing three processors (A, B, and C) [38]. Processors A and B are primary ones sharing the same cold standby processor C. The processor subsystem fails when no processor is operating correctly. Figure 10.5b shows the fault tree (FT) after replacing the CSP gates with sequential events; sequential event $(A \prec C)$ is generated from the

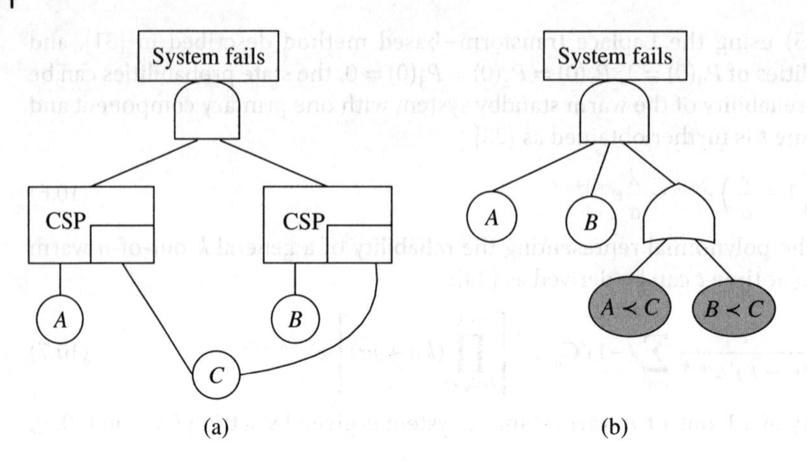

Figure 10.5 An example of a cold standby system. (a) Original DFT; (b) FT after replacement.

left CSP gate, and $(B \prec C)$ is generated from the right CSP gate in Figure 10.5a. The two sequential events are connected by an OR gate since they share the same cold standby component C.

Step 2: System SBDD generation

Similar to the traditional binary decision diagram (BDD) generation (Section 2.4.2), the SBDD model is generated from the converted FT model in a bottom-up manner using manipulation rules of (2.23).

Consider the FT model in Figure 10.5b. Assume the variable ordering is $A < B < (A \prec C) < (B \prec C)$. Figure 10.6 shows the final system SBDD model of the example cold standby processor subsystem.

Different from the traditional BDD, the SBDD model can contain sequential events, as illustrated in Figure 10.6. Note that the SBDD model generated via the bottom-up generation process may contain some *invalid* nodes (event indicated by an invalid node is in conflict with event indicated by its predecessor node) on the path from the root node of the SBDD to a sink node. The invalid nodes must be removed from the final system SBDD model. Refer to [19] for examples containing invalid nodes.

Step 3: System SBDD evaluation

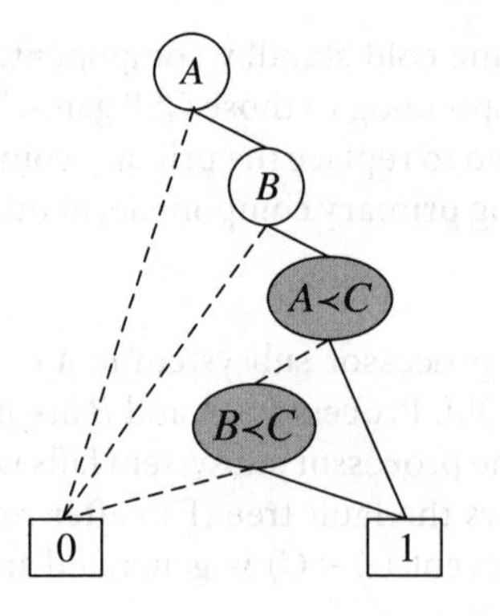

Figure 10.6 SBDD of the example cold standby system.

Similar to the traditional BDD, each path from the root node to sink node 1 in an SBDD model indicates a disjoint combination of event occurring or non-occurring that can lead to the entire system failure. The system unreliability can be obtained by adding probabilities of all the disjoint paths to sink node 1. Different from the traditional BDD evaluation, dependencies among events on the same path must be addressed during the evaluation of the path probability.

Consider the SBDD model of Figure 10.6. There are two disjoint paths leading to node 1:

$$A \Rightarrow B \Rightarrow (A \prec C),$$

$$A \Rightarrow B \Rightarrow \neg(A \prec C) \Rightarrow (B \prec C).$$

Thus, the unreliability of the example cold standby processor subsystem is:

$$UR_{system}(t) = P(A \cdot B \cdot (A \prec C)) + P(A \cdot B \cdot \neg(A \prec C) \cdot (B \prec C)). \tag{10.8}$$

The combined event $\{A{\cdot}B{\cdot}(A \prec C)\}$ in (10.8) can be reduced to $\{B{\cdot}(A \prec C)\}$ as event $(A \prec C)$ implies the occurrence of event A. The reduced event $\{B{\cdot}(A \prec C)\}$ can be further translated into $\{(A \rightarrow B){\cdot}(A \prec C)\}$ as $(A \prec C)$ implies that the standby component C replaces A after A fails, and thus B must fail after A. The symbol $\rightarrow$ denotes the order of precedence (the component to the left of the symbol fails before the component to the right of the symbol). By similar reasoning, event $\{A{\cdot}B{\cdot}\neg(A \prec C){\cdot}(B \prec C)\}$ in (10.8) can be translated to $\{(B \rightarrow A){\cdot}(B \prec C)\}$. Thus (10.8) can be rewritten as:

$$UR_{system}(t) = P((A \rightarrow B) \cdot (A \prec C)) + P((B \rightarrow A) \cdot (B \prec C)). \tag{10.9}$$

According to (8.10), given that all the n components begin to work at $t = 0$ simultaneously, the occurrence probability of a general sequential failure event of these n component, denoted as $X_1 \rightarrow X_2 \rightarrow \ldots \rightarrow X_n$ can be evaluated as:

$$P(X_1 \rightarrow X_2 \rightarrow \ldots \rightarrow X_n) = \int_0^t \int_{\tau_1}^t \cdots \int_{\tau_{n-1}}^t \prod_{k=1}^{n} f_{X_k}(\tau_k)\, d\tau_n \ldots d\tau_2 d\tau_1. \tag{10.10}$$

According to [19], the occurrence probability of the cold standby sequential event $X_1 \prec X_2 \prec \ldots \prec X_n$ can be evaluated as:

$$P(X_1 \prec X_2 \prec \ldots \prec X_n) = \int_0^t \int_0^{t-\tau_1} \cdots \int_0^{t-\tau_1-\tau_2-\ldots-\tau_{n-1}} \prod_{k=1}^{n} f_{X_k}(\tau_k) d\tau_n \ldots d\tau_2 d\tau_1. \tag{10.11}$$

$f_{X_k}(t)$ in (10.10) and (10.11) is the probability density function (*pdf*) of *ttf* of component X_k.

Note that the precedence symbol defined for CSP $\prec$ implies that the component to the right of the symbol cannot start to work until the component to the left of the symbol has failed, while the components on both sides of $\rightarrow$ can start to work at the same time.

Applying (10.10) and (10.11), one obtains

$$P(A \rightarrow B) = \int_0^t \int_{\tau_1}^t f_A(\tau_1) f_B(\tau_2) d\tau_2 d\tau_1, \tag{10.12}$$

$$P(A \prec C) = \int_0^t \left[\int_0^{t-\tau_1} f_C(\tau_3) d\tau_3 \right] f_A(\tau_1) d\tau_1. \tag{10.13}$$

Combining (10.12) and (10.13), $P((A \to B) \cdot (A \prec C))$ in (10.9) is obtained as

$$P((A \to B) \cdot (A \prec C)) = \int_0^t \int_{\tau_1}^t \int_0^{t-\tau_1} f_A(\tau_1) f_B(\tau_2) f_C(\tau_3) d\tau_3 d\tau_2 d\tau_1. \tag{10.14}$$

Similarly, $P((B \to A) \cdot (B \prec C))$ in (10.9) is obtained as

$$P((B \to A) \cdot (B \prec C)) = \int_0^t \int_{\tau_2}^t \int_0^{t-\tau_2} f_B(\tau_2) f_A(\tau_1) f_C(\tau_3) d\tau_3 d\tau_1 d\tau_2. \tag{10.15}$$

According to (10.9), the system unreliability is evaluated as

$$\begin{aligned} UR_{system}(t) = & \int_0^t \int_{\tau_1}^t \int_0^{t-\tau_1} f_A(\tau_1) f_B(\tau_2) f_C(\tau_3) d\tau_3 d\tau_2 d\tau_1 \\ & + \int_0^t \int_{\tau_2}^t \int_0^{t-\tau_2} f_B(\tau_2) f_A(\tau_1) f_C(\tau_3) d\tau_3 d\tau_1 d\tau_2. \end{aligned} \tag{10.16}$$

Example 10.3 Analysis Results: Consider two sets of failure parameters. In set 1, all the three processors fail exponentially with constant rates: $\lambda_A = 0.001$/day; $\lambda_B = 0.003$/day; and $\lambda_C = 0.0025$/day (after activation); the *pdf* in (10.16) is $f_k(t) = \lambda_k e^{-\lambda_k t}$. In Set 2, *ttf* of all the three processors follow the *Weibull* distribution with scale (λ) and shape (α) parameters: $\lambda_A = 0.001$, $\alpha_A = 1$; $\lambda_B = 0.003$, $\alpha_B = 2$; $\lambda_C = 0.0025$, $\alpha_C = 2.5$ (after activation); the *pdf* is $f_k(t) = \alpha_k (\lambda_k)^{\alpha_k} t^{\alpha_k - 1} e^{-(\lambda_k t)^{\alpha_k}}$.

Table 10.1 presents the system unreliability for different values of the mission time using the two sets of failure parameters.

10.3.2 Warm Standby System

The SBDD-based approach for reliability analysis of warm standby systems involves three steps [25]: system DFT conversion, system SBDD generation, and system SBDD evaluation, which are presented as follows:

Step 1: System DFT conversion

In this step, WSP gates in the system DFT model are replaced with sequential events. Specifically, each WSP gate, e.g. with primary component P and a spare S is replaced with two *sequential* events $(P \to S)$ and $(S \to P)$ connected by an OR gate. Sequence event $(P \to S)$ means the primary component fails first then the spare fails with a full failure rate λ_S; sequential event $(S \to P)$ represents the spare fails before the primary component with a reduced failure rate α_S.

Table 10.1 Unreliability of the example cold standby processor subsystem.

t (days)	Exponential	Weibull
300	0.062571	0.023292
600	0.249829	0.275869
1000	0.511374	0.614208

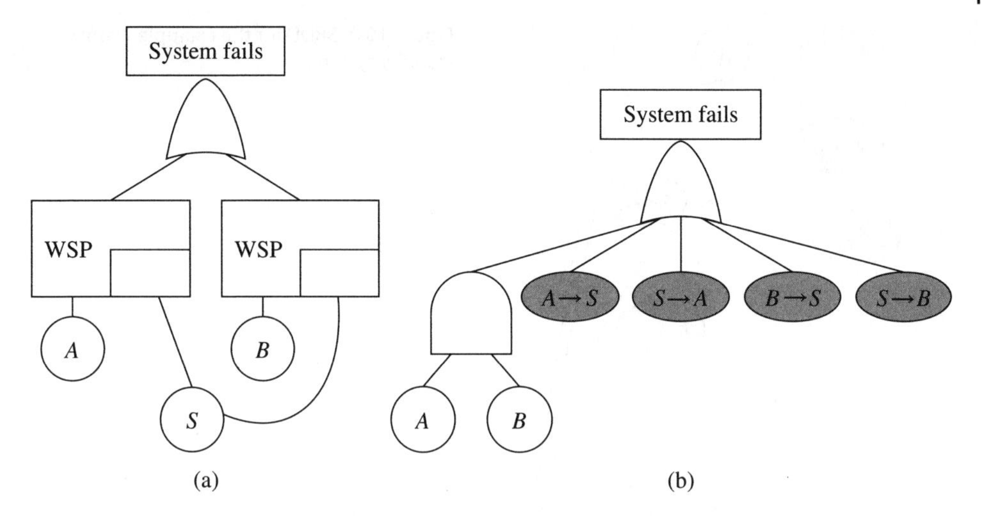

Figure 10.7 An example warm standby system. (a) Original DFT; (b) FT after replacement.

Example 10.4 Figure 10.7a shows the DFT of an example warm standby hard disk drive system containing three disks (A, B, and S) [25]. Disks A and B are primary ones sharing the same warm standby disk S. Figure 10.7b shows the FT after replacing the WSP gates with sequential events. The left WSP gate is replaced with *sequential* events ($A \to S$) and ($S \to A$); the right WSP gate is replaced with *sequential* events ($B \to S$) and ($S \to B$).

Step 2: System SBDD generation

Similar to the traditional BDD generation (Section 2.4.2), the SBDD model is generated from the converted FT model in a bottom-up manner using manipulation rules of (2.23).

Consider the FT model in Figure 10.7b. Assume the variable ordering is $A < B < (S \to A) < (A \to S) < (S \to B) < (B \to S)$. Figure 10.8 shows the final system SBDD model of the example warm standby disk system.

The two nodes with lines background in Figure 10.8 are invalid nodes because they contradict with their grandparent node B in the path from the rood note. Specifically, the left-edge of node B means B is operational, thus ($S \to B$) meaning B fails after S fails and ($B \to S$) meaning B fails before S fails should be removed from the path. Figure 10.9 shows the final SBDD after removing the two invalid nodes.

Step 3: System SBDD evaluation

The unreliability of a warm standby sparing system is obtained as the sum of probabilities of all the disjoint paths from the root node to sink node 1 in the SBDD model. Note that the sequence or temporal dependence caused by the warm standby sparing must be considered during the model evaluation, in particular the path probability evaluation in this step.

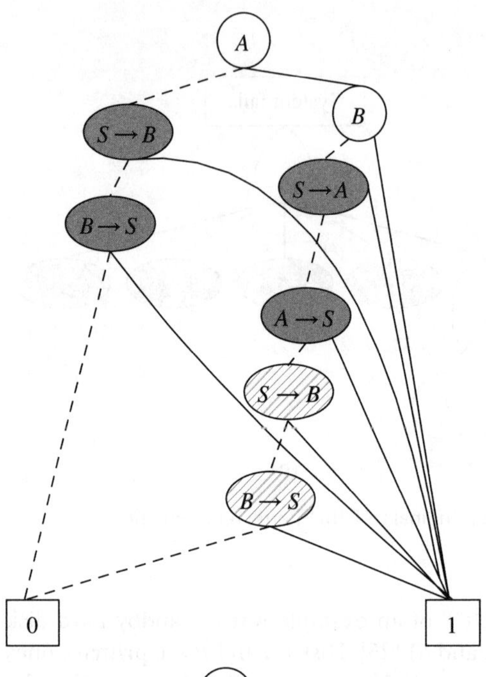

Figure 10.8 SBDD for the example warm standby system.

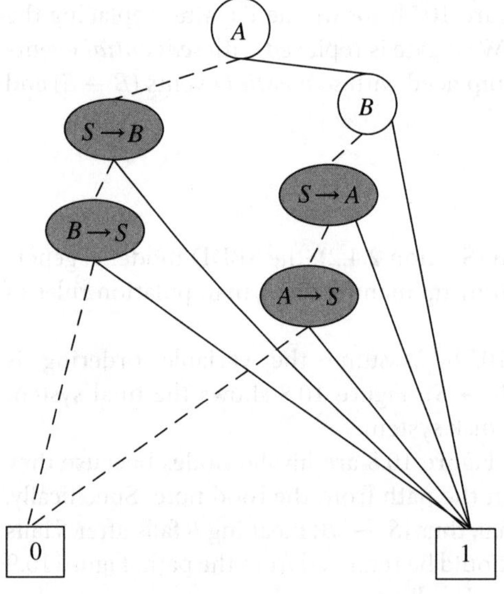

Figure 10.9 The final SBDD of the hard disk system.

Consider the SBDD in Figure 10.9. There are five paths leading to the system failure:

$$A \Rightarrow B,$$
$$A \Rightarrow \neg B \Rightarrow (S \rightarrow A),$$
$$A \Rightarrow \neg B \Rightarrow \neg(S \rightarrow A) \Rightarrow (A \rightarrow S),$$
$$\neg A \Rightarrow (S \rightarrow B),$$
$$\neg A \Rightarrow \neg(S \rightarrow B) \Rightarrow (B \rightarrow S).$$

Table 10.2 Unreliability of the example warm standby system.

t (days)	Exponential	Weibull
300	0.404359	0.183705
600	0.758792	0.716139
1000	0.938432	0.977974

Applying reductions similar to those discussed in Section 10.3.1, the unreliability of the example warm standby system is calculated as:

$$\begin{aligned} UR_{system} &= P(A \cap B) + P(\neg B \cap (S \to A)) + P(\neg B \cap (A \to S)) \\ &\quad + P(\neg A \cap (S \to B)) + P(\neg A \cap (B \to S)). \end{aligned} \tag{10.17}$$

Applying (10.10), the system unreliability in (10.17) can be evaluated as:

$$\begin{aligned} UR_{system} = & \left(\int_0^t f_A(\tau_2)d\tau_2\right)\left(\int_0^t f_B(\tau_3)d\tau_3\right) \\ & + \left(1 - \int_0^t f_B(\tau_3)d\tau_3\right)\int_0^t \int_{\tau_1}^t f_{S,\alpha}(\tau_1)f_A(\tau_2)d\tau_2 d\tau_1 \\ & + \left(1 - \int_0^t f_B(\tau_3)d\tau_3\right)\int_0^t \int_0^{t-\tau_2} f_A(\tau_2)f_{S,\lambda}(\tau_1)\left(1 - \int_0^{\tau_2} f_{S,\alpha}(\tau_1)d\tau_1\right)d\tau_1 d\tau_2 \\ & + \left(1 - \int_0^t f_A(\tau_2)d\tau_2\right)\int_0^t \int_{\tau_1}^t f_{S,\alpha}(\tau_1)f_B(\tau_3)d\tau_3 d\tau_1 \\ & + \left(1 - \int_0^t f_A(\tau_2)d\tau_2\right)\int_0^t \int_0^{t-\tau_3} f_B(\tau_3)f_{S,\lambda}(\tau_1)\left(1 - \int_0^{\tau_3} f_{S,\alpha}(\tau_1)d\tau_1\right)d\tau_1 d\tau_3. \end{aligned} \tag{10.18}$$

Example 10.4 Analysis Results: Consider two sets of failure parameters. In set 1, all the three disks fail exponentially with constant rates: $\lambda_A = 0.001$/day, $\lambda_B = 0.003$/day, $\alpha_S = 0.0015$/day (before activation), and $\lambda_S = 0.0025$/day (after activation); the *pdf* in (10.18) is $f_k(t) = \lambda_k e^{-\lambda_k t}$.

In Set 2, *ttf* of all the three disks follow the *Weibull* distribution with the same *shape* ($\alpha =$ 1.8) parameter and different *scale* parameters: $\lambda_A = 0.001$, $\lambda_B = 0.003$, $\alpha_S = 0.0015$ (before activation), and $\lambda_S = 0.0025$ (after activation); the *pdf* is $f_k(t) = \alpha_k (\lambda_k)^{\alpha_k} t^{\alpha_k - 1} e^{-(\lambda_k t)^{\alpha_k}}$.

Table 10.2 presents the system unreliability for different values of mission time using the two sets of failure parameters.

10.4 Approximation Method

This section presents fast approximation models based on the central limit theorem (CLT) for reliability analysis of 1-out-of-*n* homogeneous and heterogeneous cold standby systems. Refer to [39] for the extension of the approximate model to a more general k-out-of-n cold-standby system requiring k primary and online components. Refer to [40] for another extension of the approximate method for reliability analysis of warm standby systems.

10.4.1 Homogeneous Cold Standby System

Consider a homogeneous cold standby system with n s-identical components with A_1 being the primary component and $A_2, \ldots, A_n$ being cold standby spares, which are used in the order of their indices. Specifically, when A_1 fails at time T_1, it is replaced by A_2; when A_2 fails at time $T_1 + T_2$, it is replaced by A_3, and so on. The replacement is assumed to happen instantaneously. The time durations $T_1, T_2, \ldots$ are assumed to be independent and identically distributed (i.i.d.). The entire cold standby system fails when all of the n components have failed.

Define S_n as the time until the nth failure, i.e. the system *ttf*, which can be expressed as $S_n = T_1 + T_2 + \cdots T_n = \sum_{i=1}^{n} T_i$ [41]. Let μ denote the mean time to failure (MTTF) of each component. Based on the strong law of large numbers [31], one obtains $\frac{S_n}{n} \to \mu$ as $n \to \infty$. According to the CLT [31], $S_n = \sum_{i=1}^{n} T_i$ is asymptotically normally distributed, particularly,

$$\frac{S_n - n\mu}{\sigma\sqrt{n}} \to N(0, 1). \tag{10.19}$$

σ in (10.19) is standard deviation of T_i. Thus, the unreliability of the 1-out-of-n homogeneous cold-standby system can be approximated as [39]:

$$UR_{system} = \mathrm{P}(S_n \leq t) \approx \Phi\left(\frac{t - n\mu}{\sigma\sqrt{n}}\right). \tag{10.20}$$

$\Phi(\cdot)$ in (10.20) denotes the distribution function of the standard normal distribution $N(0,1)$.

Example 10.5 (Exponential Distribution): Consider a 1-out-of-n cold standby system with identical components failure times following an exponential distribution. The MTTF of each component is assumed to be $\mu = 1000$ hours.

Based on the approximate model in (10.20), the system unreliability in the case of exponentially distributed component *ttf* is estimated as

$$UR_{\text{exponential}} \approx \Phi\left(\frac{t - n\mu}{\sigma\sqrt{n}}\right) = \frac{1}{2}\left[1 + \mathrm{erf}\left(\frac{t - n\mu}{\mu\sqrt{2n}}\right)\right]. \tag{10.21}$$

The error function erf(·) in (10.21) is defined as $\mathrm{erf}(z) = \frac{2}{\sqrt{\pi}} \int_0^z e^{-x} dx$. For the exponential distribution, $\mu = \sigma$.

To evaluate the accuracy of the approximation model, numerical results obtained using (10.21) are compared to the exact results obtained using the CTMC–based method in Section 10.2.1. Applying (10.4) with $\lambda = 1/\mu$, the exact unreliability of the example cold standby system with $(n - 1)$ cold standby spares is

$$UR'_{\text{exponential}} = 1 - e^{-t/\mu} \sum_{i=0}^{n-1} \frac{(t/\mu)^i}{i!}. \tag{10.22}$$

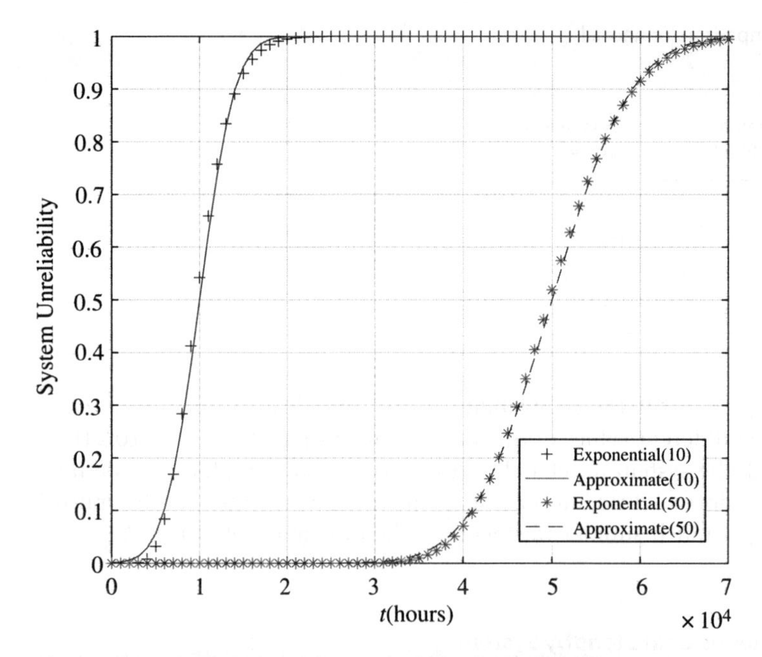

Figure 10.10 Unreliability comparison for exponential distribution.

Figure 10.10 shows the graphical comparison of unreliability results for systems with 10 components and 50 components obtained using the approximate model (10.21) and the exact results obtained using (10.22). Apparently, the approximate model serves as a very good approximation of the real system unreliability. In addition, accuracy of the approximate model increases as the number of system components n increases.

Example 10.6 (Normal Distribution): Consider a 1-out-of-n cold-standby system with identical components failure times following a Normal distribution. The MTTF of each component is assumed to be $\mu = 1000$ hours.

If $T_1, T_2, \ldots T_n$ are n i.i.d. normal *r.v.*s with mean μ and standard deviation σ, then their sum is also normally distributed [42]: $S_n = T_1 + T_2 + \cdots + T_n = \sum_{i=1}^{n} T_i \sim N(n\mu, n\sigma^2)$.

Applying (10.20), the system unreliability in the case of normally distributed component *ttf* is

$$UR_{\text{normal}} = \Phi\left(\frac{t - n\mu}{\sigma\sqrt{n}}\right) = \frac{1}{2}\left[1 + \text{erf}\left(\frac{t - n\mu}{\sigma\sqrt{2n}}\right)\right]. \quad (10.23)$$

To evaluate the accuracy of the approximation model, numerical results obtained using (10.23) are compared to the exact results obtained using (10.11). For the normal distribution, (10.11) is rewritten as

$$UR'_{\text{normal}} = \int_0^t \int_0^{t-\tau_1} \cdots \int_0^{t-\tau_1-\tau_2\cdots-\tau_{n-1}} \frac{1}{(\sigma\sqrt{2\pi})^n} e^{-\frac{(\tau_1-\mu)^2+(\tau_2-\mu)^2+\cdots+(\tau_n-\mu)^2}{2\sigma^2}} d\tau_n \cdots d\tau_2 d\tau_1. \quad (10.24)$$

Table 10.3 Results comparison for normal distribution.

t	Approximate model (10.23)	Exact method (10.24)
3000	0.0062	0.0062
3500	0.1056	0.1056
4000	0.5000	0.5000
4500	0.8944	0.8944

Using $\mu = 1000$ and $\sigma = 200$, the unreliability of the example cold standby system is evaluated for $n = 4$ at different values of the mission time using (10.23) and (10.24). As shown in Table 10.3, the system unreliability results obtained using the two methods match exactly for the normal distribution case. However, the evaluation of the approximate model (10.23) is much simpler (faster) than the evaluation of (10.24) based on multiple integrals.

10.4.2 Heterogeneous Cold Standby System

Consider a heterogeneous cold standby system with n nonidentical components with A_1 being the primary component and $A_2, \ldots, A_n$ being cold standby spares, which are used in the order of their indices in the case of the online component becoming malfunctioned. Similar to the homogeneous case, when A_1 fails at time T_1, it is replaced by A_2; when A_2 fails at time $T_1 + T_2$, it is replaced by A_3, and so on. The entire system *ttf* is $S_n = T_1 + T_2 + \cdots T_n = \sum_{i=1}^{n} T_i$.

Define μ_i as MTTF of T_i and σ_i as standard deviation of T_i. Based on the Lyapunov CLT [43], if for some $\delta > 0$, the Lyapunov's condition $\lim_{n\to\infty} \sum_{i=1}^{n} \mathrm{E}[|T_i - \mu_i|^{2+\delta}] / \left(\sqrt{\sum_{i=1}^{n} \sigma_i^2}\right)^{2+\delta} = 0$ is met, then as n goes to infinity, the sum of $(T_i - \mu_i)/\sqrt{\sum_{i=1}^{n} \sigma_i^2}$ converges in distribution to a standard normal *r.v.*, that is,

$$\frac{1}{\sqrt{\sum_{i=1}^{n} \sigma_i^2}} \sum_{i=1}^{n} (T_i - \mu_i) = \frac{S_n - \sum_{i=1}^{n} \mu_i}{\sqrt{\sum_{i=1}^{n} \sigma_i^2}} \to N(0, 1). \tag{10.25}$$

Thus, the unreliability of the 1-out-of-n heterogeneous cold standby system can be approximated as [44]:

$$UR_{system} = P(S_n \le t) \approx \Phi\left(\frac{t - \sum_{i=1}^{n} \mu_i}{\sqrt{\sum_{i=1}^{n} \sigma_i^2}}\right). \tag{10.26}$$

Same as in (10.20), $\Phi(\cdot)$ denotes the distribution function of the standard normal distribution $N(0,1)$.

Example 10.7 (Exponential Distribution): Consider a 1-out-of-n cold standby system with nonidentical components *ttf* following exponential distributions. Specifically, assume the *ttf* of the n components (when they are active) are exponentially distributed but with different failure rates, $\lambda_1, \lambda_2, \ldots, \lambda_n$. The MTTF of T_i is $\mu_i = \frac{1}{\lambda_i}$ and the variance is $\sigma_i^2 = \frac{1}{\lambda_i^2}$. It is shown in [44] that the Lyapunov's condition can be satisfied for the exponential distribution. Thus applying (10.26), the unreliability of the 1-out-of-n heterogeneous cold standby systems with different exponential component *ttf* can be approximated as

$$UR_{\text{exponential}} \approx \Phi\left(\frac{t - \sum_{i=1}^{n} \mu_i}{\sqrt{\sum_{i=1}^{n} \sigma_i^2}}\right) = \frac{1}{2}\left[1 + \operatorname{erf}\left(\frac{t - \sum_{i=1}^{n} \frac{1}{\lambda_i}}{\sqrt{2\sum_{i=1}^{n}\left(\frac{1}{\lambda_i}\right)^2}}\right)\right]. \tag{10.27}$$

To evaluate the accuracy of the approximation model, numerical results obtained using (10.27) are compared to the exact solution obtained in [46], which is

$$UR'_{\text{exponential}} = 1 - \sum_{i=1}^{n}\left(e^{-\lambda_i t} \prod_{\substack{j=1 \\ j \neq i}}^{n} \frac{\lambda_j}{\lambda_j - \lambda_i}\right). \tag{10.28}$$

Note that (10.28) can only be applicable to systems where failure rates of system components are all different while the approximate model (10.27) has no such a limitation.

Figure 10.11 shows the graphical comparison of unreliability results for systems with 5 components and 15 components obtained using the approximate model (10.27) and the exact results obtained using (10.28). μ_i is given by generating a row vector of n linearly spaced numbers between 1000 and 2000 hours. For example, MTTF in the 5 components case are 1000, 1250, 1500, 1750, and 2000 hours. Apparently, the approximate model offers a very good approximation of the real system unreliability. In addition, the accuracy of the approximate model increases as the number of system components n increases.

Example 10.8 (Normal Distribution): Consider a 1-out-of-n cold-standby system with non-identical components *ttf* following normal distributions. Specifically, assume $T_1, T_2, \ldots T_n$ are n independent normally distributed *r.v.*s with different mean μ_i and standard deviation σ_i. Thus, the sum $\sum_{i=1}^{n} T_i$ is also normally distributed with mean $\sum_{i=1}^{n} \mu_i$ and variance $\sum_{i=1}^{n} \sigma_i^2$ [46]. The *pdf* of normal distribution for *ttf* of component i is $f_i(t)_{\text{normal}} = \frac{1}{\sqrt{2\pi\sigma_i^2}} e^{-\frac{(t-\mu_i)^2}{2\sigma_i^2}}$. The unreliability of the 1-out-of-n heterogeneous cold

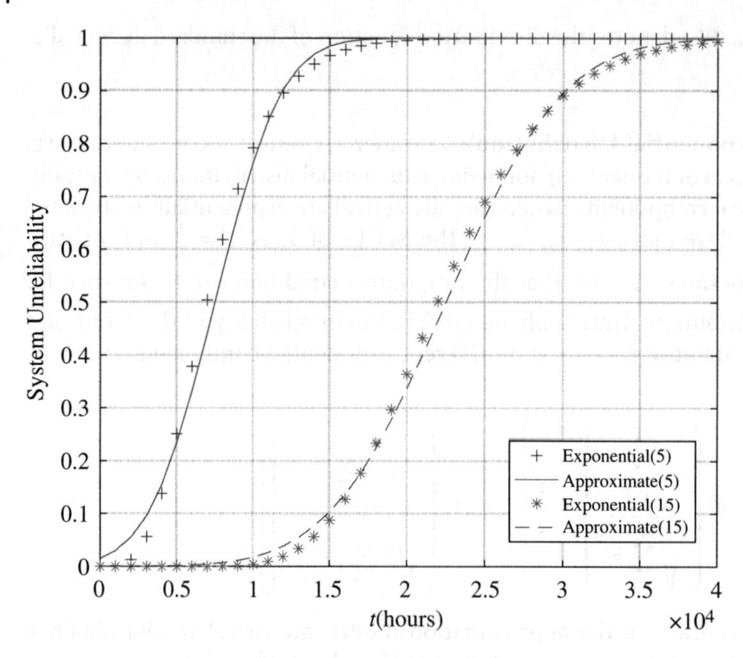

Figure 10.11 Unreliability comparison for exponential distribution.

standby systems with different normal component *ttf* can be evaluated as

$$UR_{\text{normal}} = \Phi\left(\frac{t - \sum_{i=1}^{n} \mu_i}{\sqrt{\sum_{i=1}^{n} \sigma_i^2}}\right) = \frac{1}{2}\left[1 + \text{erf}\left(\frac{t - \sum_{i=1}^{n} \mu_i}{\sqrt{2\sum_{i=1}^{n} \sigma_i^2}}\right)\right]. \tag{10.29}$$

Figure 10.12 shows the unreliability of a 1-out-of-10 cold standby system evaluated using (10.29) as a function of mission time t. Different μ_i of the 10 components are given by generating a row vector of 10 linearly spaced numbers between 1000 and 2000 hours. All components have the same standard deviation $\sigma_i = 100$ hours.

10.5 Event Transition Method

This section describes a recently emerged approach for evaluating performance and reliability characteristics of heterogeneous standby systems. The method is based on presenting the mission execution as a trajectory of parameters of random events in a state-space. It can be applied to standby systems with heterogeneous structure, random replacement time, single and multiple mission phases, and different rules of standby-operation mode transitions. The method allows building recursive numerical algorithms.

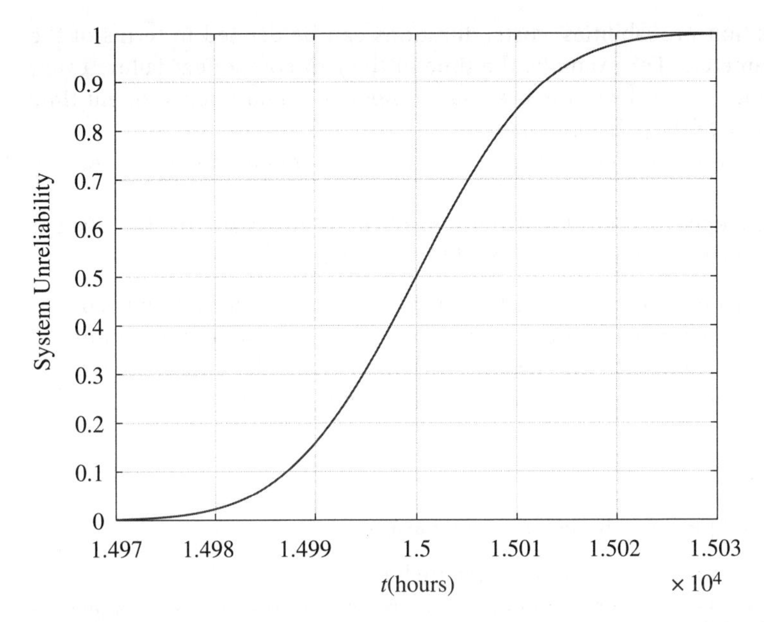

Figure 10.12 Unreliability of 1-out-of-10 cold standby system with normal distribution.

10.5.1 State-Space Representation of System Behavior

The state-space describing the mission behavior should be defined as a space of parameters that completely describe events that may happen during a mission as well as transitions among the events.

For example, for a cold standby system with random replacement time that has to operate for a certain time, the only parameter needed to define the state space is the operation time. Indeed, the *ttf* distribution of the cold standby components depends only on their operation time, and the mission failure/success is determined by the cumulative operation time of the components.

For a k-out-of-n standby system, the state-space is at least k-dimensional because operation times of all k working components affect the system stochastic behavior and the mission failure/success conditions.

The event transition method can be applied under the following assumptions.

1. A finite number of events can happen during a mission. The following standby systems meet this requirement:
   - Standby systems with a finite number of standby components, which restricts the possible number of failures.
   - Standby systems that should perform a fixed amount of work during a fixed time.
2. The mission failure/success is defined in terms of the state-space parameters. For example, the mission fails if the failure of the last standby component occurs when the total operation time does not reach a required value.

3. The event transition probabilities, costs, durations can be defined in terms of the state-space parameters. For example, the time of the j-th component failure event equals to the sum of the time of the previous component failure, replacement time and operation time of the j-th component.
4. The parameters of any state-space event depend only on parameters of the preceding event.
5. Functional dependence among *ttf* distributions of components and the times spent by the components in different modes should be known.

The method can handle any type of distributions for components' *ttf* and replacement time. Statistical dependence among components and random event parameters can exist. Any type of component performance variation (due to load variations, mode transfers, mission phase changes) can be taken into account.

10.5.2 Basic Steps

The method presumes performing the following basic steps.

Step 1: Define the problem state-space. This step includes:

- defining random events that represent the system behavior during the mission execution;
- defining random variables that fully characterize the events;
- formulating the mission success/failure criteria in terms of the random variables.

Step 2: Define the event transitions. This step includes:

- describing all possible transition scenarios;
- determining transition probabilities for each scenario;
- determining scenarios leading to the entire mission failure/success after each event.

Step 3: Determine joint state-space event parameters distribution. This step includes:

- determining joint state-space parameters distribution for the initial event (e.g., the first failure during the mission);
- determining recursive rules for obtaining distribution for state-space parameters of random event j based on distribution for state-space parameters of event $j-1$.

Step 4: Define the mission success/failure scenarios in terms of event parameters for any event and evaluate the mission performance indices for any specific last event of the mission.

Step 5: Determine the entire mission performance indices as sum of the values corresponding to mutually exclusive events of mission success/failure after any fixed number of events.

10.5.3 Warm Standby System

Assume that the order $s(1), s(2), \ldots, s(n)$ determines the activation sequence of components in a 1-out-of-n warm standby system. Consider event $\langle k \rangle$ that the last component from sequence $s(1), \ldots, s(k)$ fails during the operation mode. Let T_k be a *r.v.* representing the time when the event $\langle k \rangle$ happens, and $q_k(t)$ be the *pdf* of T_k. The time when events $T_1, \ldots, T_k$ happen constitute the (one-dimensional) event space. For $k = 1$, since only one component $s(1)$ belongs to the sequence, $q_1(t) = f_{s(1)}(t)$, where $f_{s(1)}(t)$ is the *pdf* of *ttf* for component $s(1)$.

With $q_{k-1}(t)$ and $f_{s(k)}(t)$, the *pdf* $q_k(t)$ can be derived for $k = 2, \ldots, n$ based on considering $\langle k-1\rangle \rightarrow \langle k\rangle$ event transition rules. Specifically, there exist two scenarios that can cause the failure of the last component from sequence $s(1), \ldots, s(k)$ at time t.

Scenario 1: $T_k = T_{k\text{-}1} = t$. The last component from sequence $s_i(1), \ldots, s_i(k-1)$ fails at time t; component $s(k)$ fails earlier during the standby mode. This scenario can occur with probability $F_{s(k)}(t)$, where $F_{s(k)}(t)$ is the *cdf* of *ttf* for component $s(k)$.

Scenario 2: $T_k = t$, $T_{k-1} = t - \tau$. The last component from sequence $s(1), \ldots, s(k-1)$ fails at certain time before t, e.g. $t - \tau$ for $0 \leq \tau \leq t$; component $s(k)$ fails after spending $(t - \tau)$ in the standby mode and then working for time τ in the operation mode.

Based on the two scenarios of $\langle k-1\rangle \rightarrow \langle k\rangle$ event transition, *pdf* of T_k is

$$q_k(t) = q_{k-1}(t)F_{s(k)}(t) + \int_0^t q_{k-1}(t-\tau)f_{s(k)}((t-\tau)\delta_{s(k)} + \tau)d\tau. \tag{10.30}$$

In (10.30), $0 \leq \delta_{s(k)} \leq 1$ represents a deceleration factor of component $s(k)$ during the standby mode, which is used to reflect lower stresses experienced by the component during the warm standby mode than during the operation mode under the commonly used cumulative exposure model [17]. Based on (10.30), $q_k(t)$ can be obtained iteratively for $k = 2, \ldots, n$.

The probability $p_1(t)$ that component $s(1)$ operates at time t is simply the probability that this component does not fail before time t, which is $p_1(t) = 1 - F_{s(1)}(t)$. The probability $p_k(t)$ that component $s(k)$ operates at time t can be evaluated as the probability that $T_{k-1} = t - \tau$ for any $0 \leq \tau \leq t$, and component $s(k)$ has waited for time $t - \tau$ in the warm standby mode and does not fail before spending at least time τ in the operation mode:

$$p_k(t) = \int_0^t q_{k-1}(t-\tau)[1 - F_{s(k)}(\delta_{s(k)}(t-\tau) + \tau)]d\tau. \tag{10.31}$$

As the components operate consecutively, their operation events are mutually exclusive. Therefore, the probability that the system is operational at time t (at least one of its components operates) is thus

$$R(t) = \sum_{k=1}^{n} p_k(t). \tag{10.32}$$

When each system component k has its specific performance g_k, having functions $p_k(t)$ one can obtain the system performance distribution in any time instance t. Indeed, $p_k(t)$ gives the probability that the system operates at time t with performance level g_k.

Example 10.9 Consider a 1-out-of-3 warm standby system consisting of three components characterized by Weibull *ttf* distributions. The scale (λ_j) and shape (α_j) parameters of the distributions, deceleration factor (δ_j), and nominal performance (g_j) of the system components are presented in Table 10.4.

The random system performance $G(t)$ can take four values: $G(t)\in\{0, 20, 27, 32\}$. Figure 10.13 illustrates the probabilities of the system performance levels $p_j(t) = P(G(t) = g_j)$ obtained using numerical integration in a recursive procedure based on (10.30)–(10.32) for two different component activation sequences 1, 2, 3 and 3, 2, 1. For $j = 1, 2, 3$, $p_j(t)$ represent probabilities of $G(t) = g_j$, whereas $p_0(t) = 1 - p_1(t) - p_2(t) - p_3(t)$ represents the system failure probability.

Table 10.4 Component parameters of the example standby system.

Component j	λ_j	α_j	δ_j	g_j
1	280	1.5	0.2	20
2	250	1.1	0.4	27
3	180	2	0.2	32

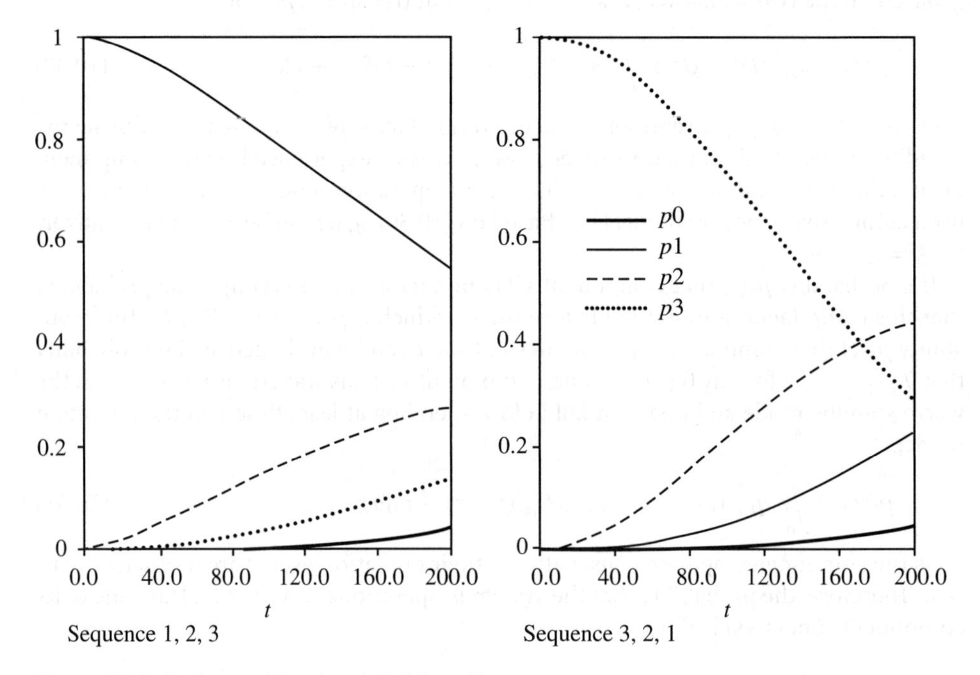

Figure 10.13 Performance level probabilities $p_j(t)$.

Figure 10.14 shows the cumulative system performance distribution $P(G(t) \geq x)$ under the two activation sequences. It can be observed that the probability $P(G(t) \geq 32)$ is always larger for sequence 3, 2, 1 where component 3 with the greatest performance is activated first; the probability $P(G(t) \geq 20)$ is always slightly larger for sequence 1, 2, 3 where the most reliable component 1 is activated first.

10.6 Overview of Optimization Problems

While this chapter focuses on the reliability modeling and analysis of standby sparing systems, numerous research efforts have been made to optimize their performance [47]. A traditional optimization problem solved for standby sparing systems is the redundancy allocation problem (RAP). It determines the number of spares or redundancies to be allocated in a subsystem or system with the purpose of maximizing the entire

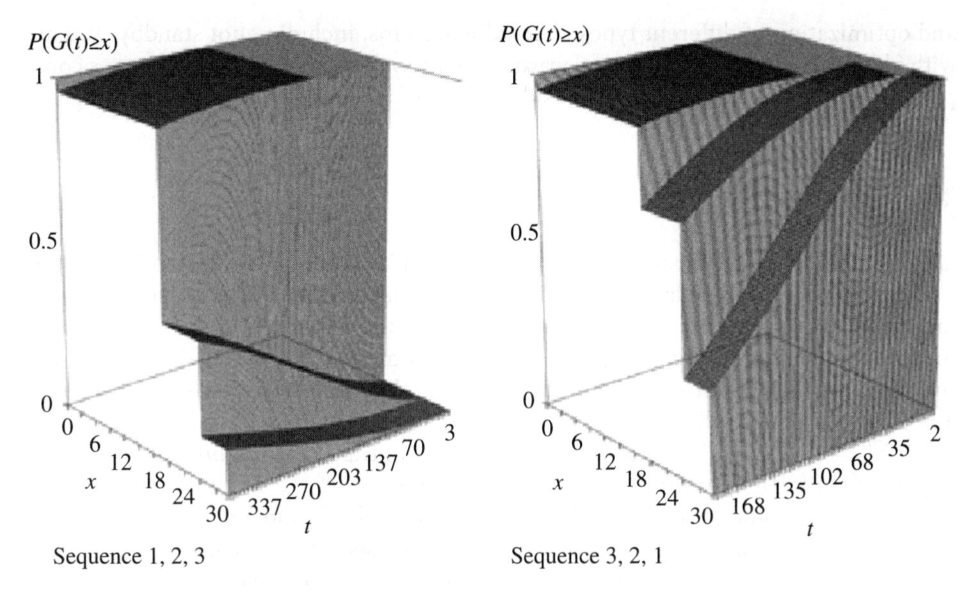

Figure 10.14 Cumulative performance distributions $P(G(t) \geq x)$.

system reliability while satisfying a certain cost constraint, or minimizing the system cost subject to a certain system reliability constraint [22, 48–56].

Recent works revealed that the component activation sequence in a standby sparing system with heterogeneous types of components can heavily affect the system performance metrics [20]. Therefore, a new type of optimization problems called optimal standby element sequencing problem (SESP) has been formulated and solved for different types of standby sparing systems, including for example cold standby systems [14, 20, 57], warm standby systems [17, 33], mixed hot and cold standby systems [58, 59], standby systems with standby mode transfers [16, 60–62], standby systems with checkpointing or backup schemes [63–65], and standby multi-phase systems [66]. Given the desired redundancy level and available component choices, the objective of the SESP is to select initiation sequence of system components to maximize the system reliability, or to minimize the expected system cost while providing a desired level of system reliability. The joint redundancy allocation and sequencing problem in which the list of different spares is chosen for each subsystem, and sequence of spares activation is determined is considered in [67].

For standby computing systems with checkpoints or backups assisting efficient system recovery in the case of a component failure occurring, backup distribution can have a nonmonotonic effect on the system performance, leading to the optimal backup distribution problem. Readers may refer to [68–73] for the formulation and solution of this optimization problem for standby systems with different types of backup schemes.

For standby systems used in life-critical applications, being able to survive the system can be more critical than accomplishing the specified mission due to factors related to safety of personnel and environments or cost. Therefore, a mission abort policy may be implemented to enhance the system survivability under those circumstances [74–76]. Readers may refer to [45, 77, 78] for modeling the mission abort policy in the analysis and optimization of different types of standby systems, including hot standby systems with identical components [45], and warm standby systems having heterogeneous components with [77] or without [78] considering state-dependent component loading.

10.7 Summary

Components in standby sparing systems, particularly in cold and warm standby sparing systems, exhibit dynamic failure behaviors. In this chapter, four different methods were presented to address these dynamic component failure behaviors in system reliability modeling and analysis, including the CTMC-based method, the SBDD-based method, the approximation method based on the CLT, and the event transition-based method. The methods presented assume the standby sparing system has a perfect switching mechanism. Refer to [33] for an extension of the event transition method for addressing imperfect fault detection and switching mechanisms.

Monte Carlo simulations [79], Bayesian networks [80], and universal generating function-based techniques [20] have also been proposed for the reliability analysis of standby sparing systems. Interested readers may refer to the relevant references for details of these methods.

References

1 Johnson, B.W. (1989). *Design and Analysis of Fault Tolerant Digital Systems*. Addison-Wesley.

2 Kilmer, W.L. (1994). Failure distributions for local vs. global and replicate vs. standby redundancies. *IEEE Transactions on Reliability* 43 (3): 476–483.

3 ANSI/IEEE 446-1995 (1995). *IEEE Recommended Practice for Emergency and Standby Power Systems for Industrial and Commercial Applications (IEEE Orange Book)*. IEEE.

4 Pandey, D., Jacob, M., and Yadav, J. (1996). Reliability analysis of a powerloom plant with cold standby for its strategic unit. *Microelectronic and Reliability* 36 (1): 115–119.

5 Zhang, T., Xie, M., and Horigome, M. (2006). Availability and reliability of k-out-of-(M+N): G warm standby systems. *Reliability Engineering & System Safety* 91 (4): 381–387.

6 Elerath, J.G. and Pecht, M. (2009). A highly accurate method for assessing reliability of redundant arrays of inexpensive disks (RAID). *IEEE Transactions on Computers* 58 (3): 289–299.

7 Hsieh, C. and Hsieh, Y. (2003). Reliability and cost optimization in distributed computing systems. *Computers and Operations Research* 30: 1103–1119.

8 Luo, W., Qin, X., Tan, X.-C. et al. (2009). Exploiting redundancies to enhance schedulability in fault-tolerant and real-time distributed systems. *IEEE Transactions on Systems, Man and Cybernetics, Part A: Systems and Humans* 39 (3): 626–639.

9 Pham, H., Phan, H.K., and Amari, S.V. (1995). A general model for evaluating the reliability of telecommunications systems. *Communications in Reliability, Maintainability, and Supportability—An International Journal* 2: 4–13.

10 Sterritt, R. and Bustard, D.W. (2002). Fusing hard and soft computing for fault management in telecommunications systems. *IEEE Transactions on Systems, Man, and Cybernetics, Part C: Applications and Reviews* 32 (2): 92–98.
11 Johnson, B.W. and Julich, P.M. (1985). Fault-tolerant computer system for the A129 helicopter. *IEEE Transactions on Aerospace and Electronic Systems* 21 (2): 220–229.
12 Sklaroff, J.R. (1976). Redundancy management technique for space shuttle computers. *IBM Journal on Research and Development* 20 (1): 20–28.
13 Sinaki, G. (1994). Ultra-reliable fault tolerant inertial reference unit for spacecraft. In: *Proceedings of the Annual Rocky Mountain Guidance and Control Conference*, 239–248. San Diego, CA: Univelt Inc.
14 Levitin, G., Xing, L., and Dai, Y. (2013). Sequencing optimization in k-out-of-n cold-standby systems considering mission cost. *International Journal of General Systems* 42 (8): 870–882.
15 Misra, K.B. (1992). *Reliability Analysis and Prediction: A Methodology Oriented Treatment*. Elsevier.
16 Levitin, G., Xing, L., and Dai, Y. (2015). Reliability and mission cost of 1-out-of-N: G systems with state-dependent standby mode transfers. *IEEE Transactions on Reliability* 64 (1): 454–462.
17 Levitin, G., Xing, L., and Dai, Y. (2013). Optimal sequencing of warm standby elements. *Computers & Industrial Engineering* 65 (4): 570–576.
18 Amari, S.V., Pham, H., and Misra, R.B. (2012). Reliability characteristics of k-out-of-n warm standby systems. *IEEE Transactions on Reliability* 61 (4): 1007–1018.
19 Xing, L., Tannous, O., and Dugan, J.B. (2012). Reliability analysis of non-repairable cold-standby systems using sequential binary decision diagrams. *IEEE Transactions on Systems, Man, and Cybernetics, Part A: Systems and Humans* 42 (3): 715–726.
20 Levitin, G., Xing, L., and Dai, Y. (2013). Cold-standby sequencing optimization considering mission cost. *Reliability Engineering and System Safety* 118: 28–34.
21 Van Gemund, A.J.C. and Reijns, G.L. (2012). Reliability analysis of k-out-of-n systems with single cold standby using Pearson distributions. *IEEE Transactions on Reliability* 61 (2): 526–532.
22 Coit, D.W. (2001). Cold-standby redundancy optimization for non-repairable systems. *IIE Transactions* 33: 471–478.
23 She, J. and Pecht, M.G. (1992). Reliability of a k-out-of-n warm-standby system. *IEEE Transactions on Reliability* 41 (1): 72–75.
24 Yun, W.Y. and Cha, J.H. (2010). Optimal design of a general warm standby system. *Reliability Engineering & System Safety* 95 (8): 880–886.
25 Tannous, O., Xing, L., and Dugan, J.B. (2011). Reliability analysis of warm standby systems using sequential BDD. In: *Proceedings of the 57th Annual Reliability & Maintainability Symposium*. FL, USA.
26 Eryilmaz, S. (2013). Reliability of a k-out-of-n system equipped with a single warm standby component. *IEEE Transactions on Reliability* 62 (2): 499–503.
27 Li, X., Yan, R., and Zuo, M.J. (2009). Evaluating a warm standby system with components having proportional hazard rates. *Operations Research Letters* 37 (1): 56–60.
28 Ruiz-Castro, J.E. and Fernández-Villodre, G. (2012). A complex discrete warm standby system with loss of units. *European Journal of Operational Research* 218 (2): 456–469.
29 Misra, K.B. (2008). *Handbook of Performability Engineering*. Springer-Verlag.

30 Gulati, R. (1996). A modular approach to static and dynamic fault tree analysis. *M. S. Thesis*, Electrical Engineering, University of Virginia.

31 Rausand, M. and Hoyland, A. (2003). *System Reliability Theory: Models, Statistical Methods, and Applications*, 2e. Wiley Inter-Science.

32 Xing, L., Li, H., and Michel, H.E. (2009). Fault-tolerance and reliability analysis for wireless sensor networks. *International Journal of Performability Engineering* 5 (5): 419–431.

33 Levitin, G., Xing, L., and Dai, Y. (2014). Mission cost and reliability of 1-out-of-N warm standby systems with imperfect switching mechanisms. *IEEE Transactions on Systems, Man, and Cybernetics: Systems* 44 (9): 1262–1271.

34 Zhai, Q., Peng, R., Xing, L., and Yang, J. (2015). Reliability of demand-based warm standby systems subject to fault level coverage. *Applied Stochastic Models in Business and Industry, Special Issue on Mathematical Methods in Reliability* 31 (3): 380–393.

35 Tannous, O., Xing, L., Peng, R., and Xie, M. (2014). Reliability of warm-standby systems subject to imperfect fault coverage. *Proceedings of the Institution of Mechanical Engineers, Part O: Journal of Risk and Reliability* 228 (6): 606–620.

36 Zhai, Q., Peng, R., Xing, L., and Yang, J. (2013). BDD-based reliability evaluation of k-out-of-(n+k) warm standby systems subject to fault-level coverage. *Proceedings of the Institution of Mechanical Engineers, Part O: Journal of Risk and Reliability* 227 (5): 540–548.

37 Zhai, Q., Xing, L., Peng, R., and Yang, J. (2015). Multi-valued decision diagram-based reliability analysis of k-out-of-n cold standby systems subject to scheduled backups. *IEEE Transactions on Reliability* 64 (4): 1310–1324.

38 Xing, L., Shrestha, A., Meshkat, L., and Wang, W. (2009). Incorporating common-cause failures into the modular hierarchical systems analysis. *IEEE Transactions on Reliability* 58 (1): 10–19.

39 Wang, C., Xing, L., and Amari, S.V. (2012). A fast approximation method for reliability analysis of cold-standby systems. *Reliability Engineering & System Safety* 106: 119–126.

40 Tannous, O. and Xing, L. (2012). Efficient analysis of warm standby systems using central limit theorem. In: *Proceedings of The 58th Annual Reliability & Maintainability Symposium*. Reno, NV, USA.

41 Amari, S.V. and Dill, G.A. (2009). A new method for reliability analysis of standby systems. In: *Proceeding of the Annual Reliability and Maintainability Symposium*, 417–422. Fort Worth, TX, USA.

42 Modarres, M., Kaminskiy, M., and Krivtsov, V. (2010). *Reliability Engineering and Risk Analysis: A Practical Guide*. CRC Press.

43 Koralov, L.B. and Sinai, Y.G. (2007). *Theory of Probability and Random Processes*. Springer.

44 Wang, C., Xing, L. and Peng, R. (2014). Approximate reliability analysis of large heterogeneous cold-standby systems. In: *Proceedings of International Conference on Quality, Reliability, Risk, Maintenance, and Safety Engineering (QR2MSE 2014)*. Dalian, Liaoning, China.

45 Myers, A. (2009). Probability of loss assessment of critical k-out-of-n: G systems having a mission abort policy. *IEEE Transactions on Reliability* 58 (4): 694–701.

46 Amari, S.V. and Misra, R.B. (1997). Closed-form expressions for distribution of sum of exponential random variables. *IEEE Transactions on Reliability* 46 (4): 519–522.
47 Kuo, W. and Wan, R. (2007). Recent advances in optimal reliability allocation. *IEEE Transactions on Systems, Man and Cybernetics, Part A: Systems and Humans* 37 (2): 143–156.
48 Boddu, P. and Xing, L. (2013). Reliability evaluation and optimization of series-parallel systems with k-out-of-n: G subsystems and mixed redundancy types. *Proceedings of the Institution of Mechanical Engineers, Part O: Journal of Risk and Reliability* 227 (2): 187–198.
49 Amari, S.V. and Dill, G. (2010). Redundancy optimization problem with warm-standby redundancy. In: *Proceedings of the Annual Reliability and Maintainability Symposium (RAMS)*, 1–6. San Jose, CA, USA.
50 Coit, D.W. (2003). Maximization of system reliability with a choice of redundancy strategies. *IIE Transactions* 35 (6): 535–544.
51 Misra, K.B. and Sharma, U. (1991). An efficient algorithm to solve integer programming problems arising in system-reliability design. *IEEE Transactions on Reliability* 40 (1): 81–91.
52 Chia, L.Y. and Smith, A.E. (2004). An ant colony optimization algorithm for the redundancy allocation problem (RAP). *IEEE Transactions on Reliability* 53 (3): 417–423.
53 Misra, K.B. (1972). Reliability optimization of a series-parallel system. *IEEE Transactions on Reliability* R-21 (4): 230–238.
54 Onishi, J., Kimura, S., James, R.J.W., and Nakagawa, Y. (2007). Solving the redundancy allocation problem with a mix of components using the improved surrogate constraint method. *IEEE Transactions on Reliability* 56 (1): 94–101.
55 Tannous, O., Xing, L., Peng, R. et al. (2011). Redundancy allocation for series-parallel warm-standby systems. In: *2011 IEEE International Conference on Industrial Engineering and Engineering Management*, 1261–1265. Singapore.
56 Zhao, R. and Liu, B. (2005). Standby redundancy optimization problems with fuzzy lifetimes. *Computers & Industrial Engineering* 49 (2): 318–338.
57 Boddu, P., Xing, L., and Levitin, G. (2014). Energy consumption modeling and optimization in heterogeneous cold-standby systems. *International Journal of Systems Science: Operations & Logistics* 1 (3): 142–152.
58 Levitin, G., Xing, L., and Dai, Y. (2015). Effect of failure propagation on cold vs. hot standby tradeoff in heterogeneous 1-out-of-N: G systems. *IEEE Transactions on Reliability* 64 (1): 410–419.
59 Levitin, G., Xing, L., and Dai, Y. (2014). Cold vs. hot standby mission operation cost minimization for 1-out-of-N systems. *European Journal of Operational Research* 234 (1): 155–162.
60 Levitin, G., Xing, L., and Dai, Y. (2014). Optimization of predetermined standby mode transfers in 1-out-of-N: G systems. *Computers & Industrial Engineering* 72: 106–113.
61 Dai, Y., Levitin, G., and Xing, L. (2017). Optimal periodic inspections and activation sequencing policy in standby systems with condition based mode transfer. *IEEE Transactions on Reliability* 66 (1): 189–201.

62 Levitin, G., Xing, L., and Dai, Y. (2015). Optimal design of hybrid redundant systems with delayed failure-driven standby mode transfer. *IEEE Transactions on Systems, Man, and Cybernetics: Systems* 45 (10): 1336–1344.

63 Levitin, G., Xing, L., and Dai, Y. (2016). Cold-standby systems with imperfect backup. *IEEE Transactions on Reliability* 65 (4): 1798–1809.

64 Levitin, G., Xing, L., and Dai, Y. (2015). Heterogeneous 1-out-of-N warm standby systems with dynamic uneven backups. *IEEE Transactions on Reliability* 64 (4): 1325–1339.

65 Levitin, G., Xing, L., Johnson, B.W., and Dai, Y. (2015). Mission reliability, cost and time for cold standby computing systems with periodic backup. *IEEE Transactions on Computers* 64 (4): 1043–1057.

66 Levitin, G., Xing, L., and Dai, Y. (2017). Reliability vs. expected mission cost and uncompleted work in heterogeneous warm standby multi-phase systems. *IEEE Transactions on Systems, Man, and Cybernetics: Systems* 47 (3): 462–473.

67 Levitin, G., Xing, L., and Dai, Y. (2017). Optimization of component allocation/distribution and sequencing in warm standby series-parallel systems. *IEEE Transactions on Reliability* 66 (4): 980–988.

68 Levitin, G., Xing, L., Zhai, Q., and Dai, Y. (2016). Optimization of full vs. incremental periodic backup policy. *IEEE Transactions on Dependable and Secure Computing* 13 (6): 644–656.

69 Levitin, G., Xing, L., and Dai, Y. (2015). Optimal backup distribution in 1-out-of-N cold standby systems. *IEEE Transactions on Systems, Man, and Cybernetics: Systems* 45 (4): 636–646.

70 Levitin, G., Xing, L., and Dai, Y. (2017). Preventive replacements in real-time standby systems with periodic backups. *IEEE Transactions on Reliability* 66 (3): 771–782.

71 Levitin, G., Xing, L., Dai, Y., and Vokkarane, V.M. (2017). Dynamic checkpointing policy in heterogeneous real-time standby systems. *IEEE Transactions on Computers* 66 (8): 1449–1456.

72 Levitin, G., Xing, L., and Dai, Y. (2017). Optimal distribution of non-periodic full and incremental backups. *IEEE Transactions on Systems, Man, and Cybernetics: Systems* 47 (12): 3310–3320.

73 Levitin, G., Xing, L., and Dai, Y. (2018). Heterogeneous 1-out-of-N warm standby systems with online checkpointing. *Reliability Engineering & System Safety* 169: 127–136.

74 Levitin, G. and Finkelstein, M. (2018). Optimal mission abort policy for systems operating in a random environment. *Risk Analysis* 38 (4): 759–803.

75 Levitin, G. and Finkelstein, M. (2018). Optimal mission abort policy for systems in a random environment with variable shock rate. *Reliability Engineering & System Safety* 169: 11–17.

76 Levitin, G. and Finkelstein, M. (2018). Optimal mission abort policy with multiple shock number thresholds. *Proceedings of the Institution of Mechanical Engineers, Part O: Journal of Risk and Reliability, https://doi.org/10.1177/1748006X17751496.*

77 Levitin, G., Xing, L., and Dai, Y. (2018). Co-optimization of state dependent loading and mission abort policy in heterogeneous warm standby systems. *Reliability Engineering & System Safety* 172: 151–158.

78 Levitin, G., Xing, L., and Dai, Y. (2018). Mission abort policy in heterogeneous non-repairable 1-out-of-N warm standby systems. *IEEE Transactions on Reliability* 67 (1): 342–354.

79 Long, W., Zhang, T.L., Lu, Y.F., and Oshima, M. (2002). On the quantitative analysis of sequential failure logic using Monte Carlo method for different distributions. In: *Proceedings of Probabilistic Safety Assessment and Management* 391–396.

80 Boudali, H. and Dugan, J.B. (2005). A discrete-time Bayesian network reliability modeling and analysis framework. *Reliability Engineering & System Safety* 87 (3): 337–349.

Index

a

Aerospace 4, 27
As good as new 12
Availability 22, 55, 58, 201
Axioms of probability 7

b

Backup 205, 221
Basic event 12–14, 16, 64, 65, 76, 78, 84, 87, 115
Bayesian network 222
Bayes' theorem 9
Binary decision diagram (BDD) xv, 7, 16, 17, 22, 29–33, 35, 36, 40–44, 52, 56, 61, 66, 69, 84, 85, 87, 88, 90, 91, 93, 109–112, 114–116, 118, 120–124, 128, 131, 133, 134, 136, 139, 140, 142, 146, 147, 151, 153, 155–158, 176–180, 186–189, 196, 206, 207, 209
  0-edge 17
  1-edge 17
  canonical 17
  compact 17
  depth-first search 18
  disjoint paths 19, 207, 209
  else edge 17, 18
  evaluation 18, 20
  generation 17–19
  heuristics 18
  if-then-else (ite) format 17, 18
  index 17, 18
  input variable ordering 18
  isomorphic 18
  operation rules 18
  ordered BDD (OBDD) 17, 18
  recursive evaluation 18
  reduced OBDD (ROBDD) 17–20
  reduction rules 18
  Shannon's decomposition 17
  sink node 17–19
  then edge 17, 18
  useless node 18
Binary-state system 2, 28, 29, 34, 44
Body sensor network (BSN) xv, 182–185, 190
Boolean algebra 18, 65, 78, 87, 88, 131, 136, 142, 172
Bridge network 36, 37

c

Canonical 17
Cascading effect 73
Cascading failure 150
Central limit theorem (CLT) 211, 212, 214, 222
Checkpointing 221
Combinatorial phase requirement (CPR) 15, 51
Common-cause failure (CCF) xi, xii, xiii, xv, 2, 4, 83–87, 89–94, 97, 104
  common cause (CC) xv, 83–87, 89, 90, 92, 93, 104, 107–119, 121–124
  external cause 83
  internal cause 83
  mutually exclusive 86, 87, 90, 92, 93, 104, 112, 113, 115
  s-dependent 89, 90, 92, 93, 104, 112, 113, 115

Dynamic System Reliability: Modeling and Analysis of Dynamic and Dependent Behaviors, First Edition. Liudong Xing, Gregory Levitin and Chaonan Wang.

Common-cause failure (CCF) (*contd.*)
 s-independent 89, 90, 92, 93, 96, 104, 108, 109, 112, 115–117, 121
 deterministic CCF xii, 83, 104, 124
 probabilistic CCF (PCCF) xi, xii, xvi, 2, 4, 107–112, 114–119, 121–125
Communication system 27, 28
Compact 17, 70, 72, 79, 89, 159
Competing failure xi, xii, xiii, 2–4, 127, 129, 135, 136, 141, 150, 152, 153, 158, 159, 169, 170, 198, 199
 deterministic xii, 2–4, 80, 127, 169, 173, 180, 184, 190
 failure isolation xii, 3, 127, 130, 131, 133, 134, 142, 150, 152, 165, 166, 169, 173, 176, 180, 186–188, 190, 196, 198
 failure propagation xii, 2, 3, 127, 129, 130, 134, 136, 142, 150, 152, 155, 158, 160, 164, 166, 169–172, 177, 187, 190–192, 196
 local failure (LF) 128–133, 135, 137, 138, 143, 144, 150–152, 154–160, 162, 164, 165, 169–172, 174–193, 195, 198
 multi-phase system xii, 127, 150, 158, 166
 PFGE method 128–131, 133, 137, 142, 143, 151, 176–179, 191, 192, 195
 probabilistic xi, xii, xiii, xvi, 2–4, 169–173, 177, 187, 190, 191, 194, 196, 198, 199
 single-phase system xii, 127–129, 135, 141, 150, 166, 170, 190, 198
Computation complexity 66, 123, 124
Computer network 2, 61
Computer system 9, 16, 52, 53, 56, 61, 62, 67, 91, 92, 109, 110, 112, 114, 131, 164, 201, 205
Conditional probability 7, 8, 28, 33, 50, 56, 65, 98, 116, 154, 186, 190
Condition monitoring 172, 173
Continuous-time Markov chain (CTMC) 20, 22, 55–57, 63, 67, 70, 72, 79, 80, 158–166, 201–203, 212, 222
 initial state 20, 158–165
 Laplace transform 21, 159, 203
 state probability 20, 159, 162–165
 state transition 20, 21, 202–204
 state transition diagram 20, 21, 203, 204
 system state 20–22, 202–204
Cumulative distribution function (cdf) xv, 9–11, 109, 137, 219
Cutset 13, 16, 22

d

Data storage 27
Decision diagram xii, xiii, xv, 7, 16, 17, 35, 51, 52, 61, 83, 84, 89, 109, 116, 128, 201, 205, 206
 binary decision diagram (BDD) 7, 16–20
 multi-state multi-valued decision diagram (MMDD) xvi, 35–38
 multi-valued decision diagram 116
 sequential binary decision diagram (SBDD) xvi, 153, 205–210, 222
 ternary decision diagram 51
Degradation 3, 16
Dependence xi, xii, xiii, xv, 1–4, 8, 14, 37, 38, 41, 61, 63, 64, 86, 127, 132, 135, 169, 170, 209, 218
 cascading failure 150
 common-cause failure xi, xii, xiii, xv, 2, 4, 83, 107, 127
 functional dependence ix, xi, xii, xiii, xv, 1, 2, 3, 5, 14, 64, 86, 127, 169, 170, 218
 negative dependence 132
 positive dependence 132, 135
Dependent behavior xi, 1, 3, 22
Deterministic common-cause failure (deterministic CCF) 83
 common cause (CC) 83–87, 89, 90, 92–94, 104
 common cause event (CCE) 86–94
 common cause group (CCG) 86, 87, 90–92
 decision diagram (DD) 89–94
 efficient decomposition and aggregation (EDA) 83, 85–87, 89, 90, 104
 expanded fault tree (expanded FT) 84, 85
 explicit method 83–85, 89
 implicit method 83, 85

universal generating function (u-function) 83, 94–100, 103
Directed acyclic graph 17
Disjoint event 7, 111, 119, 141, 145, 151, 155
Distribution 9–12, 20, 22, 43, 63, 67, 68, 70, 79, 94, 104, 109, 115, 125, 129, 131, 135, 137, 145, 154, 158, 173, 174, 183, 194, 197, 202, 205, 208, 211–221
exponential 10–12, 20, 22, 43, 63, 67, 68, 70, 73, 76, 79, 131, 137, 145, 154, 158, 183, 202–205, 208, 211–213, 215, 216
normal 212–215, 217
Weibull 22, 43, 154, 173, 183, 184, 194, 197, 208, 211, 219
Distribution function 9, 212, 215
Divide-and-conquer 172
Dynamic behavior xi, 1, 3, 4, 15, 20
Dynamic system xi, xiii, 1, 13, 21

e

Event 7, 8
certain event 7
disjoint events 7
impossible event 7
independent events 8
mutually exclusive events 8
Event tree analysis 22
Explicit method 29, 44, 83–85, 89, 107, 108, 110, 115, 116, 118, 122–125

f

Failure competition event (FCE) xv, 170–172, 175–177, 179, 180, 182, 185–187, 190–192, 194–197
Failure function 10, 11, 38, 40, 50, 52
Failure isolation xii, 3, 127, 130, 131, 133, 134, 142, 150, 152, 165, 166, 169, 173, 176, 180, 186–188, 190, 196–198
Failure mode 2, 3, 12, 16, 174
Failure propagation xii, 2, 3, 83, 94, 99, 127, 129, 130, 134, 136, 142,.150, 152, 155, 158, 160, 164, 166, 169–172, 177, 187, 190–192, 196
Failure rate 2, 10, 11, 13, 14, 20, 21, 56, 68, 73, 76, 132, 137, 138, 145, 154, 160, 164, 201–204, 208, 215
Failure symptom 12
Fault injection 29
Fault recovery 2, 27, 29, 49
Fault tolerance 3, 201
Fault tolerant system (FTS) xii, xv, 1, 2, 27, 28, 43
Fault trees (FT) xii, xv, 1, 7, 12–20, 22, 31, 32, 42, 43, 51–54, 57, 58, 61, 65, 66, 69–72, 74, 75, 77–80, 83–85, 87–91, 93, 107–112, 114–119, 121–124, 130–134, 136–140, 142, 144, 146, 147, 149, 151–153, 155–159, 163, 164, 169, 171–173, 176–180, 182–184, 186, 188, 189, 192, 193, 195–198, 205, 206, 209
binary-state 16
construction 12
dynamic fault tree (DFT) xv, 13, 14, 21, 22, 61–65, 67, 68, 70, 73, 77, 202–206, 208, 209
event 12–14, 16
basic 12–14, 16
intermediate 12, 14, 16
TOP 12–16
undesired 12
gate 12–14, 16, 52, 61, 64–66, 70, 73–76, 78, 84, 107, 108, 110, 114, 116, 118, 130, 131, 133, 136, 142, 144, 151, 169–172, 176, 182, 186, 192, 195, 202, 205, 206, 208, 209
AND gate 14, 65, 76, 116, 118
cold spare (CSP) 13, 14, 21, 202, 203, 205–207
dynamic gate 13
functional dependence (FDEP) gate 14, 61, 64–66, 70, 73–75, 78, 107, 130, 131, 133, 142, 144, 151
hot spare (HSP) xiii, 14, 202
K-out-of-N 13, 15, 16
OR gate 61, 84, 151, 182, 186, 192, 195, 205, 206, 208
PCCF gate 107, 108, 110, 114

Fault trees (FT) (*contd.*)
PFD gate 170–172, 176, 186, 192, 195
priority AND (PAND) xvi, 14
sequence enforcing (SEQ) xvi, 14
Vote 13
warm spare (WSP) 14
multistate FT (MFT) xvi, 15, 16, 36
phased-mission FT 14, 15
phased-mission system FT (PMS FT) 42, 43, 116, 118, 119, 121, 122, 152, 153
static fault tree (SFT) xvi, 13, 61, 70, 72, 73
subtree 22, 51, 52
Fault tree analysis 1
independent subtrees 22
qualitative analysis 13
quantitative analysis 13, 16
analytical 12, 13, 20
combinatorial xii, xiii, 13, 15, 16, 22, 32, 33, 40, 51, 61, 63, 66–70, 72, 74, 76, 78–80, 104, 115, 129, 132, 135, 141, 150, 166, 170, 174, 181, 190, 201, 205
modularization 13, 22
simulation 13, 40, 222
state space based 13, 20, 40, 63
Flight control 27, 201
Functional dependence 61, 64
cascading 63–67, 70, 72–74, 76, 77, 79, 80
combinatorial algorithm 63, 67, 70, 72, 74, 76, 78–80
dependent component 61–63, 67, 72, 76, 80
dual event 65, 76, 77, 80
logic OR replacement method 61–63, 67, 69–72, 76, 79
trigger 61, 63–68, 72, 73, 76, 78, 80

g

Global effect 2, 127, 150, 169

h

Hazard rate 11
Hierarchical system (HS) xii, xv, 2, 49, 51–56, 58
nonrepairable 49, 51, 58
repairable 49, 51, 55, 58
Human error 83

i

Imperfect coverage (IPC) xvi, 2, 27, 28, 37, 43, 49, 51, 58, 62, 63, 66, 67, 79, 127
BDD expansion method (BEM) xv, 29–32, 34, 40
binary-state system 28, 29, 34, 44
covered failure 30, 34, 35, 55, 56, 80
element level coverage (ELC) xv, 27–34, 37, 38, 40, 42–44
fault level coverage (FLC) xv, 27, 28, 43
modular imperfect coverage 2, 49, 51, 58
multi-fault model 28
multi-state system 28, 34, 44
performance dependent coverage (PDC) xvi, 27, 28, 43
phased-mission system 37–44, 51
simple and efficient algorithm (SEA) xvi, 29, 32–34, 37, 38, 42–44, 51
single-fault model 27
single-point failure 28, 29, 50
uncovered failure (UF) xvi, 2, 29–40, 49–52, 55, 56, 58
Imperfect coverage model (IPCM) xvi, 28–33, 35, 37, 43, 49, 58, 63
coverage factor 28, 29, 31, 32, 38, 43, 56, 68, 70, 73, 77
disjoint exits 29
entry point 28, 49
modular IPCM (MIPCM) xvi, 49–53, 58
permanent coverage 28, 50
single-point failure 28, 29, 50
transient restoration 28, 50
Implicit method xii, 29, 44, 83, 85, 107, 110, 112, 115, 119, 121, 123–125
Inclusion-exclusion (I-E) 16, 22
Internet of Things (IoT) xv, 1, 198
Isolation factor 172–174, 184, 194, 197
Isolation factor group (IFG) xv, 173, 174, 180, 181, 184, 190

j

Jamming attack 169, 172, 173, 181, 193

l

Life-critical 27, 201, 221
Loading 28, 222
Local failure (LF) xvi, 97, 108, 112–116, 118, 121, 128–133, 135, 137, 138, 143, 144, 150–152, 154–160, 162, 164, 165, 169–172, 174–193, 195, 198
Lyapunov's condition 214, 215

m

Markov process xii, 7, 20
Mean residual life (MRL) xvi, 10, 12
Mean time between failures (MTBF) xvi, 12
Mean time to failure (MTTF) xvi, 10–12, 212–215
Mean time to repair (MTTR) xvi, 12
Memoryless property 10, 12
Memory system 67, 70, 137, 138, 144, 145
Minimal cutset 13
Mission abort 221
Mission-critical 27
Modular imperfect coverage 2, 49
  conditional probability 50
  entry point 49
  exit 49, 50
  modular imperfect coverage model (MIPCM) 49–53, 58
  uncovered failure (UF) 49–51
Monte Carlo simulation 13, 222
Multi-phase system xii, 3, 4, 28, 107, 115, 125, 127, 150, 158, 166, 221
Multi-state system (MSS) ix, xii, xvi, 3, 4, 16, 28, 34–36, 44
Mutually exclusive 7, 8, 52, 64, 86, 87, 90, 92, 93, 104, 112, 113, 115, 129, 171, 173, 174, 181, 185, 186, 218, 219

n

Nuclear plant 27

p

Parallel system 31, 32, 34
Performance 3, 15, 16, 27, 28, 34–36, 94–97, 99, 100, 169, 180, 201, 216, 218–221
  degradation 3, 16
  level 3, 15, 28, 34–36, 219, 220
  sharing 3
Phased-mission system (PMS) xii, xvi, 4, 14, 15, 22, 37,38–44, 51, 93, 115, 116, 118–125, 128, 150–159, 162–165, 199
  combinatorial phase requirement (CPR) xv, 15, 51
  mini-component concept 37, 39
  nonrepairable 37, 39, 42, 93
  phase duration 153, 160, 162
  phased-mission FT 14, 15
  phase-OR requirement 15
  PMS BDD 40–44, 116, 118, 120–123, 151, 155–158
  PMS FT 42, 43, 116, 118, 119, 121, 122, 152, 153
  PMS SEA 37, 38, 42, 43
  reliability analysis 15, 37, 40, 42, 44, 93, 115, 150, 158, 166, 199
PMS BDD method 40, 42, 116, 118, 121, 151, 155, 156
  phase dependent operation (PDO) xvi, 40, 41
  PMS BDD 40–44, 116, 118, 120–123, 151, 155–158
  variable ordering 40
    backward 40, 41, 122
    forward 40
Polynomial 96, 99, 204, 205
Probabilistic common-cause failure (PCCF) xi, xii, xvi, 2, 4, 107–112, 114–119, 121–125
  explicit method 107, 108, 110, 115, 116, 118, 122–125
  impact 124
  implicit method 107, 110, 112, 115, 119, 121, 123–125
  multi-phase system 107, 115, 125
  probabilistic common-cause event (PCCE) xvi, 111–115, 119, 121, 122
  probabilistic common-cause group (PCCG) xvi, 107–109, 115–117

Probabilistic common-cause failure (PCCF) (*contd.*)
  single-phase system 107, 119, 125
Probabilistic-dependent (PDEP) xvi, 169–172, 175, 176, 180, 182, 184, 185, 191, 192, 194
Probabilistic functional dependence (PFD) xvi, 169–173, 176, 182, 184–186, 191–193, 195, 198
Probability density function (pdf) xvi, 9–11, 94, 99, 109, 128, 130, 131, 137, 154, 174, 183, 184, 207, 208, 211, 215, 218
Probability mass function (pmf) xvi, 9, 95, 96, 100
Propagated failure (PF) xvi, 83, 97–100, 103, 107, 127, 141, 143, 148–150
  propagated failure with global effect (PFGE) xvi, 2, 127–133, 135–145, 148, 150–155, 157–162, 164, 166, 169–172, 174–186, 188–196, 198
  propagated failure with selective effect (PFSE) xvi, 141–146, 148, 150, 158, 159, 162, 164, 166
Propagation time xii, 83, 94, 190

r
Random experiment 7
Random variable (r.v.) xvi, 9
  continuous 9–11
  cumulative distribution function (cdf) xv, 9–11
  discrete 9
  expected value 9, 10
  exponential r.v. 10
  mean 9, 10
  normal r.v. 213, 214
  probability density function (pdf) xvi, 9–11
  probability mass function (pmf) xvi, 9
  standard deviation 10
  variance 10
Recursive algorithm 19, 116
  exit condition 19, 41
  recursive evaluation 18, 41
Redundancy xi, 1, 22, 27, 29, 83, 127, 220, 221
Redundancy allocation problem (RAP) xvi, 220
Relayed wireless communication xii, 3, 198
Reliability analysis xi, xii, xiii, 1, 3, 7, 15–17, 20, 22, 28, 29, 32, 35, 37, 40, 42, 44, 49, 51, 53, 58, 61–63, 66, 79, 83, 84, 87, 93, 94, 107, 109, 115, 127, 135, 141, 150, 158, 166, 169, 170, 190, 199, 205, 208, 211, 212, 222
Reliability block diagram 1, 22
Reliability function 10, 11
Reliability measures xii, 7, 10
  failure function 10, 11
  failure rate 10, 11
  hazard rate 11
  mean residual life (MRL) xvi, 10, 12
  mean time to failure (MTTF) xvi, 10–12
  reliability function 10, 11
  survival function 11
Reliability software 22
  BlockSim 22
  FaultTree+ 22
  Galileo 22
Repair 3, 12, 20, 55, 56, 202
  repair rate 20, 56
  time-to-repair 12, 202
Replacement method xii, 2, 61, 63, 67, 69–72, 76, 79

s
s-relationship 87, 104, 112, 115, 171, 181, 191, 193
  mutually exclusive 7, 8, 52, 64, 86, 87, 90, 92, 93, 104, 112, 113, 115, 129, 171, 173, 174, 181, 185, 186, 218, 219
  s-dependent 89, 90, 92, 93, 104, 112, 113, 115, 128, 132, 135, 171, 180, 191
  s-independent 38, 39, 89, 90, 92, 93, 96, 104, 108, 109, 112, 115–117, 121, 128, 130, 132, 133, 135, 137, 144, 150, 151, 158, 171, 181, 182, 186, 191
Safety xi, 3, 12
Sample point 7, 8
Sample space 7–9
Selective effect xii, 94, 127, 141
Sequence or temporal dependence 209
Series-parallel system 83, 94–96, 100, 104
Series system 32–34
Shannon's decomposition theorem 17

Simple and efficient algorithm (SEA) xvi, 29, 32–34, 37, 38, 42–44, 51, 63, 65, 75, 78, 128
Simulation 13, 40, 222
Single-phase system xii, 4, 38, 107, 115, 119, 125, 127–129, 135, 141, 150, 166, 170, 190, 198
Single-point failure 28, 29, 50
Smart home xiii, 192–194, 198
  energy management system (EMS) 192, 193
  power generation system 192–194
  relay node 192
  sensor 192, 193, 195–198
Space data gathering system 15
Space mission 201
Standby element sequencing problem (SESP) xvi, 221
Standby sparing 2–4, 13, 14, 201, 202, 209, 220–222
  approximation method xiii, 201, 222
  cold 201–208, 211–217, 221, 222
  CTMC-based method 201, 202, 212, 222
  decision diagram-based method 205
  event transition method 201, 216, 217, 222
  exact method 214
  hot 201, 202, 221, 222
  optimization problem 220–222
  redundancy allocation problem (RAP) 220
  standby element sequencing problem (SESP) 221
  standby mode transfer 221
  warm 201–205, 208–212, 218, 219, 221, 222
State indicator variable 34
State probability 20, 35–37, 159, 162–165
State-space explosion 22, 63
Structure function 4, 95, 96, 107
Sum of disjoint products (SDP) 16, 22
Survivability 221
System failure probability 86, 112, 121, 128, 137, 143, 145, 151, 152, 219

t

Temporal dependence 209
Ternary decision diagram 51
Time-domain competition 3, 169
Time to failure (*ttf*) xvi, 10–12, 20, 22, 43, 63, 67, 68, 70, 79, 99, 104, 109, 115, 125, 129, 130, 135, 154, 158, 202, 205, 207, 208, 211–219
Total probability law 8, 9, 44, 52, 55, 64, 80, 112, 121, 128, 131, 136, 141–143, 151, 152, 171, 172, 182, 191
Trigger event xii, 2, 3, 14, 52, 64, 65, 67, 68, 72, 73, 76, 78, 80, 107, 110, 114, 127, 129, 130, 136, 169, 170

u

u-function xvi, 94–100, 103
Unavailability 13, 55–58
Uncovered failure 2, 29–40, 49, 50, 62
Universal generating function xii, xvi, 83, 94, 222
Unreliability 2, 11, 13, 17–20, 22, 30–34, 38, 41, 42, 54, 62, 63, 65–67, 69–72, 76, 79, 83–87, 89, 91–93, 110, 112, 115, 118, 121, 122, 124, 125, 128, 131, 135–138, 141, 143, 145, 149, 150, 153, 154, 158, 159, 164, 165, 170, 172, 174, 180–182, 190, 192, 194, 197, 198, 203, 207–209, 211–217

v

Variable ordering 17–20, 40, 206, 209

w

Weibull distribution 43, 154, 173, 183, 194, 197, 208, 211
  scale parameter 154, 173, 183, 194, 211
  shape parameter 154, 173, 183, 194, 211
Wireless sensor network (WSN) xvi, 1, 3, 116–118, 120, 121, 123–125, 127, 169, 172–175, 180, 181, 192–194, 196–198
  application communication 117
  application communication phase (ACP) xv, 117–119
  base station 116, 117
  communication reliability 116, 117
  infrastructure communication 117
  infrastructure communication phase (ICP) xv, 117–119
  sink node 3, 192